Optical Microscopy of Fiber-Reinforced Composites

Brian S. Hayes
and
Luther M. Gammon

ASM International®
Materials Park, Ohio 44073-0002
www.asminternational.org

First printing, November 2010
Second printing, February 2019

Prepared under the direction of the ASM International Technical Book Committee (2009–2010), Michael J. Pfeifer, Chair.

ASM International staff who worked on this project include Scott Henry, Senior Manager, Content Development and Publishing; Eileen De Guire, Senior Content Developer; Ann Britton, Editorial Assistant; Bonnie Sanders, Manager of Production; Madrid Tramble, Senior Production Coordinator; and Diane Whitelaw, Production Coordinator.

Library of Congress Control Number: 2010937088

ISBN-13: 978-1-61503-044-6 (print)
ISBN-10: 0-61503-044-1 (print)
ISBN: 978-1-61503-143-6 (pdf)

SAN: 204-7586

ASM International®
Materials Park, OH 44073-0002
www.asminternational.org

Printed in the United States of America

All proceeds of this book are donated to charity.

Contents

Preface

This book is designed as an instructional reference for preparing fiber-reinforced polymeric-matrix composite materials for examination by optical microscopy and the techniques of optical microscopy used for analysis. It is also meant to be a teaching tool for those who want to learn more about the microstructure of these heterogeneous and anisotropic materials. The content is also appropriate for experienced microscopists or metallurgists who become involved in the preparation and analysis of polymeric composites. This book begins with an introduction to fiber-reinforced polymer-matrix composite materials that focuses on the microstructure and morphology of these unique materials. In the following chapters, the authors explain the materials, equipment, and procedures of how to prepare composite samples, followed by the illumination and contrast techniques of optical microscopy. Included in these chapters are the methods and reagents that are used to bring out distinct features in composite materials, such as different phases and areas of degradation or damage. Also included are details of how to prepare special composite materials having vast differences in hardness and material properties. The remaining chapters present various topics of studies of fiber-reinforced polymeric composite materials that have been performed by using optical microscopy.

These studies include a majority of the microstructural information that is of primary interest when working with composite materials. Insight into processing effects, toughening approaches, damage mechanisms, and environmental effects on the microstructure of composite materials is presented. Through these chapters and the micrographs throughout this book, it will be evident that optical microscopy is one of the most valuable tools for analyzing fiber-reinforced polymeric-matrix composites. Unfortunately, with all of the nonoptical methods of analysis available today, optical microscopy is often overlooked or not used to the fullest extent. The contents of this book were developed after many years of the authors presenting on this subject and teaching short courses about optical microscopy of composite materials. It is hoped that this text is found useful to everyone who desires to further increase their knowledge of optical microscopy and the microstructure of fiber-reinforced polymeric composite materials.

Acknowledgments

The authors wish to express their gratitude to The Boeing Company for allowing the publication of this work. Also, this work could not have been accomplished without the help of many people at The Boeing Company. Specifically, the authors wish to thank Kurt W. Batson for his contributions to Chapter 4, D. Jean Ray for her contributions to Chapter 6, and Anthony Falcone for his contributions to Chapter 15.

CHAPTER 1

Introduction—Composite Materials and Optical Microscopy

Composite Materials

The unique and diverse characteristics of composite materials have caused an increase in their utilization worldwide. From featherweight fly fishing rods to high-performance airplane parts, the use of fiber-reinforced composite materials is becoming more popular due to their high strength-to-weight ratio combined with easy manufacturing methods. Fiber-reinforced polymeric-matrix composites consist of reinforcing fibers and a polymer resin. The fibers are considered as the principal load-carrying constituent of the composite, while the role of the polymer matrix is to transfer the load between fibers as well as provide corrosion resistance, damage tolerance, and thermal and environmental stability (Ref 1). Fiber-reinforced polymeric composites are developed from thermoplastic or thermoset resins combined with either discontinuous or continuous unidirectional fibers or woven fabrics (Fig. 1.1a to 1.1c). Typical reinforcements consist of glass, carbon, or aramid fibers, but other materials, such as boron, ceramic, and thermoplastic fibers, may also be used for specific applications (Fig. 1.2a to 1.2c). Some fiber types may be the same chemical makeup and have similar mechanical properties but are vastly different in structure, depending on the manufacturer (Fig. 1.3a, b).

Many methods are used to manufacture fiber-reinforced composites, including hand lay-up of prepreg materials, automated tape lay-up of prepreg materials, resin transfer molding, vacuum-assisted resin transfer molding, resin film infusion, wet lay-up, filament winding, pultrusion, and compression molding of sheet molding or bulk molding compound. While

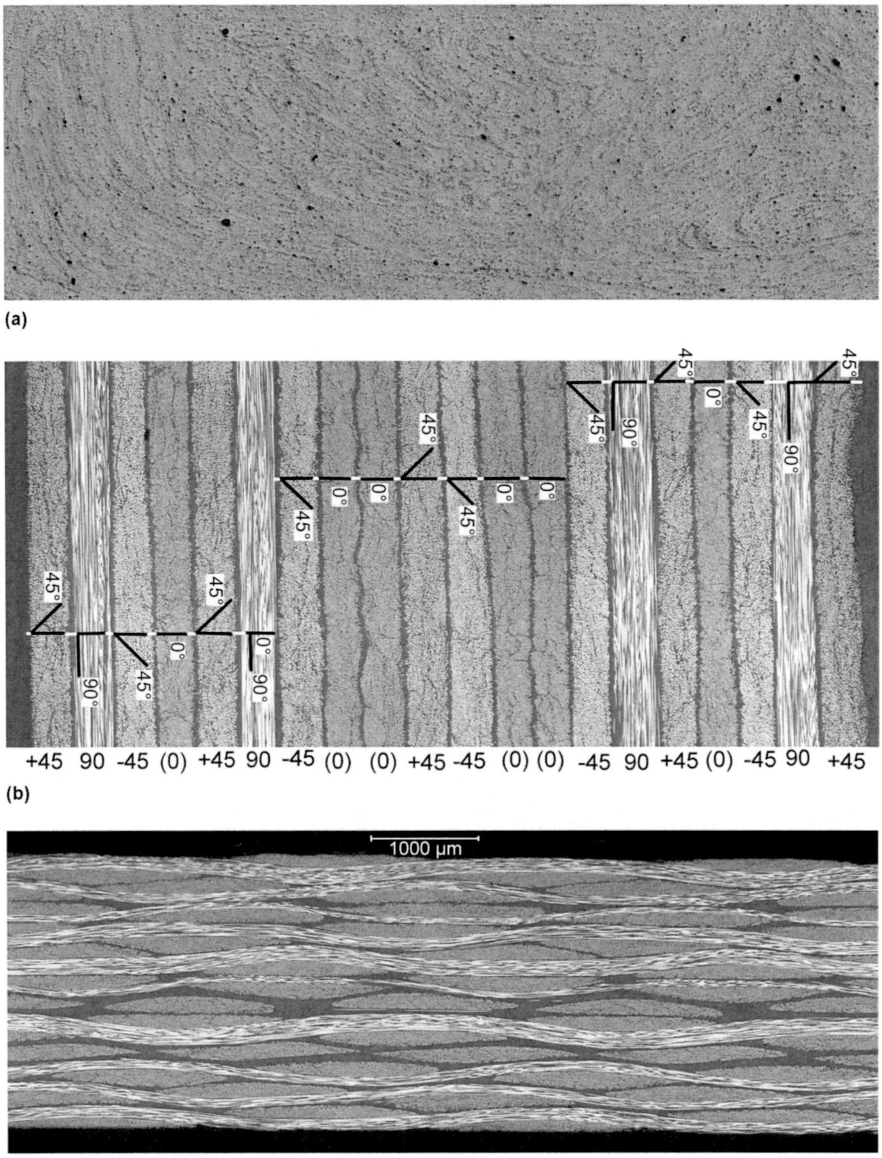

Fig. 1.1 Composite cross sections. (a) Sheet molding compound made from carbon-black-filled epoxy resin and chopped glass fiber. Bright-field illumination, 65 mm macrophotograph montage. (b) Quasi-isotropic unidirectional prepreg laminate. Slightly uncrossed polarized light, 5× objective, montage. (c) 3k-70 woven carbon fabric laminate. Bright-field illumination, 5× objective, montage

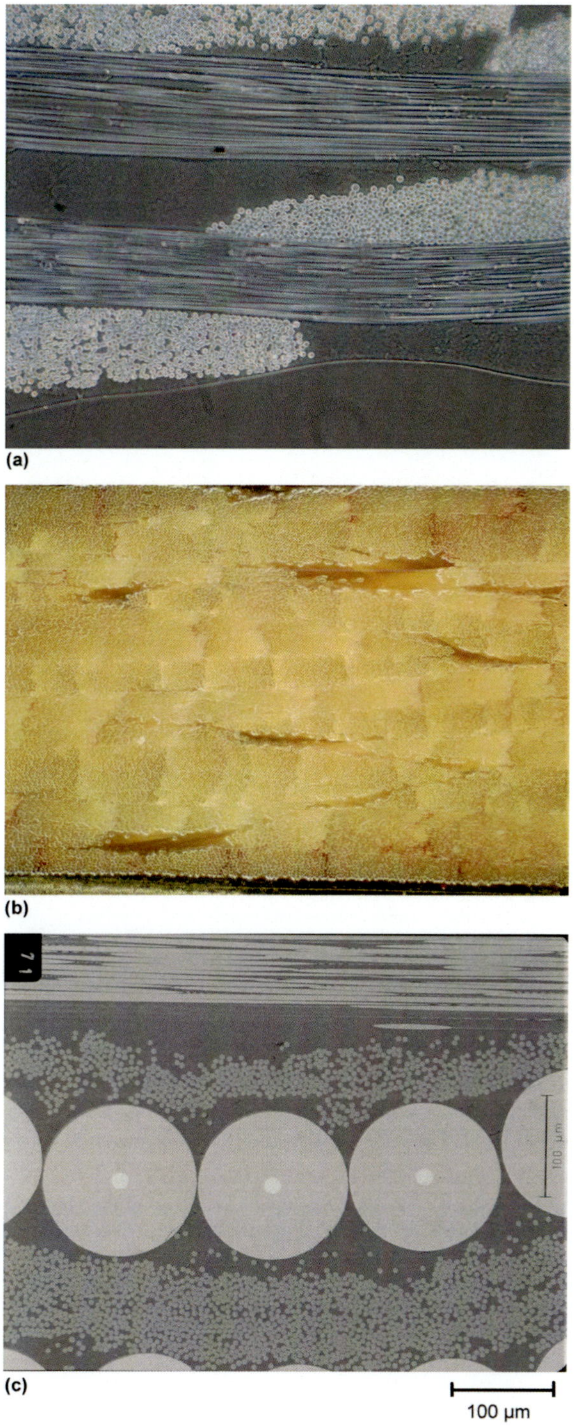

Fig. 1.2 Composite materials made from different types of fibers. (a) Woven glass fiber fabric composite revealing a multiphase-matrix morphology. Ultrathin section, transmitted-light phase contrast, 20× objective. (b) Kevlar (E.I. du Pont de Nemours and Company) fabric composite cross section. Darkfield illumination, 25× objective. (c) Boron fiber polymeric-matrix composite cross section. Bright-field illumination, 50× objective (200× original magnification)

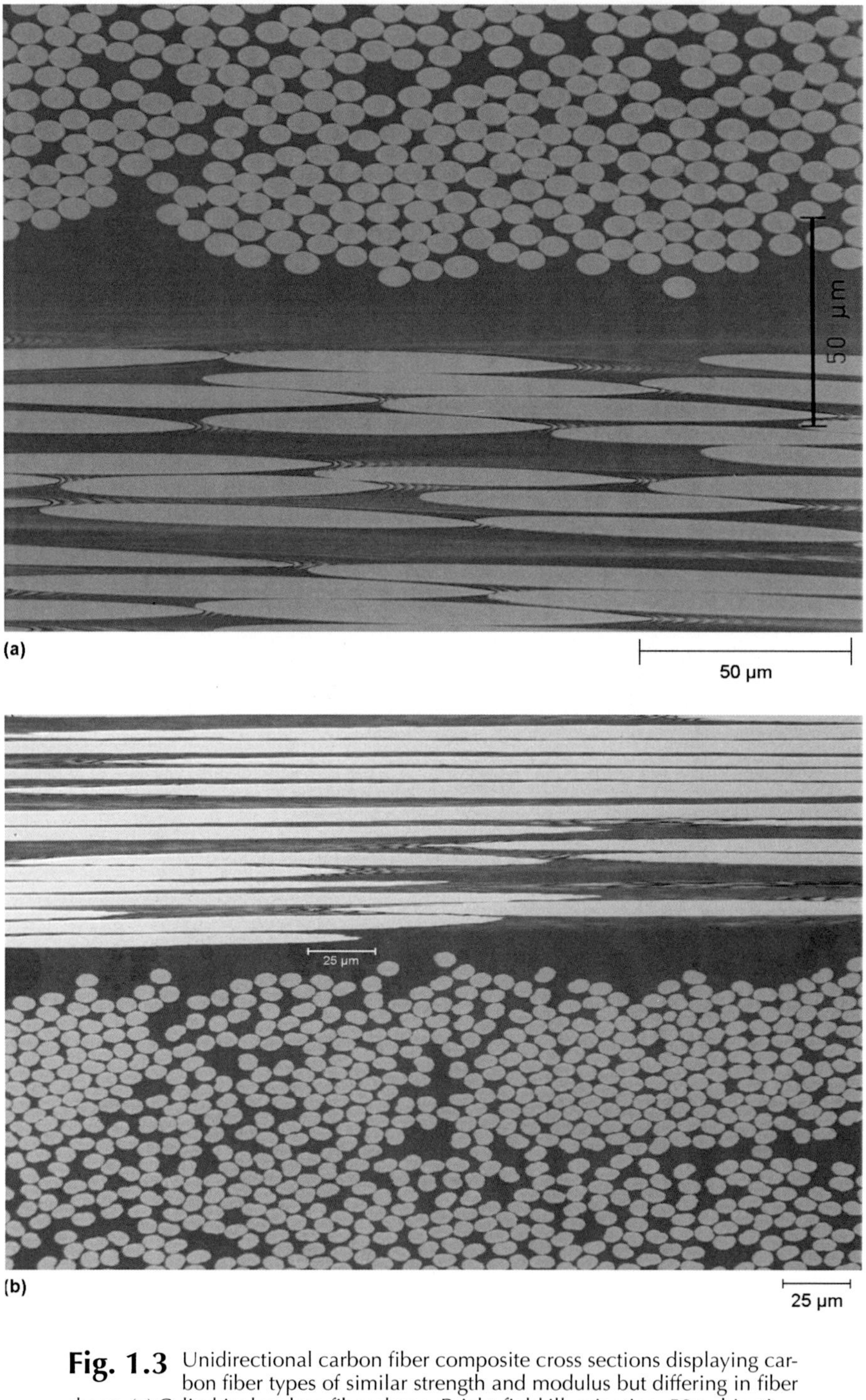

Fig. 1.3 Unidirectional carbon fiber composite cross sections displaying carbon fiber types of similar strength and modulus but differing in fiber shape. (a) Cylindrical carbon fiber shape. Bright-field illumination, 50× objective. (b) Irregular bean-shaped fibers. Bright-field illumination, 25× objective

these processes are general in description, the actual process of combining the fibers and matrix resin is unique and can be different with each resin-fiber system and between manufacturers. The processes described and the manufacturing parameters can lead to differences in morphology of the uncured and cured composite structures.

Polymer Matrices

Polymer matrices used for composite materials can be generally described as either thermoplastics or thermosets. Thermosets are polymer resins that crosslink and form a three-dimensional structure when cured. Once cured, the network structure is irreversible and cannot be reshaped or made to flow below its decomposition temperature (Ref 2). In contrast, thermoplastics, which consist of high-molecular-weight linear or branched polymer chains (not crosslinked), can be reshaped with the application of heat and pressure (Ref 2). In relation to composite materials, the distinction between these types of matrices is that they are reactive (thermosets) and nonreactive (thermoplastics) polymers. Most thermoplastic-matrix composites are developed with their polymerization complete. As a result, thermoplastic fiber-reinforced composites are generally more difficult to produce due to high viscosity resulting from the high-molecular-weight polymer chains. These materials usually require high temperatures, pressures, or the use of solvents for processing that must be removed after manufacturing. An added complexity of processing thermoplastics exists in the ability for some thermoplastics to form a semicrystalline structure (Fig. 1.4a to 1.4c) (Ref 3). This is very important to realize, if the material used is amorphous or semicrystalline, because the cooling and heating rates can affect the crystallinity of the matrix and hence the final composite properties (Ref 4). Current commercial high-performance composite matrices may contain an engineering thermoplastic in combination with a thermoset, thereby taking advantage of the different properties.

There are many types of thermosetting and thermoplastic materials that are used as matrices in fiber-reinforced composites. The selection of the matrix material and fiber type is dependent on the physical and mechanical properties that are required for the designed part. Thermosetting matrices commonly used in composite materials include polyesters, epoxy vinyl esters, epoxies, cyanate esters, bismaleimides, and other higher-temperature resins. Some common thermoplastic matrices used in composites include polyethylene terephthalate, polyamides (nylon), polyetheretherketone, and polyphenylene sulfide.

Prepreg Materials

Prepregs are the most widely used materials for manufacturing high-performance composites (Fig. 1.5). The manufacturing of prepreg is usu-

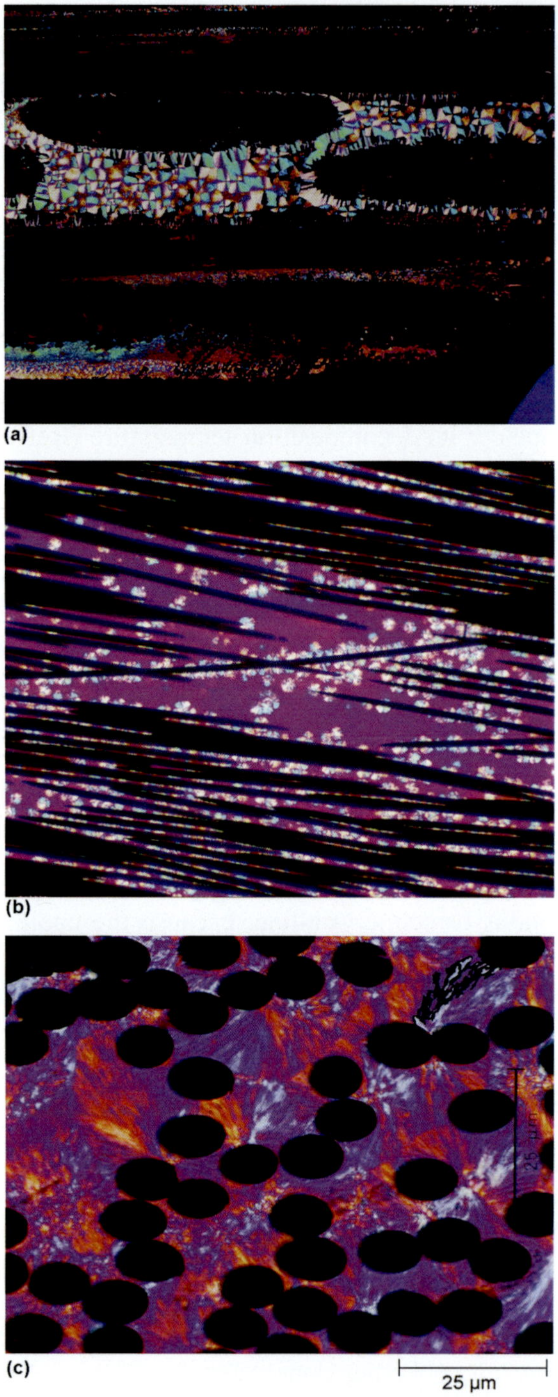

Fig. 1.4 Crystallinity in thermoplastic-matrix carbon fiber composites. (a) Crystalline region in the center area of a woven carbon fabric composite cross section. Ultrathin section, transmitted polarized light with a full wave plate (540 nm), 20× objective. (b) Fiber-induced spherulite growth. Ultrathin section, transmitted polarized light with a full wave plate (540 nm), 20× objective. (c) Large spherulitic growth in a carbon fiber composite. Ultrathin section, transmitted polarized light with a full wave plate (540 nm), 100× objective

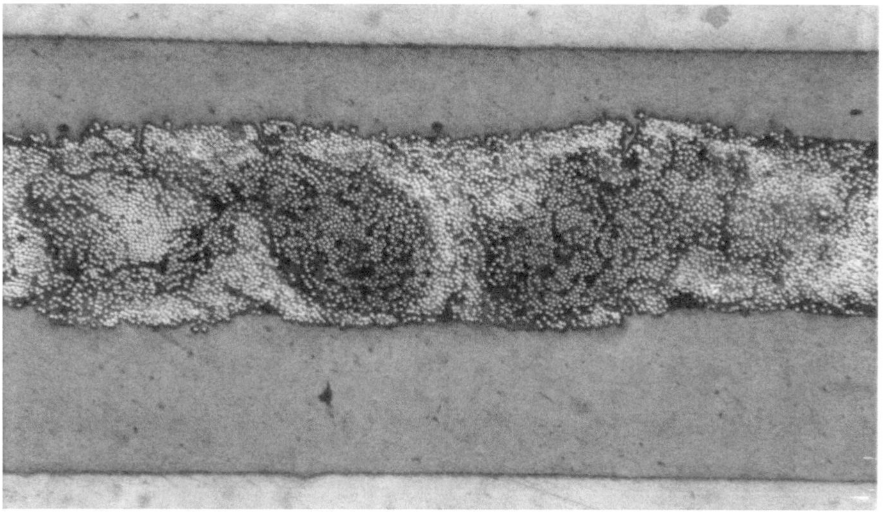

Fig. 1.5 Cross section of uncured prepreg material showing unimpregnated areas (dark) and impregnated areas (gray). Bright-field illumination, 25× objective

ally accomplished through either solvent or hot-melt impregnation of the matrix resin into the continuous unidirectional or woven fiber fabric materials (Ref 5, 6). These pre-engineered laminating materials have a discrete resin/fiber ratio that requires further lay-up of the continuous fiber plies to achieve the final composite. The prepreg lay-up is subjected to elevated temperature and pressure to cure and consolidate the part. The method of pressure application is dependent on the part and configuration, but the use of an autoclave is most common for high-performance parts.

The physical difference between thermosetting and thermoplastic polymers imparts an extreme differentiation in the handling characteristics of prepreg materials in the form of tack and drape. Tack can be defined as the ability of prepreg plies to stick together, while drape is the ability of the prepreg to conform to different contours (Ref 7). The high-viscosity characteristic of thermoplastic-matrix prepregs provides limited tack-and-drape capabilities. While this limits the use of thermoplastic materials for some applications, the inherent toughness of these materials is superior in most cases when compared to unmodified thermosets. This desirable aspect, along with short manufacturing times, has provided an increasing market for these materials. However, the compaction pressure and temperature required for consolidation of thermoplastic prepregs are usually substantially higher than what is typically used for thermosetting-matrix prepreg materials. Thermoset prepreg materials are attractive because they have desirable handling characteristics and a wide range of properties. If the tack of thermoset prepreg material is too high, it may enable greater air entrapment between the prepreg plies. Entrapped air in composites that is not removed during manufacturing can lead to voids in the cured composite (Fig. 1.6). Also, care must be taken to remove water and solvent contained in the matrix materials and prepregs, which may cause voids in the cured composite, resulting from an increase in the solvent vapor pressure with cure temperature (Fig. 1.7) (Ref 8 to 10).

Infusion Processes

Fiber-reinforced composites can be developed by omitting the intermediate impregnation or "prepreg" step and directly infusing the dry fabric

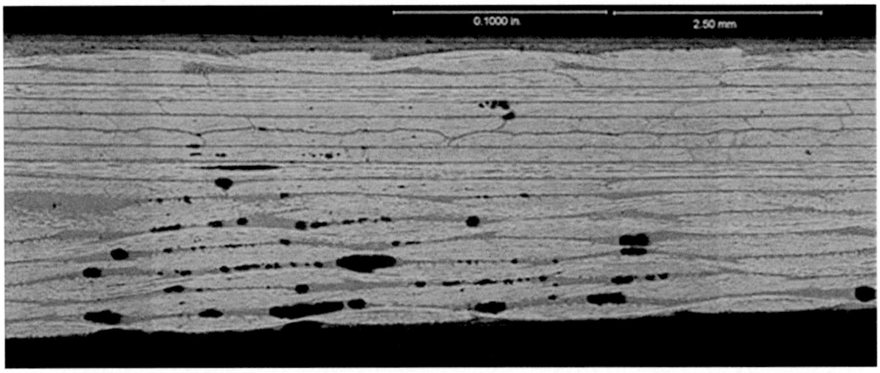

Fig. 1.6 Entrapped air in a composite part made from unidirectional carbon fiber prepreg and woven fabric prepreg. Voids (dark areas) are shown mainly in the interply regions of the cross section. Bright-field illumination, 65 mm macrophotograph montage

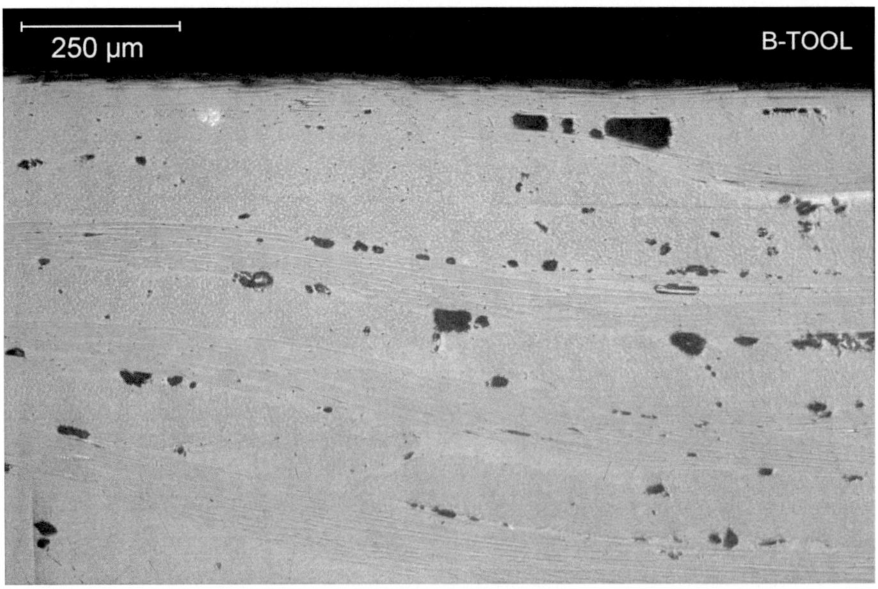

Fig. 1.7 Voids found in a glass fiber composite cross section due to solvents from manufacturing. Bright-field illumination, 10× objective

reinforcement or fibrous preform with the matrix resin. Infusion processes are low-cost composite manufacturing methods and are being readily adopted for replacement of some prepreg composite parts. Of the many infusion methods and processes that have been developed, the major techniques are variations of resin transfer molding (RTM), vacuum-assisted resin transfer molding (VARTM), and resin film infusion (RFI). In comparison to prepreg matrices, infusion resins must be homogeneous and single phase before impregnation, or the particles may become filtered by the fibers. This limits the use of some standard prepreg curing agents and modifiers, because many are in particle form (Fig. 1.8). Also, because the

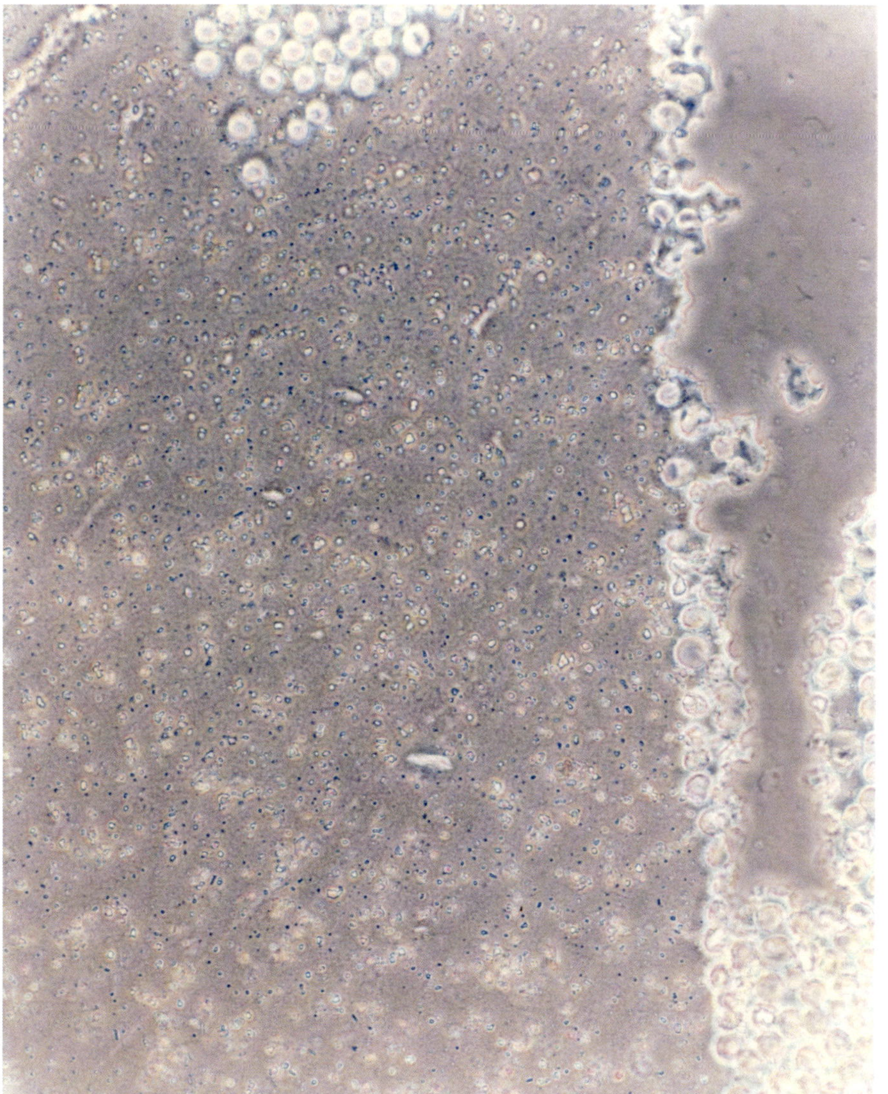

Fig. 1.8 Residual curing agent particles in a thermoset-matrix glass fiber composite. Reflected-light phase contrast, 40× objective

resin is not preimpregnated, many conventional toughness modifiers, such as thermoplastics and high-molecular-weight elastomers, cannot be used in these systems, since the viscosity is increased to an unacceptable level. If the viscosity of these systems is too high, long infusion times may be required, and the resin may gel before complete infusion. This may also result in a thicker part and improper fiber wet-out. The physical difference of the resins used for RFI versus VARTM or RTM is that the resin is usually in sheet or film form. Resins for VARTM or RTM may be one-component systems, containing the resin and catalyst/curing agent combined from the supplier, or, more generally, are two-part systems where the catalyst/curing agent is separate from the resin.

Although it is not defined as an infusion process, hand wet lay-up of continuous fabrics or preforms is very cost-effective and highly controllable if performed correctly. These wet prepreg materials commonly are consolidated under vacuum or compression cured with or without the application of heat to create the final part. Many of the parts that are developed with wet prepreg materials also include wood or foam cores to enhance the stiffness. These types of processes are currently being used for developing many high-performance composites.

Composite- and Matrix-Toughening Methods

The desire to use fiber-reinforced composites in more damage-prone environments has created a need for tougher, more damage-tolerant polymer-matrix composites. One method to increase composite toughness is through the use of more damage-tolerant fibers made from aramid or glass; however, other performance requirements may not allow the use of these materials for the design. Work also continues on the development of stitched fibrous preforms (Fig. 1.9) to improve toughness and provide easier manufacturing. While stitching may work very well for some applications, problems may arise, such as microcracking, due to the differences in the properties of the stitching materials.

Dispersed-Phase Toughening

There are many methods used to increase the toughness of thermosetting resin systems. However, the methods used must be selected so as not to reduce the processing, handling characteristics, or the mechanical and/or physical properties. Extensive research has been conducted over the past several decades involving the modification of high-performance thermosetting resins and fiber-reinforced composites. As a result of this effort, many materials and techniques have been and are currently being developed for improving the toughness of these materials. One of the most widely researched techniques involves the development of a multiphase structure with the use of either rubber or thermoplastic modifiers (Fig.

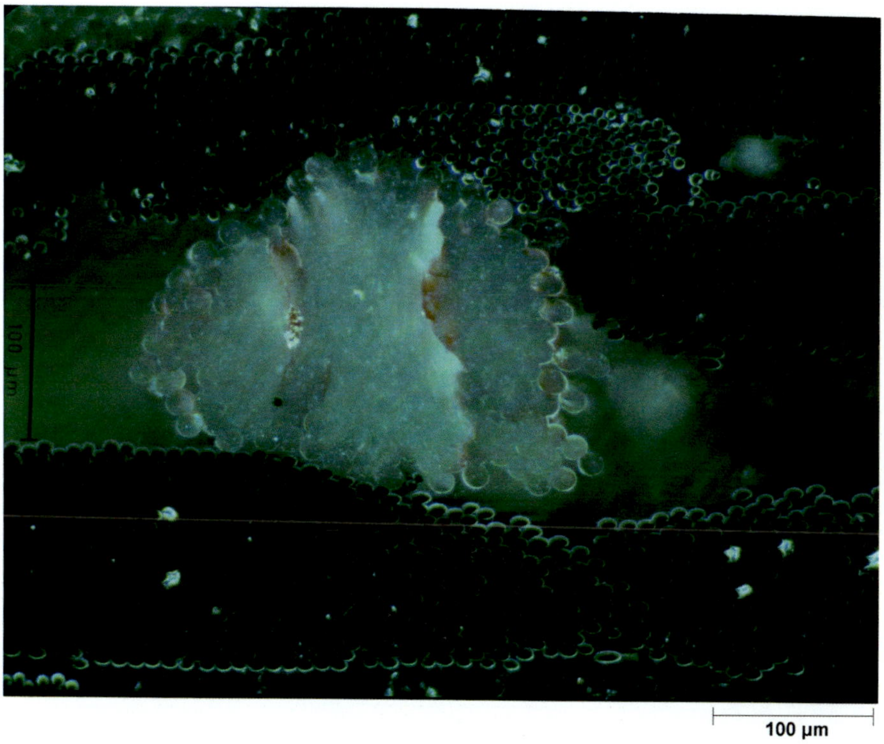

Fig. 1.9 Thermoplastic stitch in carbon fiber composite material. Note the microcracks in the center of the stitch. Epi-fluorescence, 390–440 nm excitation, 25× objective

1.10a to 1.10c) (Ref 11, 12). The most common rubber materials used for toughening thermoset matrices are based on butadiene-acrylonitrile and are often functionalized for co-reacting with the thermoset matrix (Ref 13, 14). Thermoplastic materials have also found significant use as modifiers for thermosetting-matrix resins. Engineering thermoplastics that commonly are used in high-performance prepreg matrices include polyethersulfone and polyetherimide (Ref 15 to 17). Some other thermoplastics that have been used for modifying epoxy matrices include polyvinyl formal, polyvinyl butyral, and polyether-blockamides (Ref 18). As with the rubber modifiers, phase separation is a complex function of the matrix formulation and processing conditions. The dispersed phases for these particles generally range from 0.3 to 10 μm and may be spherical or have an irregular shape. In some cases, the modifier may be designed to remain in the continuous phase of the resin as opposed to phase separation.

There are essentially two methods used to generate a multiphase morphology within a thermosetting resin, with a third method leading to true nanostructured phase separation. The first method involves the formation of an initially soluble solution of the modifier and uncured thermoset in which phase separation of the modifier occurs upon cure, forming distinct domains (Ref 19). The second technique involves the addition or forma-

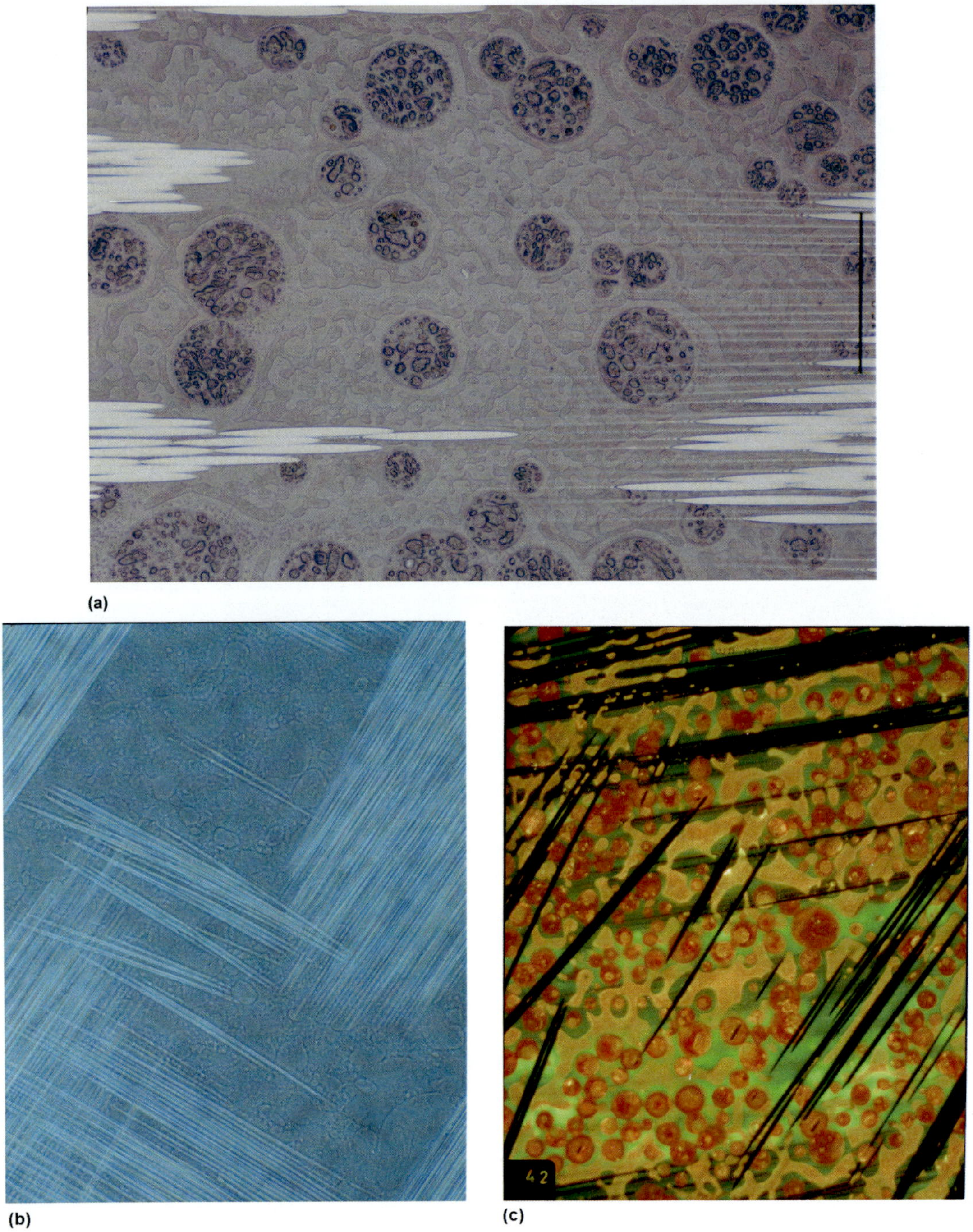

Fig. 1.10 Multiphase thermosetting-matrix composites. (a) Carbon fiber composite cross section. Etched and viewed using reflected-light phase contrast, 25× objective. (b) Glass fiber composite ultrathin section. Transmitted light, 20× objective. (c) Carbon fiber composite cross section having a multiphase-matrix structure. Rhodamine B dyed. Epi-fluorescence, 390–440 nm excitation, 25× objective

tion of preformed particles in the uncured resin that are initially distinct particles and remain so after cure (Ref 20). These materials, often core-shell modifiers, have a matrix-compatible shell with a rubber or thermoplastic core. Common particle sizes range from 50 nm to 3 μm. Dispersed nanostructured phases in thermosetting resins have been developed using block copolymers in which phase sizes less than 100 nm have been found in many resin systems (Ref 21 to 23). Due to the resolution of optical microscopy, these nanomorphologies cannot be observed, because these phases are smaller than the wavelength of light. Therefore, nanometer-scale phases or inclusions are best viewed using a technique such as transmission electron microscopy.

Interlayer-Toughened Composites

The need to improve high-performance thermosetting composite toughness further than is possible through chemistry has resulted in an engineering approach to toughen composite materials (Fig. 1.11a, b). Without improving the toughness of the resin itself, the addition of a modified interlayer can impose large increases in final composite damage tolerance (Ref 24). This modified interlayer reduces the interply delamination, which has been found to be the life-limiting factor in most composites (Ref 25, 26). To date, the most efficient and highest damage-tolerant composites arise from the use of interlayer-toughening concepts. Interlayer toughening comprises the placement of a thin toughened resin between each ply of the composite structure. The first interlayer-modified composites consisted of two resins of different composition, with the interlayer resin being much tougher than the resin used in the intraply region (3M patent) (Ref 27). Materials frequently used in the interlayer were thin thermoplastic sheets. However, prepreg handling problems, including tack and drape as well as hot-wet performance, limited the use of these types of composite systems. As a result, the present state of the art-toughened resins that were developed for the interlayer consisted of the same or similar base resin but were toughened using preformed thermoplastic or rubber particles (Fig. 1.12) (Ref 20, 28 to 30). Particles used in these applications have ranged in size from 1 to 100 μm but usually have averaged between 20 and 50 μm, so that the particles remain in the composite interlayer and do not penetrate the fiber tows (Ref 20). The most common method of interlayer toughening is through adding preformed particle-modified resin on the surface of the prepreg, which, when consolidated and cured, results in an interlayer-toughened composite. However, this same method of toughening has recently been extended to infusion-type processing, wherein the fabric plies are sprayed or coated with a tackifier containing preformed particles (Ref 31). When the resin is infused, the preformed particles that were placed on the fabric surfaces are "locked in" and produce an interlayer-modified composite.

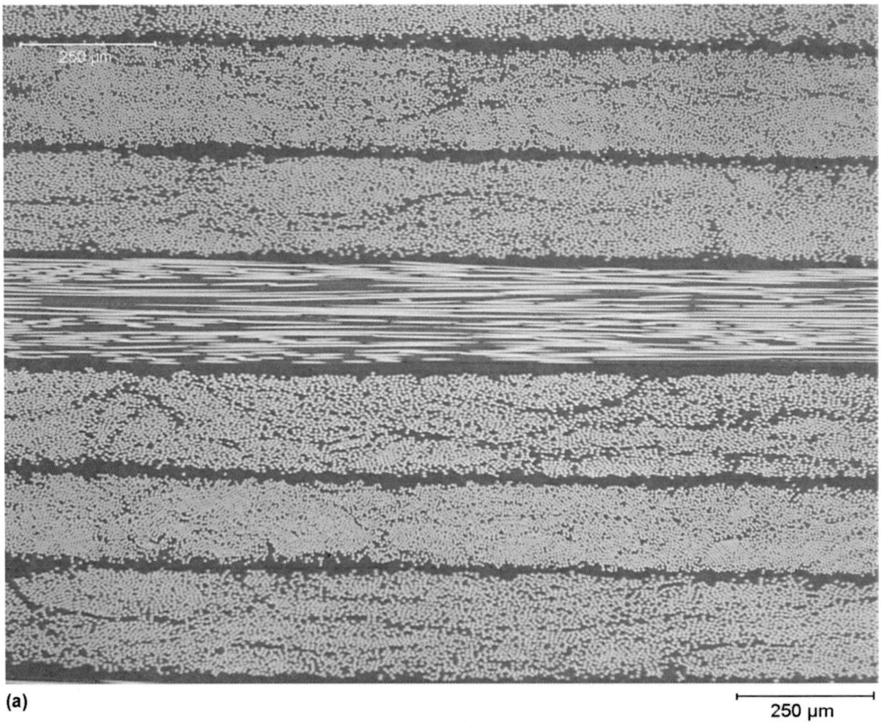

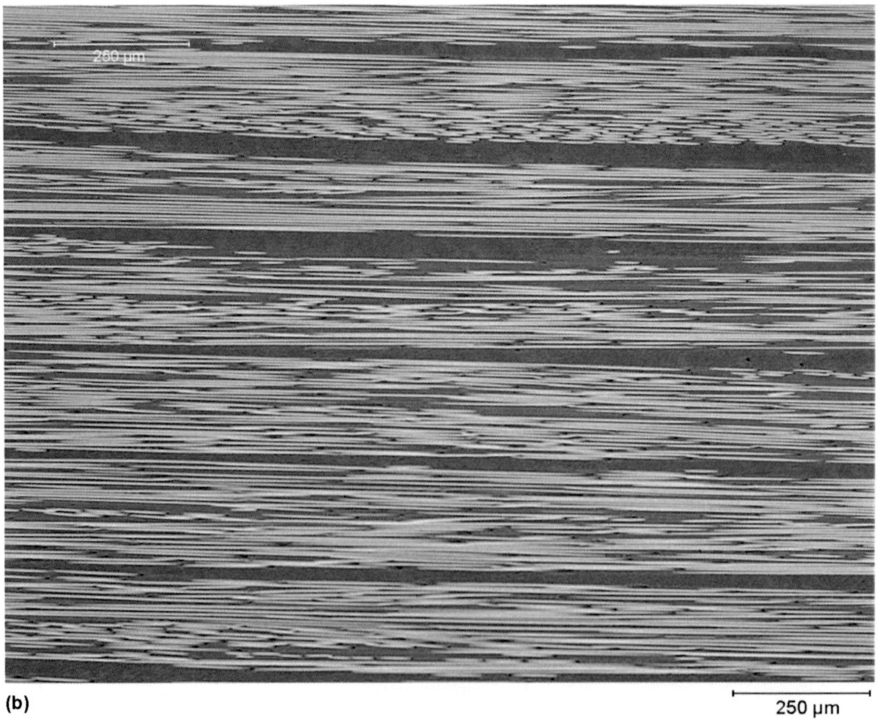

Fig. 1.11 Cross sections of interlayer-modified composite materials. (a) Cross section showing a middle ply at 90°. Bright-field illumination, 10× objective. (b) Cross section taken parallel to the fiber direction. Bright-field illumination, 10× objective

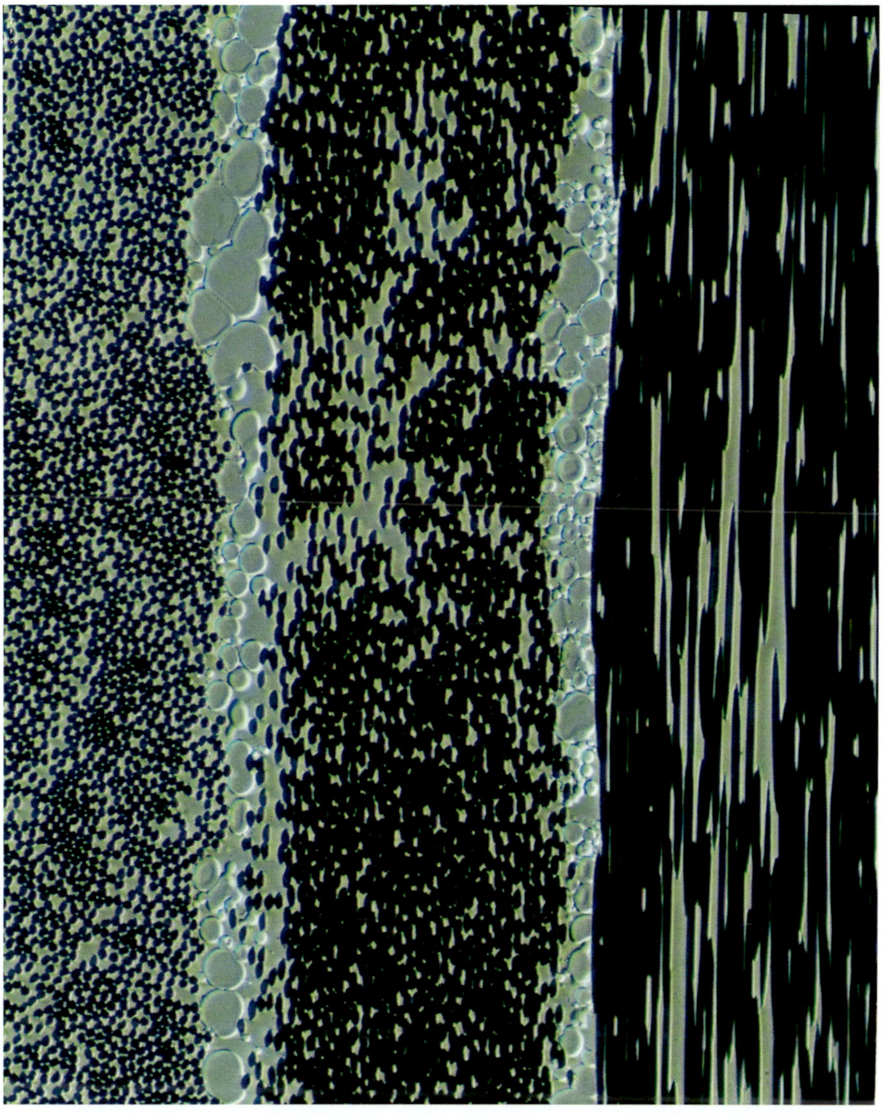

Fig. 1.12 Ultrathin section of a particle-modified interlayer-toughened composite material. Transmitted-light Hoffman modulation contrast, 20× objective

Honeycomb/Foam Structure Composite Materials

To further reduce the already lightweight composite materials and increase stiffness, structures have been developed whereby the composite layers are separated by a lightweight material, making a sandwich structure. Typical materials used for the sandwich materials are honeycomb core or foam materials. Honeycomb core materials are usually based on Nomex (E.I. du Pont de Nemours and Company), glass fiber, aluminum, or carbon, depending on the application (Fig. 1.13a, b). The high strength-

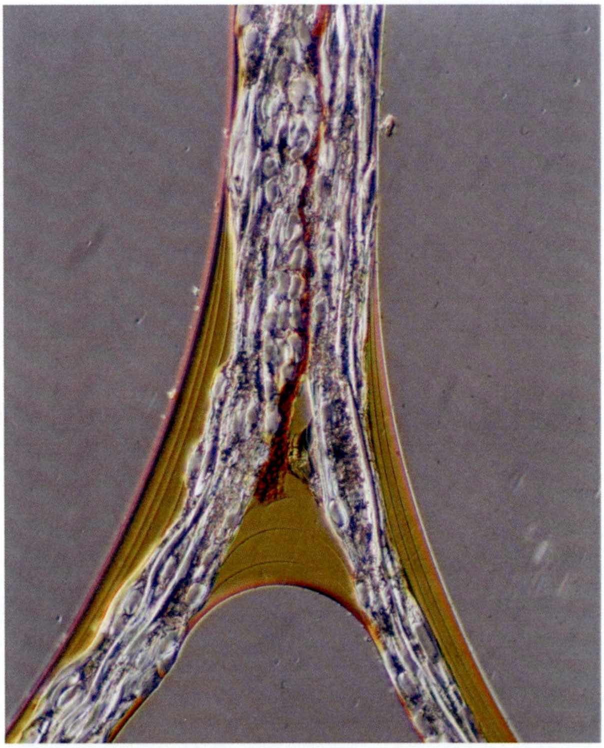

(a)

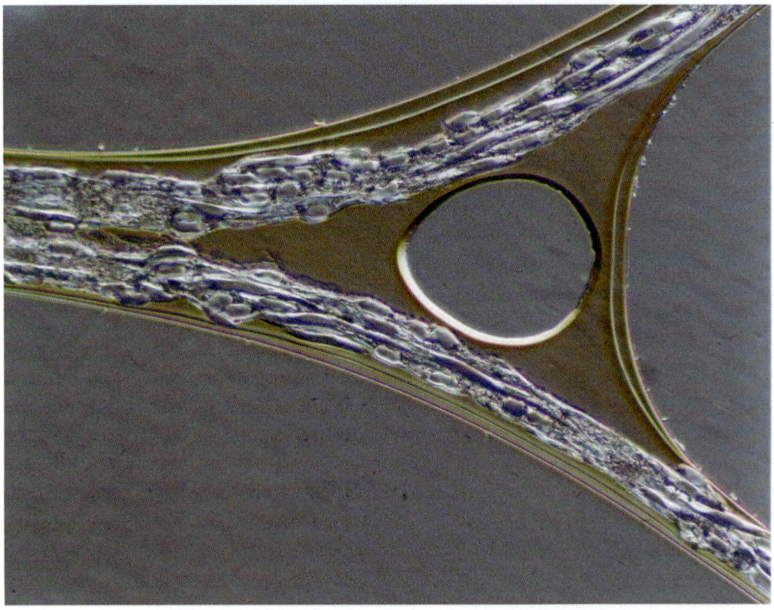

(b)

Fig. 1.13 Cross sections of honeycomb node areas showing the number of phenolic resin dip coats. (a) Transmitted light, 100× objective. (b) Void in the node area. Transmitted-light phase contrast, 100× objective

to-weight ratios of these structures are especially attractive to the airplane industry, where stiff, lightweight parts are required. In the airplane industry, composite skin honeycomb structures are used for such parts as radomes, fairings, nacelles, internal flooring, storage compartments, and elevators (Ref 32, 33). In addition to aerospace applications, many other composite structures are developed using honeycomb materials, which are found in boats, sporting goods, and infrastructure applications. Thermosetting resins are usually used as the prepreg matrices for composite honeycomb structures because of the necessity of the appropriate tack-and-drape characteristics. The fiber reinforcement usually consists of unidirectional and/or woven fabric developed from glass, Kevlar (E.I. du Pont de Nemours and Company), or carbon fibers. Prepregs with different reinforcing fibers may be used together to make hybrid parts that take advantage of the different mechanical properties supplied by different fibers. If the prepreg material does not satisfy the skin-to-core adhesive characteristics, a film adhesive must be used in combination with the prepreg, adding weight to the overall structure (Fig. 1.14).

Although not generally considered as high performance as honeycomb materials, foam (Fig. 1.15a, b) and wood cores (Fig. 1.16) have also found extensive use in the composites industry for making sandwich structures. These materials have the distinct advantage of not having large open cells and therefore can be used effectively with infusion methods or wet lay-up processing methods. The majority of foam- and balsa-cored composites are found in the marine, sporting goods, and infrastructure industries.

Optical Microscopy of Composite Materials

Optical microscopy is a valuable tool in materials investigations related to problem solving, failure analysis, advanced materials development, and quality control. Microscopy has been used for many decades to provide insight into the micro- and macrostructure of fiber-reinforced composites.

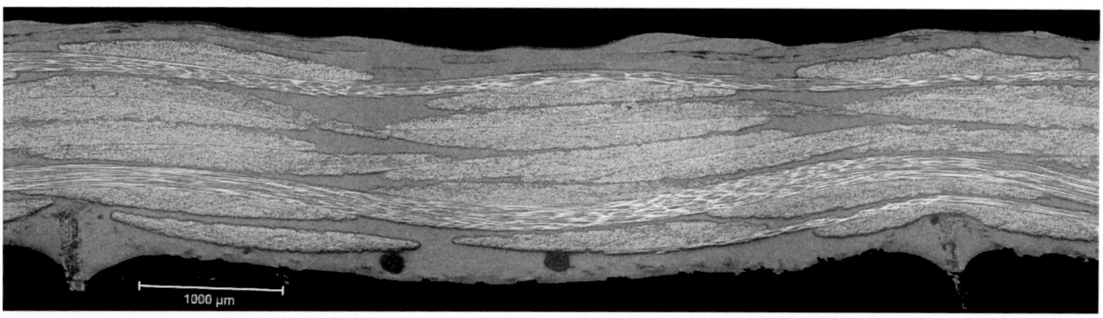

Fig. 1.14 Cross section of a carbon fiber prepreg skin-film adhesive co-cured honeycomb composite showing two fillet regions. A few voids are shown in the adhesive areas. Bright-field illumination montage, 5× objective

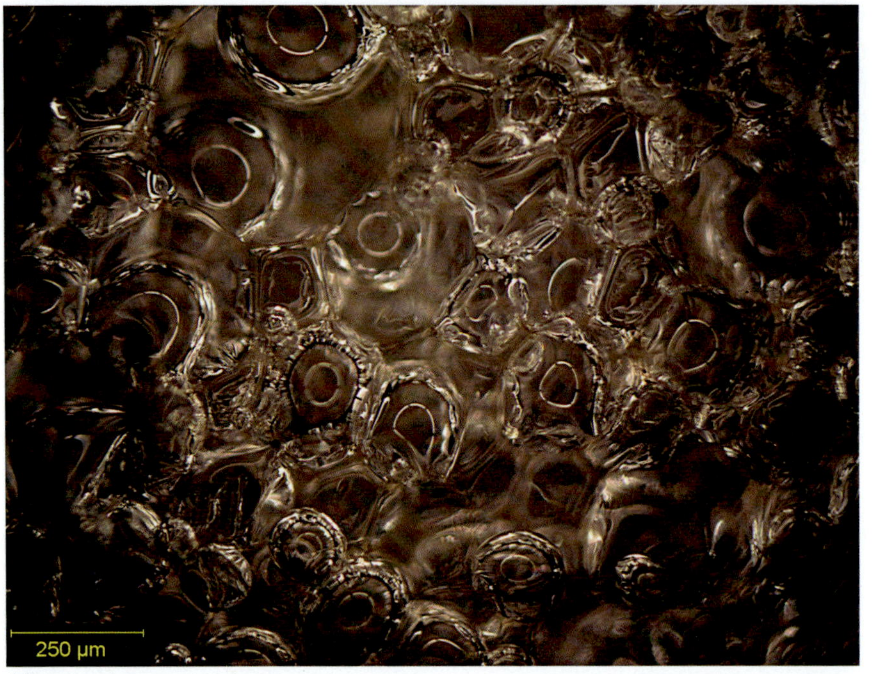

(a)

(b)

Fig. 1.15 Unprepared cross sections of structural foams. (a) Dark-field illumination, 10× objective. (b) Bright-field illumination, 65 mm macrophotograph

Fig. 1.16 Ultrathin section showing the microstructure of spruce wood. Transmitted-light phase contrast, 40× objective

The most widespread use of optical microscopy for composites is determining void content, ply counts, and fiber orientations. While this makes up the majority of analysis, the investigation of failure mechanisms and microstructural analysis is also common. Furthermore, insight into fiber morphology, matrix modifiers, fillers, and the effect of processing parameters used for manufacturing composite materials are also elucidated using optical microscopy techniques.

For the majority of cases, reflected-light microscopy provides most of the necessary information one would desire. In some cases, however, etchants, stains, or dyes may be required for further clarification of the morphology or crack identification. If reflected techniques do not yield the required information, transmitted-light optical microscopy can provide insights into the microstructures of these materials that would otherwise remain hidden when using standard bulk metallographic preparation techniques and reflected illumination. Because many thermoset materials are inert to metallographic etchants, often the sample is best observed with transmitted polarized light and various contrast media to enhance the differences in refractive index of discrete phases in the composite.

Although an array of different types of composite materials is in use today, the utilization of both thermosetting- and thermoplastic (polymer)-

matrix fiber-reinforced composites continues to dominate the field in terms of both volume and applications. This is due to easy processing, a wide range of materials and properties, and a much lower cost than other composite materials, such as metal-matrix composites and ceramic-matrix composites. While not to discount the limited use of other types of composites, all of these heterogeneous and anisotropic materials have unique properties and characteristics, lending their use in specific applications.

Over the years, many methods for polishing composite materials have been used, most with varying success and quality. This continues today and is frustrating when there is no standard to reference. Throughout this book, the easiest, most cost-effective, and reproducible techniques the authors have found for sample preparation, polishing, and analysis are described. The most common types of polymeric-matrix fiber-reinforced composite materials are the primary focus of the book, but also included is how to prepare more exotic composite materials and hybrids. Most of the materials that are described in this book are continuous thermosetting-matrix fiber-reinforced composites, but all techniques that are explained herein are applicable to short, discontinuous fiber composites and thermoplastic-matrix materials.

REFERENCES

1. J.C. Halpin, *The Role of the Polymeric Matrix in the Processing and Structural Properties of Composite Materials,* L. Nicolais and J.C. Seferis, Ed., Plenum Press, New York, 1983
2. F. Rodriguez, *Principles of Polymer Systems,* Hemisphere Publishing Co., New York, 1989
3. F.W. Billmeyer, *Textbook of Polymer Science,* 2nd ed., John Wiley & Sons, Inc., New York, 1971
4. L. McKague, Thermoplastic Resins, *Composites,* Vol 21, *ASM Handbook,* ASM International, 2001
5. W.J. Lee, J.C. Seferis, and D.C. Bonner, Prepreg Processing Science, *SAMPE Q.,* Vol 17, 1986, p 58
6. K.J. Ahn and J.C. Seferis, Prepreg Processing Science and Engineering, *Polym. Eng. Sci.,* Vol 33, 1993, p 1177
7. K.J. Ahn, J.C. Seferis, T. Pelton, and M. Wilhelm, Analysis and Characterization of Prepreg Tack, *Polym. Compos.,* Vol 13, 1992, p 197
8. J.L. **Kardos,** M.P. Dudukovic, and R. Dave, Void Growth and Resin Transport during Processing of Thermosetting—Matrix Composites, *Epoxy Resins and Composites IV, Advances in Polymer Science,* Vol 80, Springer Berlin/Heidelberg, 1980, p 102–122
9. B.S. Hayes, J.C. Seferis, and R.R. Edwards, Self-Adhesive Honeycomb Prepreg Systems for Secondary Structural Applications, *Polym. Compos.,* Vol 19 (No. 1), Feb 1998, p 54–64

10. R. Dave, J.L. Kardos, S.J. Choi, and M.P. Dudukovic, Autoclave vs. Non-Autoclave Composite Processing, *32nd International SAMPE Symposium,* April 6–9, 1987, p 325–337
11. R.Y. Ting and R.J. Moulton, Fracture Properties of Elastomer Toughened Epoxies, *12th National SAMPE Technical Conference,* 1980, p 265
12. M. Akay and J.G. Cracknell, Epoxy Resin-Polyethersulphone Blends, *J. Appl. Polym. Sci.,* Vol 52, 1994, p 663–685
13. W.D. Bascom, R.J. Moulton, E.H. Rowe, and A.R. Siebert, The Fracture Behavior of an Epoxy Polymer Having Bimodal Distribution of Elastomeric Inclusions, *Org. Coat. Plast. Preprints,* Vol 39, 1978, p 164
14. D. Verchere, H. Sautereau, J.-P. Pascault, C.C. Moschiar, S.M. Richardi, and R.J.J. Williams, Rubber-Modified Epoxies: Analysis of the Phase-Separation Process, *Toughened Plastics I,* C. K. Riew and A. J. Kinloch, Ed., American Chemical Society, Washington, D.C., 1993, p 335–363
15. C.B. Bucknell and I.K. Partridge, Addition of Polyethersulphone to Epoxy Resins, *Br. Polym. J.,* Vol 15, 1983, p 71
16. Z. Zhang, J. Cui, S. Li, K. Sun, and W. Fan, Effect of Hydroxyl-Terminated Polyethersulfone on the Phase Separation of Polyetherimide-Modified Epoxy Resin, *Macromol. Chem. Phys.,* Vol 202, 2001, p 126–132
17. B.S. Hayes and J.C. Seferis, Variable Temperature Cure Polyetherimide Epoxy-Based Prepreg Systems, *Polym. Eng. Sci.,* Vol 38 (No. 2), 1998, p 357–370
18. M.A. Boyle, C.J. Martin, and J.D. Neuner, Epoxy Resins, *Composites,* Vol 21, *ASM Handbook,* ASM International, 2001, p 78–89
19. L.T. Manzione, J.K. Gillham, and C.A. McPherson, Rubber Modified Epoxies, Part II: Morphology and Mechanical Properties, *J. Appl. Polym. Sci.,* Vol 26, 1981, p 907
20. B.S. Hayes and J.C. Seferis, Modification of Thermosetting Resins and Composites through Preformed Polymer Particles: A Review, *Polym. Compos.,* Vol 22 (No. 4), 2001, p 451–467
21. E. Girard-Reydet, J.-P. Pascault, A. Bonnet, F. Court, and L. Leibler, A New Class of Epoxy Thermosets, *Macromolecular Symposium*, Vol 198, 2003, p 309–322
22. S. He et al., Studies of the Properties of a Thermosetting Epoxy Modified with Block Copolymers, *Polym. Int.,* Vol 54, 2005, p 1543–1548
23. A. Bonnet et al., Commercial Nano Impact Modifiers for Epoxy Thermosets, *Fifth World Congress—Nanocomposites,* Aug 22–24, 2005 (San Francisco), p 1–10
24. S. Singh and I.K. Partridge, Mixed-Mode Fracture in an Interleaved

Carbon-Fibre/Epoxy Composite, *Compos. Sci. Technol.,* Vol 55, 1995, p 319

25. K.R. Hirschbuehler, An Improved 270 F Performance Interleaf System Having Extremely High Impact Resistance, *SAMPE Q.,* Vol 17, 1985, p 46
26. S.S. Wang, Fracture Mechanics for Delamination Problems in Composite Materials, *J. Compos. Mater.,* Vol 17, 1983, p 211
27. R.A. Frigstad, U.S. Patent 3,472,730, Minnesota Mining and Manufacturing Company, 1969
28. N. Odagiri et al., T800H/3900-2 Toughened Epoxy Prepreg System: Toughening Concept and Mechanism, *Proceedings of the American Society of Composites—Sixth Technical Conference,* 1991, p 43
29. M.A. Hoisington and J.C. Seferis, Process-Structures-Property Relationships for Layered Structured Composites, *Proceedings of the American Society of Composites—Sixth Technical Conference,* 1991, p 53–62
30. M.R. Groleau, Y.-B. Shi, A.F. Yee, J.L. Bertram, H.J. Sue, and P.C. Yang, Mode II Fracture of Composites Interlayered with Nylon Particles, *Compos. Sci. Technol.,* Vol 56, 1996, p 1223–1240
31. R.W. Hillermeier et al., Interlayer Toughening of Resin Transfer Molded Composites, *Compos. Part A: Appl. Sci. Manuf.,* Vol 32, 2001, p 721
32. G.E. Woodley, Light Aircraft Structural Design in Non-Metallics—Use of Composite Honeycomb for Light Aircraft, *Aeronaut. J.,* Vol 85, 1981, p 332
33. J.L. Corden and T.N. Bitzer, Honeycomb Materials and Application, *32nd International SAMPE Symposium,* April 6–9, 1987, p 68

CHAPTER 2

Sample Preparation and Mounting

SPECIMEN PREPARATION is the first step that determines the quality of the microstructural information that can be obtained using optical microscopy. Without high-quality preparation techniques, much of the information that is desired from a sample will be lost. The sample preparation methods that are described here are applicable to most types of composite materials containing short discontinuous or continuous fibers (Ref 1). These composites may have one or multiple different fiber types, including carbon, glass, Kevlar (E.I. du Pont de Nemours and Company), or polymer fibers. The matrix composition can also vary widely in morphology, hardness, and mechanical properties. There are, however, a few cases that require special mounting techniques combined with nonstandard grinding and polishing steps. The special mounting, grinding, and polishing techniques are discussed in detail in Chapter 4, "Special Sample Preparation and Polishing," in this book.

The first question to ask when analyzing a composite material with optical microscopy is what type of information is desired. This question determines how the composite should be sectioned, documented, and labeled. In most analyses, the composite is sectioned through the *z*-axis and viewed normal to the *zx* or *zy* plane or an angle in between. In other cases, the composite must be viewed normal to the *xy* plane or at an angle through the thickness (Fig. 2.1). Typically, there are many more analyses performed normal to the *zx* and *zy* planes of composites, but this sometimes does not enable the complete understanding of the morphology. Some of the features that may require analyses in the *xy* plane are microcracks, voids, dispersed phases, and crystallinity. Whatever the case, the sample may have to be viewed normal to all three planes and angles in between. Commonly, this is not understood until the composite has been through the final polishing stage. Therefore, understanding what information is re-

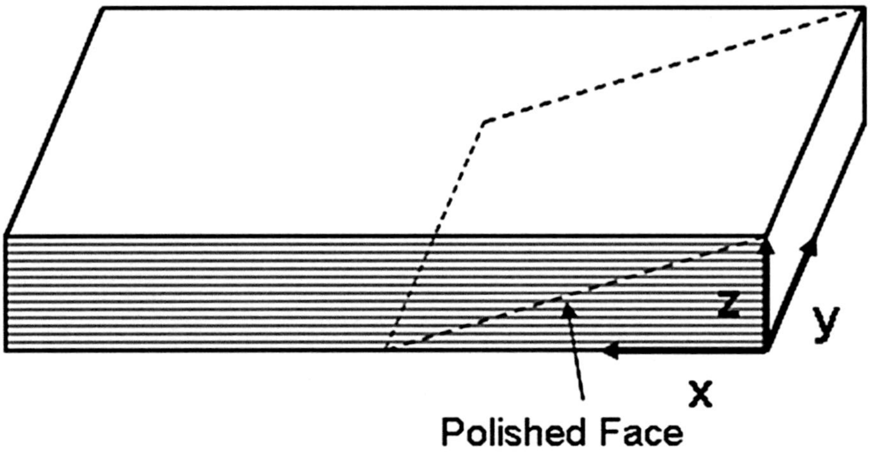

Fig. 2.1 Coordinates defined for composite material sample preparation as related to sectioning and viewing planes. Sectioning through the composite thickness on an angle helps in determining ply orientations (i.e., fibers will become elongated).

quired in the specimens before sectioning the composite often makes the entire process easier and faster and the analysis and information more complete.

Documentation and Labeling

After it is known what type of information is necessary to obtain from a composite, the part or sample must be systematically and meticulously documented and labeled before sectioning. The labeling must be easily identified and able to be carried intact through to the final analysis. Problems in identification of the labels occur after subsequent potting of the samples or grinding off the labels. Therefore, the best area to label may be the side opposite the polishing face. An easy method that can be used to document the sections is to take a digital photograph of the composite material and then organize the sections. Figure 2.2 shows a composite part that was damaged and removed from an aircraft for analysis. After the documentation, the labeling of composite materials is best performed using thin felt-tip permanent markers. It is best to use markers that show good contrast on the part. If the part is dark in color or made from carbon fibers, a silver ink permanent marker works very well (Fig. 2.3). Likewise, if the composite contains glass or Kevlar fibers, a black permanent-ink marker will work very well.

Sectioning the Composite

After labeling the samples, the composite part usually must be sectioned so that it is manageable for polishing. Some parts are small enough that no

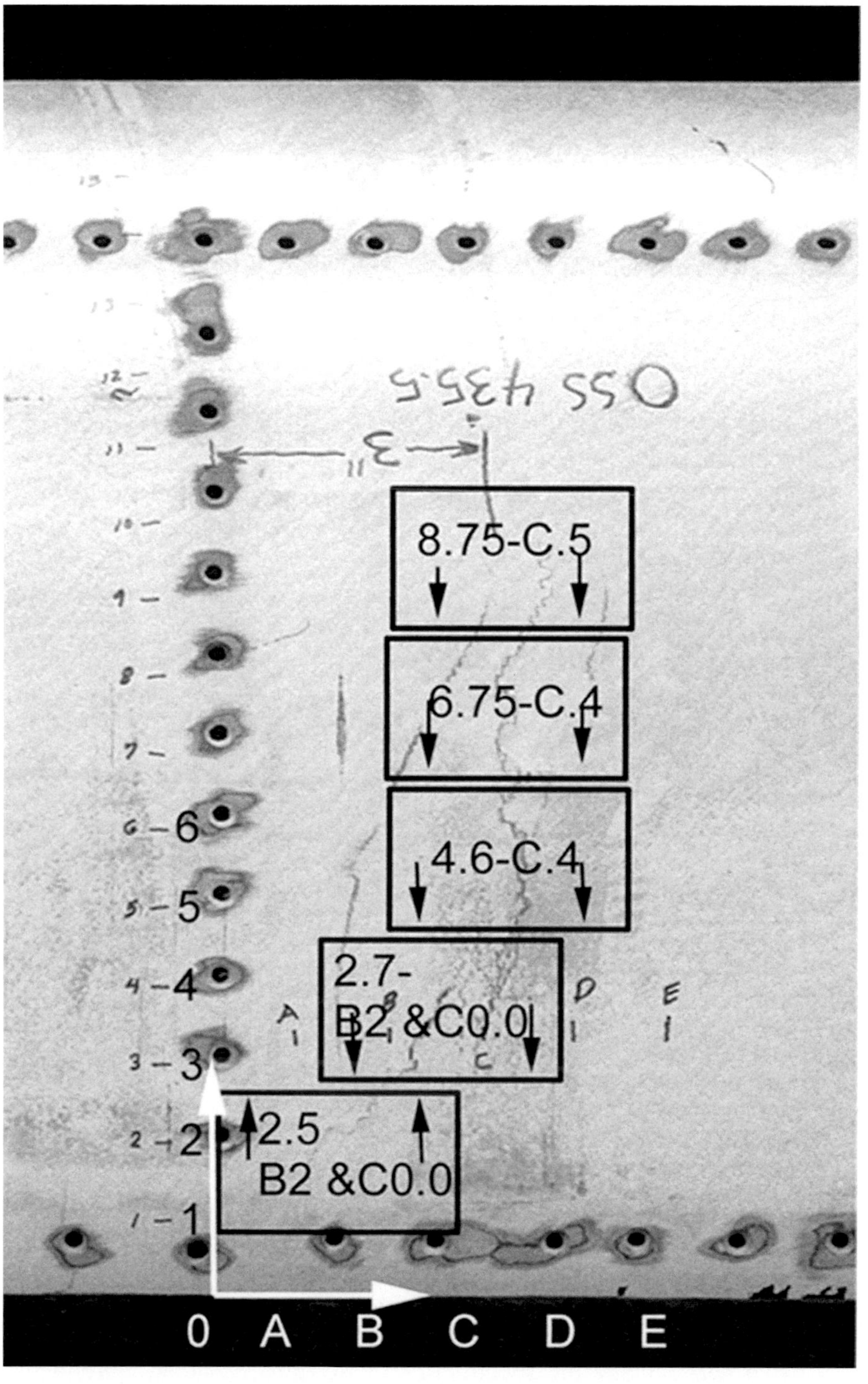

Fig. 2.2 Glass fiber honeycomb composite part submitted for failure analysis. The coordinates were established with a tape measure and a felt-tip permanent-ink marker. The starting point is the lower left corner, numbering "1" to "15" vertically (next to holes) and "A" through "E" horizontally. The 5.08 × 7.62 cm (2 by 3 in.) specimens were traced and labeled with a felt-tip permanent-ink marker. Labeling reflects the projected polishing plane and crack locations.

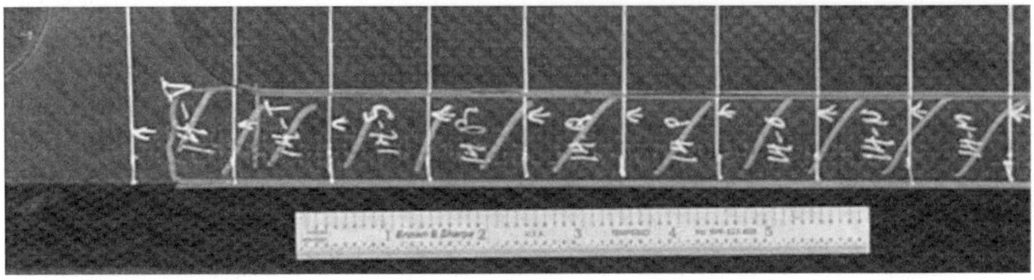

Fig. 2.3 Carbon fiber composite laminate labeled for sectioning using a silver ink felt-tip permanent marker. This sample with the corresponding section map was originally sent for nondestructive inspection.

sectioning is required and therefore are ready for grinding and polishing. It is important in the sectioning stage that no damage is done to the specimen, so that artifacts are not created. Therefore, when sectioning the part, the use of a band saw with a tooth blade is never recommended, because it may significantly damage the specimen. Instead, an abrasive band saw is best used to cut large parts down to a manageable size. Even with an abrasive band saw, some damage may be introduced to the edge of the sample (Fig. 2.4a, b). Rough cutting may be performed without lubrication or cutting fluids; however, a vacuum apparatus should be used to minimize dust. A waterjet, if available, provides the cleanest cut and the least amount of damage to the specimen (Fig. 2.5). For final composite specimen cutting, an abrasive cut-off saw with coolant should be used to minimize damage to the specimen. The damage introduced by this type of saw blade is usually less than 2500 μm (Fig. 2.6). The use of a coolant is necessary, because increases in temperature during cutting can alter the microstructure of polymer matrices. In general, thin blades reduce damage and material loss; however, they may bend, resulting in a nonplanar surface. Additionally, thin blades are more likely to break during cutting. This can be quite costly. When using thin blades, it is necessary that a coolant and lubricant are used, or the blade may bind and become damaged. Common lubricants/coolants that can be used are kerosene and emulsified oils, but newer commercial liquid lubricants are also available. Dressing the blades often may be required for the most efficient cutting and to extend blade life. Dressing removes built-up material and dull abrasive, creating a new, sharp cutting surface.

In some cases, if the sample is small enough or after rough cutting, the sample can be ground down to the required size with 80- or 120-grit silicon carbide (SiC) paper. It is important to understand that the use of 80-grit paper can damage the specimen, so care must be taken when using this for grinding. To use this technique, the ground edge of the sample must be stout enough that it is not destroyed or distorted during the grinding process. For thin materials, the specimen must be mounted first to resist deformation and then ground down to size. This grinding operation

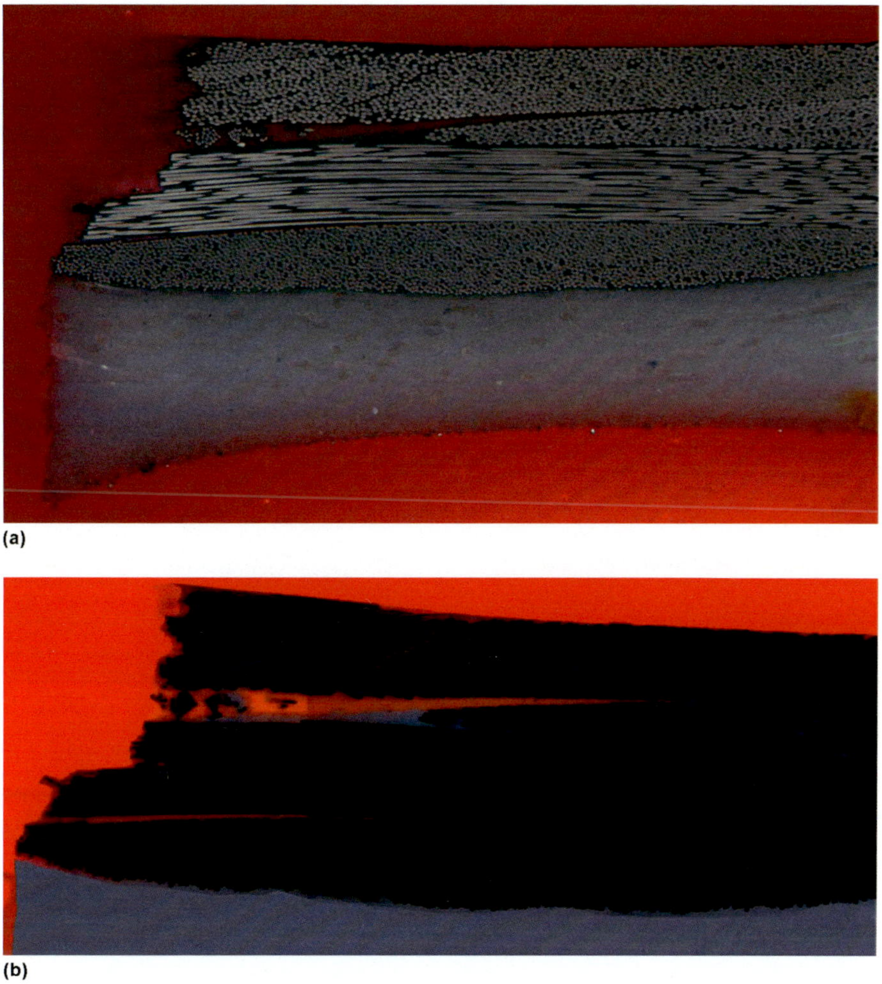

Fig. 2.4 Micrographs showing damage from cutting a carbon fiber composite part with an abrasive band saw. (a) A surfacing film can be observed on the surface of the carbon fiber composite. Polarized light, 25× objective. (b) Same view but using epi-fluorescence, 390–440 nm excitation, 25× objective

should also be done with ample cooling fluid so that the temperature of the sample is not increased.

Before mounting the specimens, it is necessary to thoroughly clean the samples to remove any particulates from the cutting or sectioning process. Also, at this point, any greases or oils (e.g., kerosene from cutting) that may have come in contact with the specimens must be removed. If not adequately cleaned, these materials can contaminate the grinding or polishing surfaces or reduce the efficiency of the process. Furthermore, if the sample requires mounting using a casting resin, a clean sample is necessary to enable optimal adhesion to the casting resin. The best method for cleaning the specimens is to use an ultrasonic bath. If this is not available,

Fig. 2.5 Composite material that was cut using a waterjet. Very little damage is observed at the cut edge of the specimen. A fluorescing dye was applied to the cut edge to determine if cracks were present. Epi-fluorescence, 390–440 nm excitation, 25× objective

Fig. 2.6 Cut edge of a composite material after sectioning with an abrasive cut-off saw. The composite was mounted using a Rhodamine-B-dyed epoxy resin and viewed using epi-fluorescence, 390–440 nm excitation, 25× objective.

the use of a fine brush and soap and water will work very well for cleaning the specimen.

Clamp-Mounting Composite Samples—Automated Polishing Heads

In all cases, "the best mount is no mount." This means that the specimen is prepared and polished without the use of a permanent mounting resin or material (Ref 2). Automated polishers have holders that allow easy mounting of the specimens by mechanical clamping or grips. The benefit is that after polishing, the specimens may be removed and directly viewed. Likewise, the original sample identifications are easier to discern. An automated polishing head having three 35 by 75 mm (1.38 by 2.95 in.) rectangular openings works very well for both laminate and honeycomb composite materials (Fig. 2.7). With this type of mount, it is necessary to use scrap backing pieces to protect the sample edges. After polishing, these scrap pieces can be easily removed. For carbon-fiber-reinforced polymeric (CFRP) composite samples, CFRP backup pieces are recommended, because the material-removal rates will be well matched. The same is true with other composite materials: backing pieces of similar materials and hardness provide superior results. The placement of the sam-

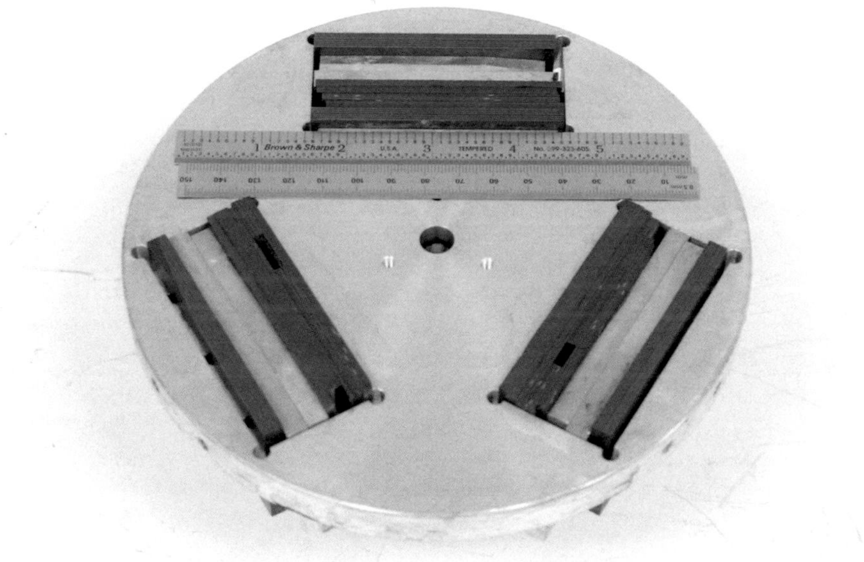

Fig. 2.7 Automated grinding/polishing head showing composite specimens mounted in the three rectangular openings. After the specimens are cut to shape, they can be clamped in the openings of the head for polishing and then easily removed for analysis. All composite samples to be polished are located near the center (hub) of the polishing head. This is done to balance the head.

ples in the rectangular openings is best located closest to the center of the head, with the backing pieces on the outside of the samples (Fig. 2.8). Using this mounting method, specimens for void analysis, ply orientation, and ply-count analysis may be prepared quickly, with excellent results.

Mounting Composite Samples in Casting Resins

Fragile materials, multiple samples, samples with surface features, and thin samples require support to ensure a flat planar surface. Also, in many cases, the area that must be analyzed requires the use of a resin to impregnate and preserve the features of the damaged area. In these cases, polishing unmounted specimens may cause fiber pullout or further damage and create artifacts. Some of these types of specimens include areas that have been impacted or contain fracture areas. In these specimens and others, the use of a contrast dye may also be incorporated into the mounting resin so that there is greater contrast between the composite matrix and the mounting medium. For all samples that require mounting, a mounting medium is necessary, such as epoxy, polyester, or acrylic resin. Fiber-reinforced composites are usually mounted with a casting resin in contrast to a powdered compression-mounting resin. Of these resins, an epoxy casting resin is preferred because it has the least shrinkage during cure and excellent adhesion. With the many options for epoxy casting resins, products with a higher hardness usually provide better results. A high hardness

Fig. 2.8 Filler composite pieces are used on the outer edges of the samples to aid clamping. A filler block of a different material hardness is placed below the sample surface so it will not contact the grinding and polishing surfaces.

will correlate with most high-performance composite matrices and provide excellent edge retention. Room-temperature curing resins are often ideal; however, they frequently need to be cured for at least 12 hours or longer at 20 °C (68 °F). It is not advised to oven cure the mounted specimens at elevated temperature. Although this increases the cure speed, it also leads to greater cure shrinkage of the mounting resin. To ensure complete impregnation of the specimens, the mold can be placed in a vacuum chamber while infusing the resin (Fig. 2.9). This is essentially a bell jar with a vacuum and resin port and a pinch valve. The use of moderate vacuum to remove entrapped air, followed by applying pressure during cure, is highly effective for sample preparation and creating good adhesion (Fig. 2.10). It must be mentioned that the greatest recurrent cost in sample preparation is the mounting resin.

Fig. 2.9 Vacuum infusion apparatus used for removing entrapped air in the mounting resin before it is added to the mold. In this apparatus, vacuum is maintained in the chamber. The mold is placed under the chamber, and a tube for resin transport is placed above the sample that connects to the container with the mixed mounting resin. Atmospheric pressure pushes the resin into the tube, and it is dripped onto the specimen in the mold. When the mold is filled, it is removed and cured under pressure.

Addition of Contrast Dyes to Casting Resins

In some analyses, there is need to distinguish the specimen matrix from the mounting polymer (Ref 3). This is best accomplished by using contrast dyes added to the mounting resin. Some of the dyes that can be added to the mounting resin to increase contrast are shown in Table 2.1, but this is by no means complete. In addition to these materials, commercial suppliers of consumables have their own set of dyes that can be used for this purpose. Caution: Read the material safety data sheets to handle appropriately. In Table 2.1, the dye that finds the most use is Rhodamine B, due to its function and cost. This dye will luminesce under polarized light as well as fluoresce in the appropriate optical wavelength range. In contrast, Rhodamine G6 does not luminesce but will fluoresce under the appropriate wavelength range but is not as efficient. In some composite samples, the use of multiple dyes having different colors may be beneficial to provide contrast. An example of a situation where this may be of use is when mi-

Fig. 2.10 Pressure chamber that is used to ensure good wet-out and reduce the size of remaining entrapped air, providing high-quality mounted specimens

Table 2.1 Contrast dyes for addition to epoxy mounting resins*

Rhodamine B: Basic Violet 10; dye/laser dye
Rhodamine G6: Basic Red 1; dye/laser dye
Coumarin 35: 7-(diethylamino)-4-(trifluoromethyl) coumarin; laser dye
Coumarin 151: 7 amino-4-(trifluoromethyl) coumarin; laser dye

*Caution: These dyes are hazardous. Read material safety data sheets before use!

crocracks that emanate from the surface must be isolated and contrasted with damage that occurred from the interior of the sample. In this case, Rhodamine B can be used with Coumarin 35 (Fig. 2.11). The use of this combination of dyes will provide significant differences in contrast, with Rhodamine B fluorescing red and Coumarin 35 fluorescing blue-green. Coumarin 35 should not be used as the primary dye and should only be used to provide contrast with another dye, such as Rhodamine B, due to its high cost. Coumarin 151 can also be used in place of Coumarin 35, but it is more visually transparent. Most of these dyes will fluoresce over a large range of wavelengths, but most polymeric matrices will only fluoresce in the shorter wavelengths, lower than 450 nm.

To use the dyes listed in Table 2.1, they first must be dissolved and mixed into the mounting resin, so that the dye will fluoresce after the specimen mount is prepared. If the dyes are not completely dissolved and any particles remain, the color will bleed out of the sample during polishing. This not only creates a mess but also is hazardous. The best Rhodamine B dye concentration is prepared by dissolving 0.7 g of Rhodamine B dye in 7 mL of methanol per 100 g of the epoxy mounting resin (part A). This does not include the hardener. The methanol and dye are premixed and then mixed with the part A component (epoxy) prior to mixing with part B (curing agent) of the mounting resin. After these steps, the sample is vacuum impregnated, followed by curing under pressure with at least 275 kPa

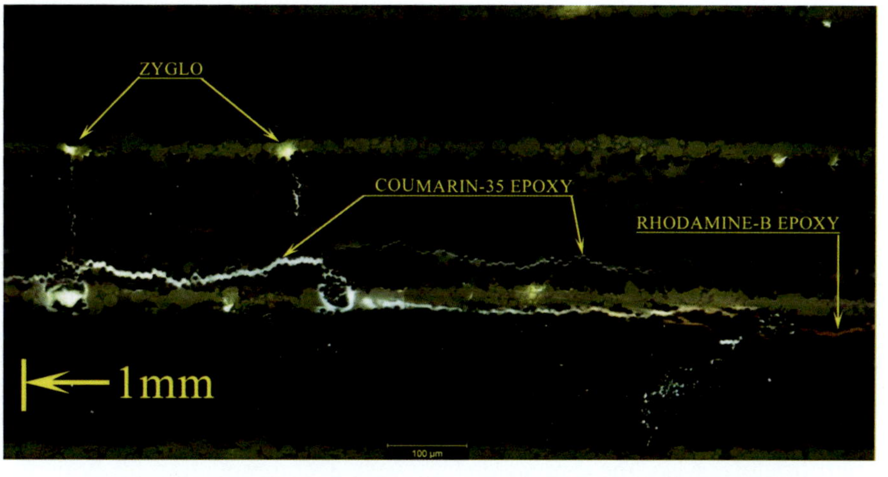

Fig. 2.11 Composite material that was subjected to a laboratory-induced lightning strike. The section shown is 1 mm (0.04 in.) away from the center of the strike. This sample was first impregnated with Rhodamine-B-dyed epoxy casting resin and then, after sectioning, mounted with Coumarin 35. Microcracks that were not impregnated by the Rhodamine-B- or Coumarin-35-dyed casting resins were dyed using a very low-viscosity penetration fluorescing dye (Magnaflux Zyglo, Magnaflux Corp.) in solvent. The fluorescing penetration dye was applied to the surface after polishing, allowed to dry, and then subjected to the last stage of polishing to remove additional dye from the surface. Epi-fluorescence, 390–440 nm excitation, 50× objective

(40 psi). The same technique is used for adding the Coumarin 35, except that 0.25 g of the dye is dissolved in 7 mL of methanol. To achieve the desired contrast, more or less of the dye may be necessary, depending on the microscopist and the capability of the microscope used. The polished cross-sectional mounts can be examined with a variety of microscopy techniques, including polarized light, bright- and dark-field illumination, and epi-fluorescence. The viewing of dyed specimens is discussed in Chapter 5, "Viewing the Specimen Using Reflected-Light Microscopy," in this book.

Molds for Mounting Composite Materials

The molds used for composite materials should provide convenient mounting dimensions and be economical and easy to use. Ideally, the molds provide mounted samples that can be effectively used for both automated and hand-polishing methods. Commercial rubber molds made from silicone or ethylene propylene diene (EPDM) are very economical and can be used to prepare multiple samples if prepped appropriately. A convenient mold size for composite specimens is 57 by 25 by 25 mm (2.25 by 1 by 1 in.) (Fig. 2.12). Of these two materials, silicone molds provide the best release and lowest buildup, but EPDM molds are more tear resistant. To extend the life of rubber molds, a good release agent should be applied between each use. Some laboratories still create custom-cavity molds for mounting composite specimens. Custom-cavity molds are commonly made from wrapping aluminum foil around a preset shape, such as previously molded samples or the end of a block of wood. This is followed by taping the mold to reinforce the structure. The object can then be removed from the aluminum foil mold. To ensure the best release from these types of molds, the aluminum foil should be release-coated. After the sample cures, the aluminum foil mold can be unwrapped from the specimen. Care must be taken in using any custom mold, because the dimensions are not as controlled as in cavity molds that are purchased to preset dimensions. Also, great care must be taken to ensure the custom molds have a flat surface, so that the final sample surface is not on an angle or requires excessive grinding. These molds are not generally recommended, because commercial rectangular-cavity rubber molds are cost-effective and can be used for making many samples if release-coated appropriately. Also, the dimensions are very accommodating for both automated and hand polishing.

Summary of Mounting Procedure

The following steps are recommended to achieve a good mounted specimen without voids or specimen pull-out:

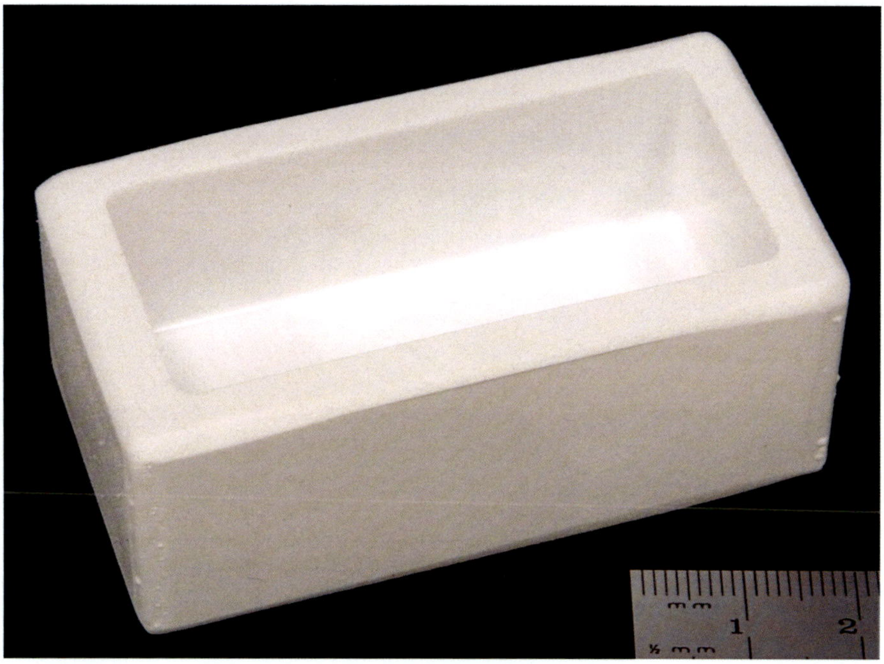

Fig. 2.12 The most useful cavity mold type for mounting composite materials. A single mold can last for many samples. Each time one is used, it should be release-coated for easy sample removal and extended life. For producing samples for transmitted-light analysis (Chapter 6), this mold geometry is necessary.

1. Select a mold to hold the specimens, and coat with a mold release agent. Silicone rubber molds are the most convenient and can be used repeatedly. A convenient-sized mold for composite specimens is 57 by 25 by 25 mm (2.25 by 1 by 1 in.) (Fig. 2.12). The resulting samples may be used in automated polishers, hand-polishing applications, and thin sections. Circular plastic molds can also be used for mounting composite samples, but they are not as convenient for the geometry of most composite materials or for polishing.
2. Wash and degrease the samples to create a clean interface to bond to the mounting resin.
3. Completely dry all the components of the sample before mounting. Vacuum desiccate at low temperature, 40 to 50 °C (105 to 120 °F). The use of a drying oven should be considered for at least 12 hours.
4. Place dry specimens in the mold, and position a strip of loosely woven glass cloth between the flat adjacent surfaces of the sample. This will ensure sufficient wetting with the resin (Fig. 2.13). The glass fabric allows the resin to impregnate and wick between the specimens no matter how tightly they are packed together. This provides the best

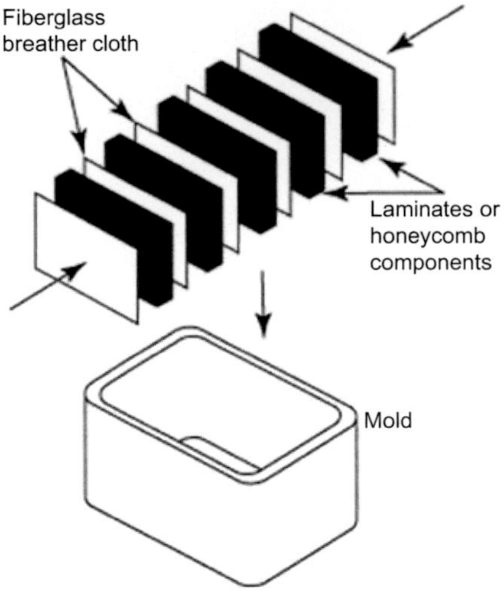

Fig. 2.13 Schematic showing the mounting of composite specimens in a rubber cavity mold. Glass fabric cloth is used for separation of the samples. The addition of a highly permeable cloth is necessary in between each of the composite specimens, or there can be areas that are not impregnated by the casting resin.

possible bonding of the specimens, so that the mounted sample does not come apart during the grinding and polishing stages.

5. When possible, excess specimens should be used for backing material within the mount to aid in polishing. As a general rule, use materials with the same mechanical properties for the best support. Also, if there is only one specimen of interest, place the specimen as close to the center of the mount as possible (in this case, do not place all backing pieces on one side of the mount).
6. Mix the necessary quantity of mounting resin to be used, based on the number of samples and volume of free space in the molds. Determine if it is necessary that the mounting resin be dyed to show contrast with the matrix resin. If so, mix the dye into the mounting resin as described previously. Do not remove the methanol that is used with the dye. While the addition of methanol will slightly plasticize the mounting resin, it will not degrade the performance. Also, some of the methanol will be removed in the following vacuum step.
7. A vacuum infusion apparatus is ideally used to remove entrained air in the mounting resin before addition to the mold (Fig. 2.9). If this type of apparatus is not available, a vacuum oven can be used to remove the air from the mixed epoxy resin before addition to the mold. In general, leave the resin in the oven at ambient temperature for five minutes under full vacuum. However, look to see if the entrapped air (air bubbles) has been removed. Although the entrapped air from mixing has been removed, care must be taken when the resin is added to the sam-

ple, because air can be entrained in this step. The advantage of a degassing apparatus is that the mounting resin is added slowly in an evacuated environment and results in little to no air entrapment.

8. After the impregnation step, it is recommended that the mount be cured under pressure of at least 275 kPa (40 psi) (preferably greater than 400 kPa, or 58 psi) (Fig. 2.10). The cure time depends on the specific epoxy mounting resin selected. Addition of pressure during cure helps reduce in size even the smallest air bubbles. If pressure is not available for cure, there may be small bubbles present in the mounting material after cure. Although these are still fine for grinding and polishing, the mount may not have as high a strength due to larger voids. When voids are present in the cured mounting resin or sample, care must be taken to clean/remove any grit or polishing compound from the previous grinding or polishing step. If the sample is not cleaned well, these particles can cause scratches on the specimen surface.
9. Remove the cured sample from the mold (Fig. 2.14). Clean the mold of any cured mounting resin before future use.

In mounting composite samples, part of the quality of the mount is dictated by the equipment that is available for the microscopist to use. It must be emphasized that while vacuum impregnation and pressure curing of the mounted specimens is optimal, specimens can be prepared without using this equipment; however, the cured mounting material often has small voids, and areas between the specimens may have entrained air. In the

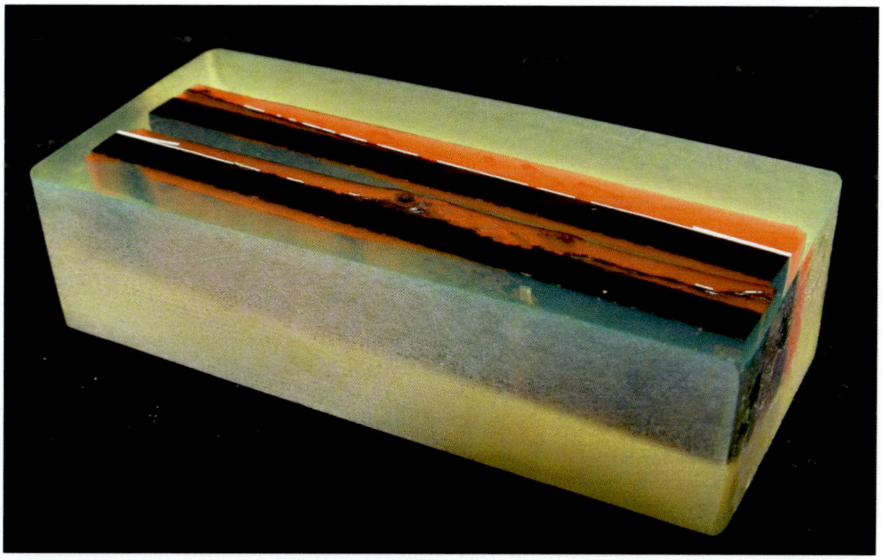

Fig. 2.14 Photograph of mounted composite materials after removal from a rubber mold. This figure shows a polished top surface.

worst case, this can lead to the loss of a sample during the grinding or polishing operation if the voids are so severe that the mount separates. This is uncommon but possible. More common is that areas that were unimpregnated, having void space, can lead to edge rounding, resulting in areas incapable of high-quality polishing. These areas also can retain polishing compound and lead to scratches in the final polish if not properly cleaned between each step.

Clamping Mounted Composite Samples in Automated Polishing Heads

The use of automated polishing heads having rectangular openings is recommended because the rectangular openings are best suited to accommodate the geometry of most composite materials (i.e., thin and long). However, in some laboratories that commonly analyze metals, the automated polishing head may have circular openings. This type of polishing head can be used for polishing composites, but the samples must be mounted in resin and cannot simply be clamped in the circular openings, as in rectangular openings. The sample size cannot be nearly as long and is much less convenient. Figure 2.15(a and b) show a comparison of mounted composite materials in automated polishing heads having circular cavities and rectangular cavities, respectively. As shown in Fig. 2.15a, spring clips or the plastic spiral from notebooks can be used to mount a single thin specimen in the upright position in a mount. However, this is not ideal and can result in significant edge rounding and pressure variation. It is always best to fill the mount with other composite materials of equal hardness to create an even pressure distribution and the best edge retention. As discussed previously, composite materials are best located in the rectangular openings closest to the center of the head.

Mounting Composite Materials for Hand Polishing

Polishing unmounted specimens by hand can be performed if the samples are thick enough, but it is often difficult because most composite materials are thin and long. The narrow nature of most composite samples makes it difficult to maintain a high-quality flat surface free of artifacts. Additionally, the edges are prone to tearing the polishing cloth. With special exceptions, this is the least favorable specimen preparation method. One exception is when the sample surface that is to be polished is very wide, generally 13 mm (½ in.) or wider. In this case, the composite can be held planar without a lot of difficulty. Samples that are less than 13 mm (½ in.) high can result in difficulties holding the sample, and polished fingertips. These samples should be less than 65 mm (2.5 in.) long for good control and preferably around 40 mm (1.5 in.).

(a)

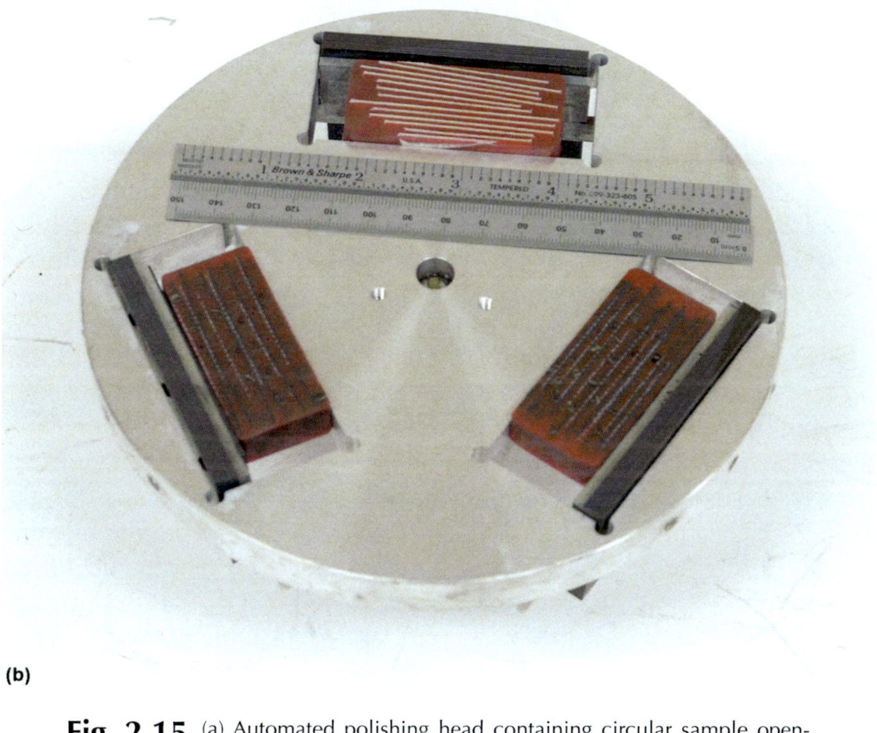

(b)

Fig. 2.15 (a) Automated polishing head containing circular sample openings. Notice the smaller sample length in the circular mounts. Also, circular cavities cannot accommodate unmounted composite specimens. The samples in this mount have been made with different mounting resins, which result in different mechanical properties. This ultimately affects the removal rate of the mounting resin. (b) Automated polishing head with rectangular sample openings containing epoxy-mounted specimens. Here, the backing pieces made from composites (located on the outside edge of the mounted sample) are below the mounted specimens so as not to contact the polishing surface. It is best to use the same mounting resin for all mounts in the head, because the removal rate will be the same. It is also best to fill the cavity molds with as much of the composite material as possible to reduce edge rounding.

While the previous mounting procedure described mounting samples in a preset cavity mold, which required impregnation of the casting resin, another type of mounting is also beneficial only for manual or hand polishing. This type of custom mounting is quite fast and economical. Samples can be mounted, polished, and viewed within 20 minutes. In many small companies and laboratories that do not have the capital to purchase automated polishing equipment, this method of mounting is essential. For the processing of one-off samples and quick analysis, even the most advanced laboratory may find this technique useful. The mount for manual polishing can be prepared as follows (Fig. 2.16):

1. Cut samples of the laminate or honeycomb specimen into strips that are 25 to 65 mm (1.0 to 2.5 in.) long by 12 to 25 mm (0.5 to 1 in.) high (Fig. 2.16a).
2. Cut sacrificial side panels that are 12 mm (0.5 in.) longer than the subject specimens, using scrap laminates of a similar material to create a custom mold (Fig. 2.16b).
3. Tightly hold and bind the sample together, making sure to press down while compressing the ends and sides. A large metal binder clip can be used effectively to help hold the mount. The open side should be opposite the side to be polished (Fig. 2.16c). In this step, "5 min epoxy" may be added to each face of the samples to be clamped together but is oftentimes not necessary.
4. Use vinyl tape to make a custom mold with the composite specimens, and fill with "5 min epoxy." After the epoxy is cured, the vinyl tape can easily be peeled off, leaving the specimen ready to polish (Fig. 2.16d). The result is a mount that can be handled easily and also maximizes the surface area of the sample.

A photograph of a mounted and polished specimen is shown in Fig. 2.16e. Notice that the backing pieces are of the same type of material, carbon fiber composite laminates, as the sample to be viewed in the middle of the mount. This provides similar sample-removal rates. If there are gaps in the interfaces between the backing pieces and the sample in the center, special care is required to clean the removed material or polishing medium from the previous step so it does not scratch the sample during the next operation. Different from making custom-cavity molds, this technique is much faster for a few samples, and the sample plane is easy to maintain.

Mounting Technique Summary

Table 2.2 compares the various mounting techniques and when each technique is best used. The categories include sample clamping—no mold; cavity mold—casting resin; and hand mount. The first category involves not mounting the specimens using an adhesive or casting resin and corre-

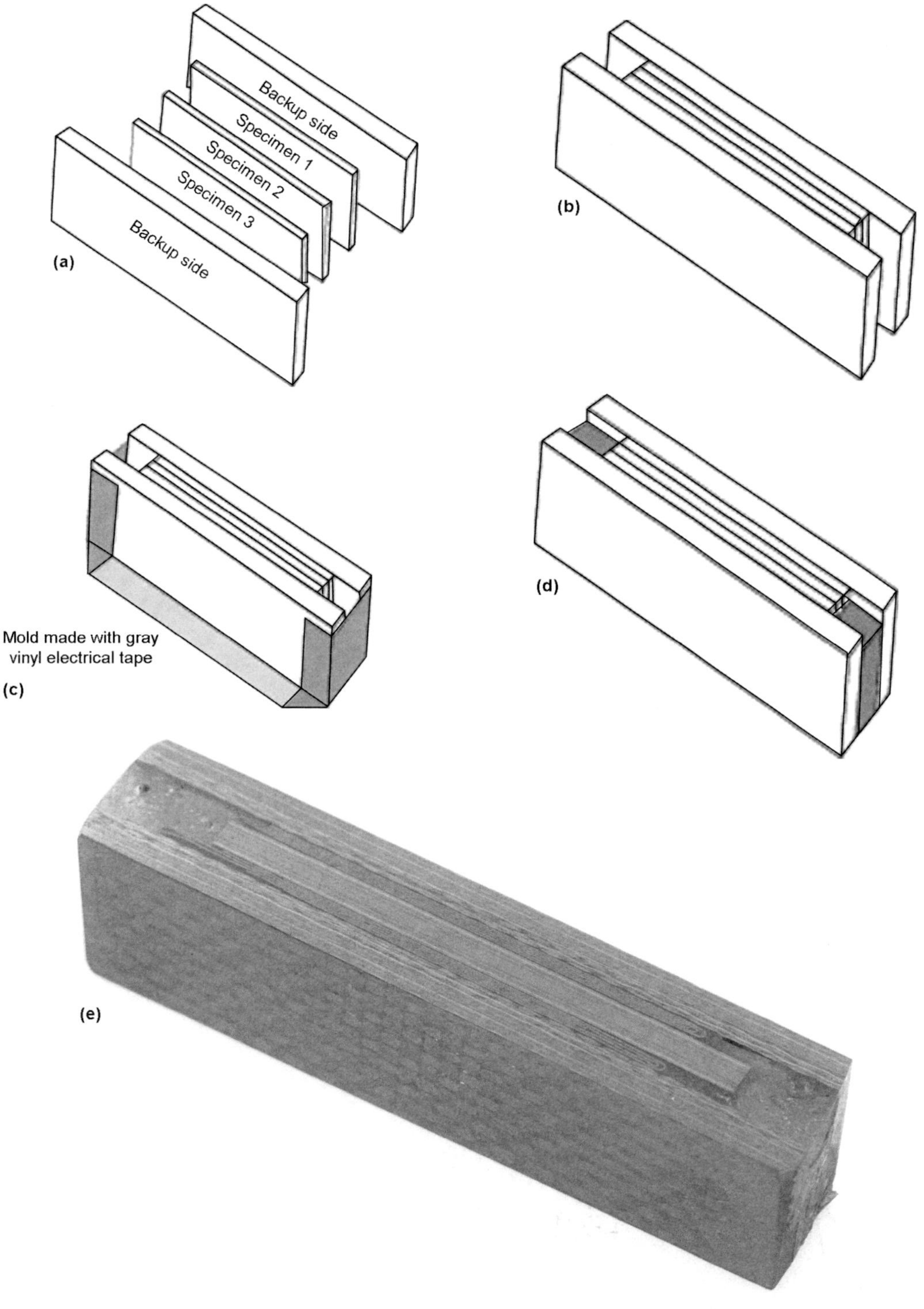

Fig. 2.16 Preparation steps for the development of a manual polishing mount. (a) Backup sides with three specimens in the center. (b) Mount before bonding with epoxy. (c) Mold with taped ends for retaining the bonding resin and holding samples while curing. (d) Mounted specimens ready for grinding and polishing. (e) Photograph of a manual mount after final polish

Table 2.2 Comparison of mounting methods and when each is advantageous

	Sample clamping—no mold	Cavity mold—casting resin	Hand mount
Automated polishing	X	X	...
Hand polishing	...	X	X
Fragile features	...	X	...
Length/volume of specimens	X	z	...
Multiple samples per mount	X	z	z
Quick preparation	X	...	X
Cost-effective sample preparation	X	...	z
One specimen	...	...	X
Mount integrity	X	X	z
Consumable cost	X	...	z
Sample identification	X	z	z
Reflected-light analysis	X	X	X
Transmitted-light analysis	...	X	...

X = best; z = acceptable

sponds to clamped samples used in automated polishing heads. Cavity molds involve mounting the composite specimen(s) using a casting resin in a preset mold. These molds are commercially available or can be made to a set dimension. Hand mounts are made from backing pieces of the same or similar material as the material that is to be mounted and bonded together. These methods were described previously.

The mounting methods that were described have been found to be the best techniques that result in the highest-quality mounted specimens. The omission of steps in this procedure may result in lower-quality mounts but may be necessary, depending on the equipment. Knowing the effect each step incurs, the microscopist can tailor the specimen and viewing area for the grinding and polishing steps so there will be no loss in the information that is desired from the specimen. The mounting techniques detailed here can be used for almost all but a few types of special composite specimens.

REFERENCES

1. B.S. Hayes and L.M. Gammon, Microscopy, *Composites,* Vol 21, *ASM Handbook,* ASM International, 2001, p 964–972
2. L.M. Gammon and B.S. Hayes, Microscopy of Composite Materials, *Structure—J. Materialogr.,* Vol 38, April 2001, p 16–18
3. L.M. Gammon, Fracture Analysis of Composites, *Microsc. Anal.,* Vol 11 (Suppl. 2), 2005, p 1588–1589

CHAPTER 3

Rough Grinding and Polishing

ROUGH GRINDING AND POLISHING of mounted specimens is required to prepare the composite sample for optical analysis (Ref 1). The rough grinding step removes the surface of the specimen to eliminate sectioning- and mounting-induced artifacts. Sectioning artifacts include large cracks and microcracks as well as induced matrix strains. Other common artifacts may be created by incomplete impregnation of fracture areas or microcracks by the mounting resin (Fig. 3.1). This can lead to rounding of the crack edges and the entrainment of polishing compound, which can scratch the sample in further processing. If this occurs, the sample surface can be further removed or can be remounted using the techniques described in Chapter 2, "Sample Preparation and Mounting," in this book. The grinding process is also used to adjust the level where the information is desired. If the sample is ground too much, the area where the analysis was desired may be lost. Rough grinding is essential to make the surface planar, but overgrinding can lead to difficulties and complications in holding or clamping. This will be dictated by automated mount head openings and the clamping apparatus or by the ability to hold and control the sample by hand. Following the rough grinding step, the sample(s) should be ready for rough polishing. Rough polishing further removes the specimen surface and subsequently results in a better-quality surface with less surface roughness after each polishing step. After the last step of rough polishing, the sample should be free of artifacts when viewed at 100× magnification. The final polishing quality will dictate the resolution of the features that can be distinguished in the analysis (Ref 2).

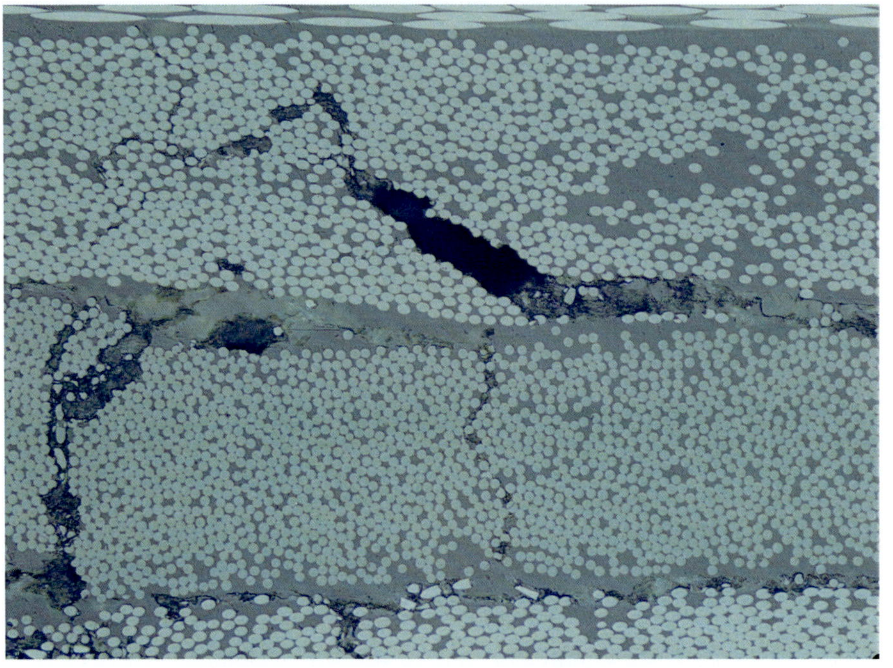

Fig. 3.1 Cross section of a polished composite showing areas of the microcracks partially filled with epoxy mounting resin. Incomplete impregnation of the mounting resin can cause edge rounding and also fill the unprotected edges with grinding and polishing debris. Unimpregnated areas in the specimens can usually be eliminated by vacuum impregnation of the sample with the mounting resin, along with a pressure cure. In this case, a dyed mounting resin would have provided better contrast with the matrix resin. Bright-field illumination, 25× objective

Grinding and Polishing Equipment and Process Variables

Over the past years, there have been significant advancements in the equipment and consumables used for preparing samples for optical microscopy analysis. More companies have entered into this market, and quality grinding and polishing equipment is becoming more affordable. In many laboratories, the use of automated polishing equipment has become the standard, while small laboratories still rely on hand polishing a limited number of samples. Many laboratories with full-time analysis requirements have banks of multiple automated polishers (Fig. 3.2). With this type of setup, very fast sample throughput can be achieved because the polishers can be preset and programmed to move from one step to the next (i.e., grinding to rough polishing to final polishing). Companies that use optical microscopy for limited analysis may have only one automated or a manual polishing wheel and require the changing of the platens, polishing media, and parameters for each step. In this case, magnetic attachment of the plat-

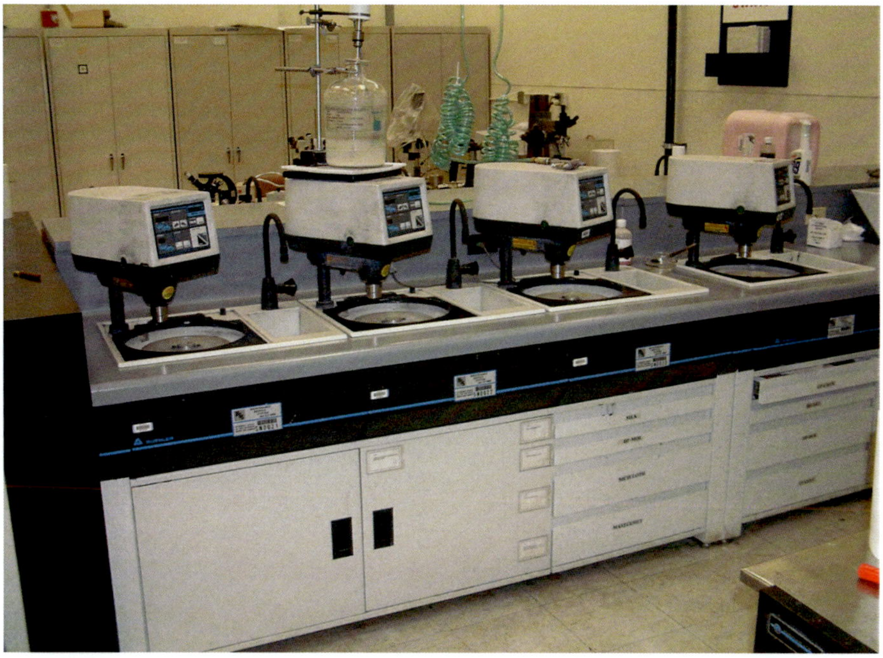

Fig. 3.2 Photograph of automated polishing equipment showing a series of four automated polishing stations. This allows easy sample movement from one automated polisher to the next without having to change platens or process parameters.

ens is essential, either on the polishing wheel or backside of the platen, so the process is more efficient (Fig. 3.3). Whatever the available equipment, the final sample quality should be the same. In addition to the new equipment, new grinding and polishing disks that can last through the production of thousands of specimens are available and becoming standard in most laboratories that process many samples. These diamond-coated disks, either continuous or discontinuous (patterned), are taking the place of the traditional silicon carbide papers, which usually last up to approximately three samples. The new diamond-coated disks are much more affordable in the long term for producing large numbers of samples.

When preparing new types of composite materials or using new equipment or consumables, it is recommended that the sample be viewed using optical microscopy after each grinding/polishing step. This will help determine the time and parameters required for optimal sample preparation as well as help in extending the life of the consumable grinding and polishing materials. It is recommended that grinding/polishing parameters be changed if any scratches or damage are evident from the previous step. If this does not solve the problems, changing the grinding/polishing consumables may be required.

The development of the sample preparation procedure depends on the equipment available as well as the material being prepared (Fig. 3.4). The

Fig. 3.3 Photograph of a manual polishing wheel with a magnetic disk adhered to the platen and the corresponding metal disk, which has the polishing pad or cloth adhered to the surface. All samples to be processed should be subjected to the first step and proceed in order, so that the platens only have to be changed once for each step for multiple samples.

type of equipment, either automated or manual, will dictate the consumable materials that are used and the process parameters. Consumable materials such as the grinding/polishing surface, abrasive type and size, and lubricant may be similar for either automated or manual grinding/polishing, with a few exceptions. However, what will change are the process variables, which include sample/platen speed, vertical force, and process time.

Processes for Sample Preparation

There are essentially two types of processes that are used for removing material in the preparation of samples for optical analysis. These processes include: 1) grinding and polishing; and 2) lapping. The process of grinding is essentially the same as polishing and therefore is shown as one process type. In this process, the abrasive is locked on the polishing surface, and the sample surface is continuously removed by the abrasive (Fig. 3.5) (Ref 3). The removal of the sample surface using coarser abrasives results in faster sample removal and deeper and wider cuts. Most of the rough grinding and rough polishing is performed on composites using either silicon carbide papers or diamond-coated (continuous and discontinuous) disks. In some cases, such as hand polishing, rough polishing may be performed

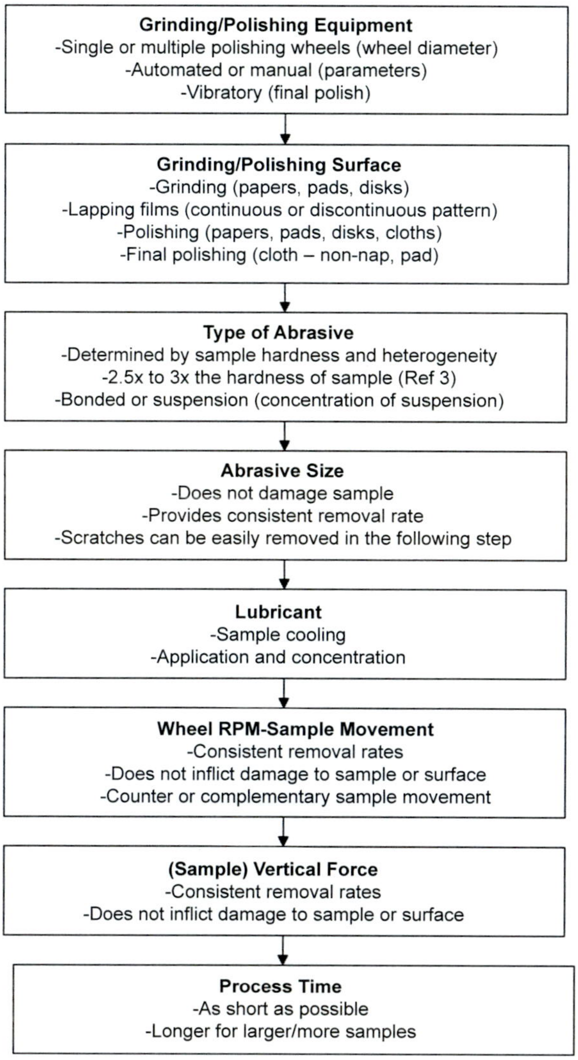

Fig. 3.4 Polishing equipment, consumables, and process variables that influence the preparation procedure

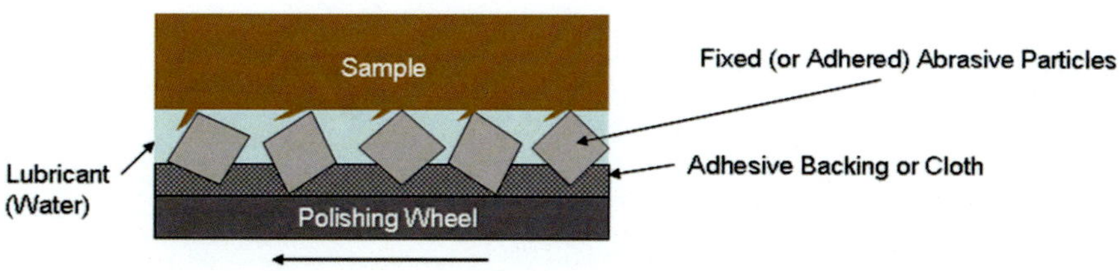

Fig. 3.5 Schematic showing the grinding process. The abrasive is either adhered or mechanically locked into the surface.

using deagglomerated alumina suspensions. In the grinding and rough polishing steps, the abrasive may be adhered to the surface of a disk, pad, or paper but can also be mechanically locked into the cloth or pad surface (Fig. 3.6). In contrast, in the final polishing process, the abrasive is typically locked into a cloth surface only by mechanical means and not actually adhered to the surface (Fig. 3.7). The type of cloth, hardness, flatness, resilience, and weave affect how the abrasive is locked in the cloth, as well as abrasive size and shape. A critical factor that is often overlooked is the amount and concentration of polishing compound in the form of deagglomerated abrasive suspensions. If there is too much of the polishing compound, or it is applied too frequently, the abrasive particles can roll on the cloth surface and not actually cut the sample. This creates features such as edge and fiber rounding and erosion of the sample. A result is that this process looks very similar to the second process that was mentioned—lapping. Different from grinding and polishing, in the lapping process the abrasive particles are not locked in place but are free to move and roll under the sample surface (Fig. 3.8) (Ref 3). This process does not actually cut the specimen and can be better described as a carving of the specimen surface by the abrasive grains. This process is not recommended for com-

Fig. 3.6 Micrograph of 240-grit alumina (Al_2O_3) showing the sharp edges and cutting surfaces of grinding and polishing compounds. Polarized light, 1/4 wave plate, 10× objective

Fig. 3.7 Micrograph of the surface of a silk polishing cloth. The interstitial areas are where the abrasive material is able to be mechanically locked into the weave. Polarized light, 1/4 wave plate, 10× objective

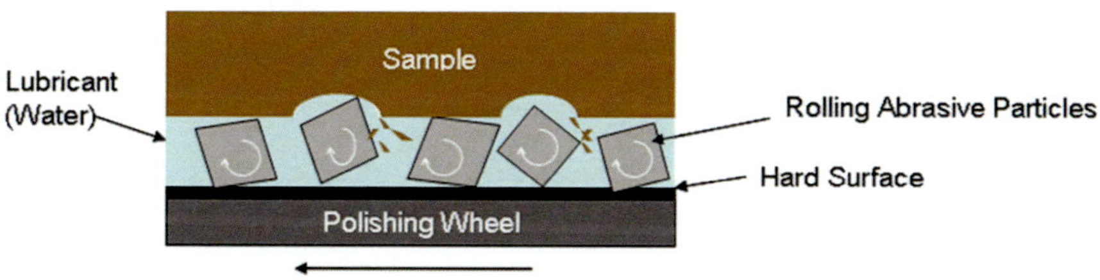

Fig. 3.8 Schematic showing the lapping process. The abrasive is applied to a hard surface and therefore is free to roll and erode the sample surface. The arrows placed on the particles indicate the rolling direction opposite to the platen direction.

posites until the final polishing step, and even at this step, it is not necessary. In the final polishing step, the lapping that may be performed is actually only partial lapping, because the abrasive particles are semifixed on the surface of a pad. A common material that can be used for the final polishing surface for preparing composite samples is a neoprene (E.I. du Pont de Nemours and Company) (foam) pad. This material is relatively soft, and through the application of pressure, the particles can grab the surface and result in both semifixed and free particles.

The predominant use of the grinding and polishing process for all of the preparation steps is due to the heterogeneous and anisotropic nature of composites. Multiple phases in the matrices as well as the fibers create a material of varying hardness and therefore are significantly affected by the preparation procedure. It is for this reason that only the grinding and polishing process is used, because lapping can erode areas of the sample plane. If lapping is performed or occurs by accident, different phases or areas of different material properties may be removed at different rates and show relief. This can create a similar situation as described for the addition of too much polishing compound. While the general rule for preparing composite specimens is not to use a lapping process, there can be exceptions when preparing hybrid materials consisting of composites and metallic components, which is discussed in Chapter 4, "Special Sample Preparation and Polishing," in this book.

Abrasive Sizing for Grinding and Rough Polishing

Abrasive or grit size determines the rate at which material is removed from the sample. Before describing the appropriate procedures, it must be mentioned that various standards have been established for grit size. The various standards can influence the sample-removal rate and sample quality. These standards are different in average grit size as well as size distribution. The most common standards used in North America are the Coated Abrasive Manufacturers Institute (CAMI) (now part of the Unified Abrasives Manufacturers' Association), Federation of European Producers of Abrasives (FEPA) P-grade, micron grading, and, to a lesser extent, Japan Industrial Standards Committee (JIS). Up to 220 grade (grit), the CAMI and FEPA grades are very similar in average sizes of grit. In grades greater than 220 (smaller grit size), more deviation is found in average grit size between two of the same grades. Table 3.1 shows the CAMI and FEPA grades in comparison to the average micron sizes of these grades. However, the particle size range for each grade is not shown. The FEPA (P-grade) standard has much tighter tolerances than the CAMI-grade abrasives and therefore a tighter grain size distribution per grade. This creates more even sample-removal rates, less variability, and fewer scratches on the specimen surfaces. Even different abrasive suppliers have a different particle size range, and therefore, consistency in suppliers for the grade of abrasive is necessary for complete consistency. In the procedures specified in this text, the abrasive standard, CAMI versus FEPA, is designated by either placing no letter after the abrasive size or by placing a P after the size, respectively. In contrast to these standards for abrasive size, the new diamond disks and pads used for grinding and polishing have tighter tolerances in particle size. This provides the ultimate in consistency in sample removal as well as repeatability, because wear is limited compared with silicon carbide papers.

Table 3.1 Coated Abrasive Manufacturers Institute (CAMI)- and Federation of European Producers of Abrasives (FEPA)-grade abrasive (grit) sizes in comparison to the average micron sizes of these grades

FEPA (P-grade)	CAMI grade	Micron scale (average)
60	...	269
...	60	268
80	...	201
...	80	192
100	...	162
...	100	141
120	...	125
...	120	116
150	...	100
...	150	93
180	...	82
...	180	78
220	...	68
...	220	66
240	...	58
...	240	53
280	...	52
320	...	46
360	...	40
...	320	36
400	...	35
500	...	30
...	360	28
600	...	25
...	400	23
800	...	21
...	500	19
1000	...	18
...	600	16
1200	...	15
1500	800	12
2000	...	10
...	1000	9
2500	...	8
...	1200	6

Rough Grinding—Sample Removal

The first stage of grinding should be aggressive enough to remove the material quickly and easily but not so aggressive as to induce damage to the sample. In all steps, water (wet grinding) should be used as a lubricant and to displace the removed sample particles. The water also cools the specimen surface and eliminates heat damage. The use of 60-grit abrasive paper, like the band saw, can induce significant damage and can leave the sample in worse condition than it was before grinding, so care must be taken if using a coarse grit. Sample removal with silicon carbide (SiC) paper will not introduce damage that cannot be removed easily if papers increasing in grade (finer grit size) are used sequentially. The most common sequence of grit size for polishing composites is 120- followed by 320- and 600-grit papers. However, because of their short useful life, as mentioned previously, SiC papers are one of the most expensive options for rough and fine grinding. Figure 3.9a, b) show cross sections of 120-grit SiC paper before and after the grinding step of one mounted specimen. There is a significant reduction in the surface roughness after only one

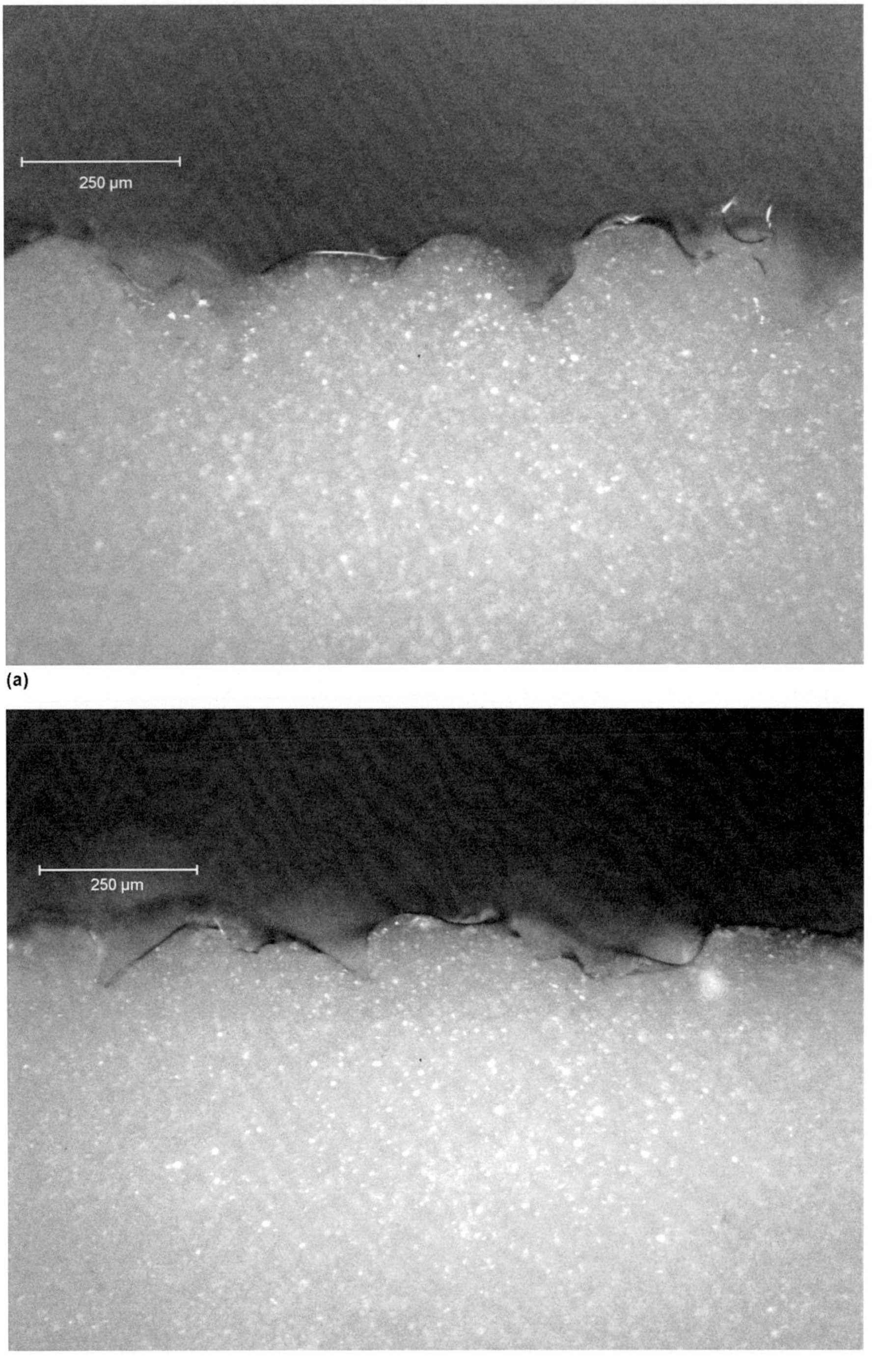

Fig. 3.9 Cross sections of 120-grit silicon carbide paper that was polished using diamond polishing compound. (a) Unused paper. Bright-field illumination, 50× objective. (b) After the preparation of one sample. A more uniform surface can be observed as compared to the surface shown in (a). Bright-field illumination, 50× objective

sample. In the first 60 seconds of grinding or polishing with abrasive papers, approximately 75 percent of the effective use is consumed (for example, 120-grit SiC, 300 rpm, counterdirection, 140 N, full three-opening head). It is important that worn or dull papers are not used, because the removal rate is decreased and may introduce damage. If there is short supply of abrasive papers, a worn paper can be used for most of the process, followed by quick use of a fresh abrasive paper of the same size. To get the greatest life out of SiC paper as well as other grinding and polishing surfaces, it is necessary that the entire surface is used across the paper radius. The sample-removal rate is also more uniform if this procedure is observed.

Rough and fine grinding using water-lubricated, diamond-coated discs or pads remains some of the most efficient means of preparing specimens for final polishing. If used only for composites, a single diamond disc can last for thousands of samples. A high-quality diamond disc can remove material at a continuous rate with less damage than SiC papers for a given grit size (120 P, or 120, diamond grit disk), making it possible to move to the next step (1200 P, or 600, diamond grit disk). A noncontinuous diamond pattern or a low-density distribution is recommended, because the water and material are easily displaced and washed away, ultimately increasing the material-removal rate (Fig. 3.10).

After the grinding substrate has been determined, the processing parameters of pressure and velocity dictate the sample-removal rate and the quality of preparation. This also ultimately affects the process time, which influences the sample-preparation economics. The processing parameters for both automated and hand sample preparation are influenced by the equipment available and, in the case of hand polishing, also the person performing the task. Movement of the specimens counter to the polishing

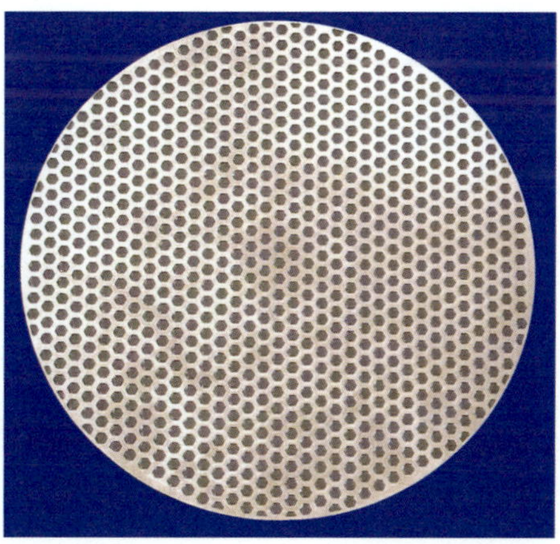

Fig. 3.10 Photograph of a diamond platen that can be used very effectively for preparing thousands of composite samples

platen rotation is used for composite sample preparation until the final polishing step (Fig. 3.11). Countermovement of the sample increases the removal rate but also provides better sample control when grinding or polishing by hand. In automated grinding and polishing, the speed is typically 300 rpm or higher for a 200 to 300 mm (7.87 to 11.81 in.) platen. In these steps, the head rotation is counter to the platen rotation, with a speed of approximately 150 rpm (Fig. 3.12). When using automated polishing equipment and the method of mounting described in Chapter 2, "Sample Preparation and Mounting," the samples should remain on the platen surface for the entire process and not travel off the edge of the platen. Higher sample-surface speed, whether increased through wheel speed, wheel diameter, or counterspecimen rotation, increases removal rates. In the case of hand grinding and polishing of composite specimens, the speed of the platen can be set as high as possible, usually not exceeding 1000 rpm, while the movement of the sample by hand in the counterdirection usually proceeds at less than 1 rpm (Fig. 3.13).

The pressure on the sample surface is dictated by the force that is applied and the area of the sample(s) surface. Higher pressure on the samples increases the sample-removal rate. However, if the pressure is too high, damage can occur to composite specimens, such as separating or tearing out dissimilar materials and frictional heating. In any event, the pressure applied to the sample(s) must be high enough to eliminate hydroplaning while also accomplishing the procedure at a fast enough rate without sample damage. A force of 40 N (9 lbf) or higher is recommended to be applied to the head, but this depends on the capability of the equipment. A fully loaded automatic head can have as much as 80 cm^2 (12.4 $in.^2$) of sample contact area with the grinding surface. It may not be possible to apply sufficient pressure without overloading the equipment if the polishing head is full. In general, for heads containing samples with large sur-

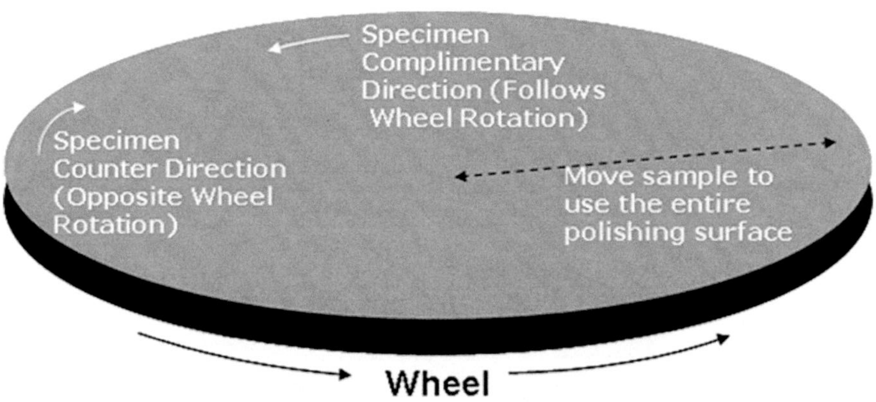

Fig. 3.11 Schematic of a polishing wheel showing complementary and counter sample movement relative to the wheel direction

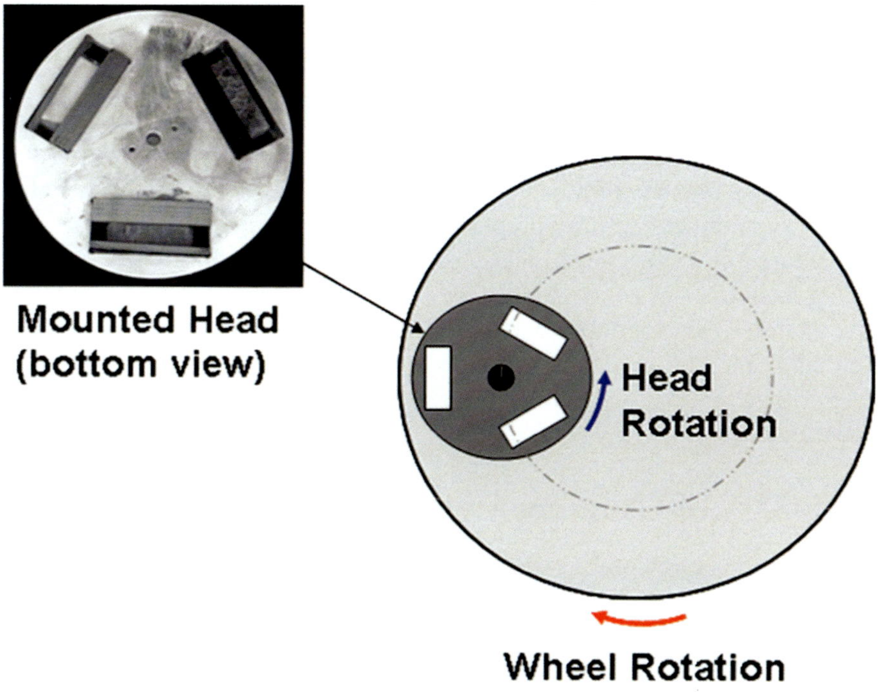

Fig. 3.12 Schematic showing the automated head movement relative to the wheel (platen) movement

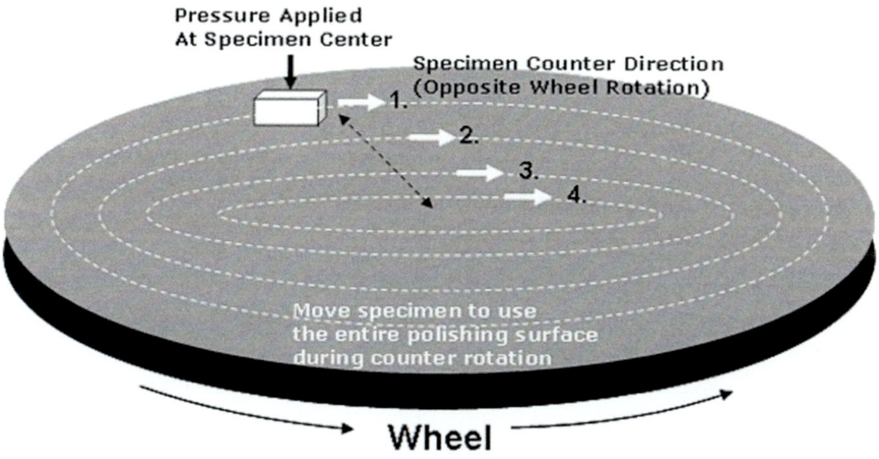

Fig. 3.13 Schematic showing specimen movement relative to the wheel (platen) movement for hand grinding/polishing composite materials. The entire grinding/polishing surface should be used to maximize sample removal and extend the life of abrasive paper. Moving the sample to use the complete radius (i.e., from the circumferences depicted as "1" through "4") of the paper will provide uniform wear rates and sample removal.

face areas as compared to smaller surface areas, the polishing time should be increased instead of implementing an increase in pressure. As a result, friction will be minimized. For hand polishing, the pressure is limited by the ability of the operator to hold the specimen under control.

The final parameter that is often not addressed is the process time for each step or for the entire sample preparation. This is critical to developing an economical sample-preparation process. Using the materials and procedures described in this text, each sample-preparation step (such as going from one abrasive paper to the next and so on) should take less than five minutes, whether automated or performed by hand. The process time for hand preparation of one sample is commonly less than two minutes per step. However, this depends on the capability of the person and the equipment that is available.

Summary of Grinding Methods

The two methods for grinding can be summarized as follows:

- *Method 1:* Hand grinding:
  a. 120 P (120): Silicon carbide paper, sample movement counterdirection to platen
  b. 400 P (320): Silicon carbide paper, sample movement counterdirection to platen
  c. 1200 P (600): Silicon carbide paper, sample movement counterdirection to platen
- *Method 2:* Automated grinding:
  a. 120 P (120): Diamond disk, sample movement counterdirection to platen
  b. 1200 P (600): Diamond disk, sample movement counterdirection to platen

As with all steps, it is necessary to thoroughly clean the samples between each polishing step so there are no residual abrasive particles that can be transferred to the next step. If this occurs, it can cause all following samples to contain scratches, because the larger particles from the previous step will contaminate the next steps that require smaller abrasive sizes. The use of a brush with an adequate amount of soapy water or an ultrasonic bath provides excellent results. After this procedure, it is best to dry the sample using clean pressurized air instead of a mechanical wipe with a cloth.

Rough Polishing

Transitioning from rough grinding to rough polishing involves a further reduction of abrasive size and the use of abrasive suspensions. Like the rough grinding steps, these steps also require the use of fixed abrasives, either adhered to a surface or mechanically locked into a cloth. For automated polishing, a 9 μm diamond lapping film is recommended for the first rough polishing step. The term *lapping film* is used only as a commercial description of this type of abrasive film but can be confusing in the sense that these films have fixed abrasive particles. Therefore, these lapping films are not associated with the fundamental lapping process, which again is not used (if at all) until the final polishing step. As with diamond-coated disks, one sheet of lapping film is capable of polishing up to 1000 specimens, and it is highly economical. The use of a lapping film is applicable to almost all composite materials for the onset of rough polishing, with a few exceptions that are described in Chapter 4, "Special Sample Preparation and Polishing," in this book. After the 9 μm lapping film, the next step transitions to polishing using mechanically fixed alumina (aluminum oxide, Al_2O_3) abrasive particles applied in suspension to a cloth surface. For automated polishing, this step proceeds directly to 0.3 μm alumina. In comparison to automated polishing, rough hand polishing is best performed using only alumina abrasive suspensions on cloth surfaces. The first step uses 15 μm alumina abrasive, which is followed by 0.3 μm alumina. It may be preferred to use an intermediate step of 5.0 μm alumina after the 15 μm alumina to reduce the work required in the last 0.3 μm alumina step. This is a personal preference and will not make a difference in the final specimen quality. It must be emphasized that when transitioning to each alumina abrasive size, different polishing cloths must be used for each abrasive size. The cloths cannot be washed after use and then used with a smaller abrasive size.

Polishing with deagglomerated alumina suspensions is inexpensive and effective when used correctly. Alumina may be used to polish most composites with fiber hardness less than that of alumina. The deagglomerated alumina powders can be purchased in many sizes, but composites can be efficiently rough polished using only two or three sizes, as mentioned previously. In some literature, fine polishing compounds are shown to be below 5 μm; however, in this text, the final polishing materials correspond to the last polishing step, below 0.3 μm. A fine balance of alumina powder in the lubricant and its proper application are necessary for efficient and quality polishing, as is the cloth type, speed of the platen, and applied pressure. Achieving this balance will lock the cutting material into the cloth, resulting in a clean cut across both the soft resin and hard fibers. The distribution of the alumina can be best controlled by adding it to the cloth in a premixed suspension in distilled water. The correct concentration of alumina is critical. There is a tendency to overconcentrate the mixture. In this case, the alumina will roll, becoming an ineffective cutting material. As mentioned previously, the rolling action will erode the resin from around the fibers, leaving the fibers rounded and the surrounding resin undercut. This condition ultimately destroys the surface. For this reason, it is better to use a less concentrated alumina solution to accomplish the rough polishing. A ratio of 12 g of 15 μm alumina to 1 L of distilled water and 5 g of 0.3 μm alumina to 1 L of distilled water has been found to be optimal. The same concentration applies to the 5.0 μm alumina as the 15 μm alumina, if it is used. Deagglomerated alumina suspensions can also be purchased; however, they usually require dilution with more distilled water to arrive at an optimal concentration for composite polishing. This is usually best accomplished by diluting one part of the deagglomerated alumina suspension with 10 parts distilled water by volume. Dry alumina powder should never be applied to the cloth. This can destroy the cloth and possibly the sample as well. It should be noted that although diamond abrasives can also be used effectively, they are not generally recommended because the cost is significantly higher. There are a few cases of sample preparation where diamond abrasives may be effectively used, and this is discussed in Chapter 4, "Special Sample Preparation and Polishing," in this book.

The purpose of a lubricant is to dissipate the heat from polishing and to act as a carrier for the abrasive material, such as alumina. The lubricant must have a low viscosity to prevent hydroplaning during polishing. Usually, the lubricant consists of distilled water, but in some special preparation cases, other lubricants may be required. Any contaminant in the lubricant can cause deep gouges, so using tap water should be avoided because it often contains abrasive particles. The introduction of the alumina suspension at the right rate—one to two drops/second—will cool the sample and supply an ample amount of abrasive. Cloths having no nap do not retain lubricant, so caution must be exercised. Damage to the samples may occur in four to ten seconds if the platen is allowed to run dry. It is advantageous to use an apparatus for applying the deagglomerated alumina suspension at the correct rate when using automated polishing. A simple apparatus that can be used effectively is shown in Fig. 3.14. This apparatus provides constant mixing and controlled release of the alumina suspension to the polishing surface (Fig. 3.15). For hand polishing, the application of the deagglomerated alumina suspension can best be applied by dripping the material from a pint bottle onto the cloth surface with one hand while holding the sample (polishing) with the other (Fig. 3.16). The alumina suspension should be shaken before use each time and also lightly during use so that the abrasive particles remain suspended.

The polishing surface that is used with the alumina suspensions is critical for obtaining a high-quality polished specimen. In all composite preparation steps, the use of nonnap cloths is recommended. If polishing cloths are used that have a nap, areas of the composite specimens may be eroded, and artifacts may be created. Cloths that provide hard polishing surfaces are recommended. Although there are many types of cloths promoted for rough polishing, such as polyester, wool, acetate, and polyamide, cloths made from silk are recommended. Woven silk cloths come in a variety of different types, synthetic or natural, and allow the fastest removal rates.

Of these, satin woven silk is preferred because it is harder and provides better edge retention. Like a good diamond disk or lapping film, a silk cloth should be able to prepare thousands of samples. Damage to cloths may result from allowing the surface to run dry, tears from inadequate sample holding or support, and contaminants that find their way into the

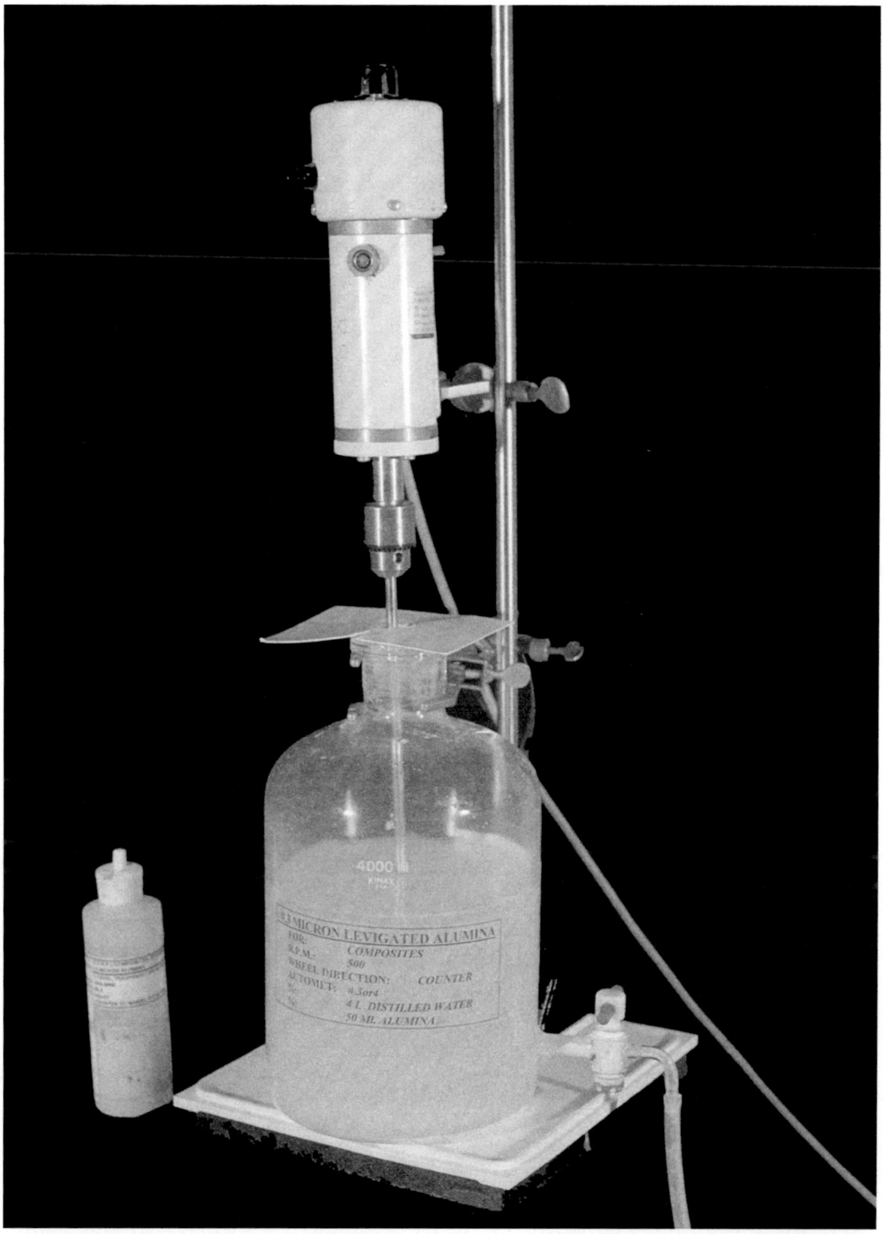

Fig. 3.14 Apparatus used to provide consistent alumina suspension concentrations and flow rate to the platen surface. Also shown is a pint bottle containing the same alumina suspension for hand polishing composite samples.

Fig. 3.15 Photograph showing the apparatus that is used to apply the deagglomerated alumina suspension mounted on top of automated polishing equipment

cloth from impurities in the abrasive suspension (in this case, due to water contaminants) or that fall onto the cloths. In regard to the latter case, it is important that all polishing cloths are covered when not in use.

The processing parameters, sample pressure and velocity, that are best used for the rough polishing steps were stated previously in the rough grinding section. These processing parameters should be similar whether used with grinding/polishing surfaces that have abrasives adhesively

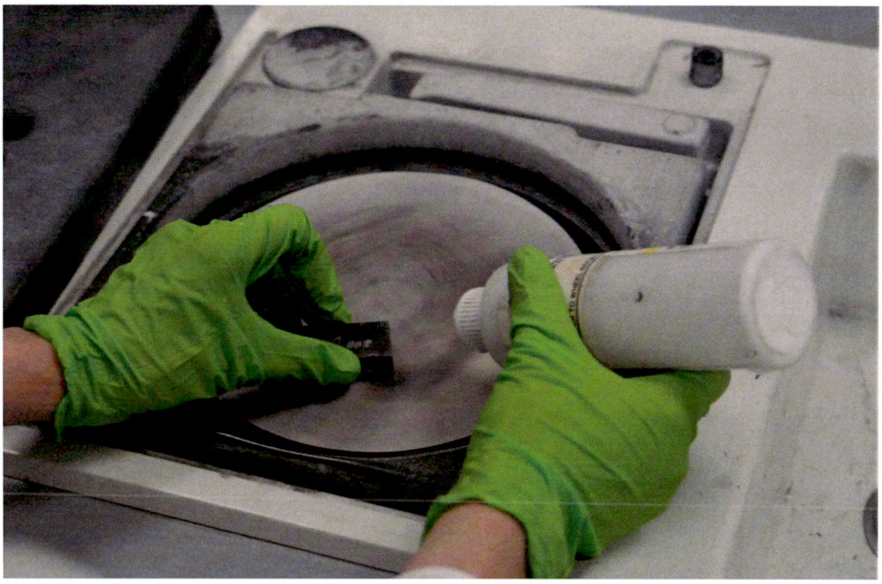

Fig. 3.16 Application of alumina polishing suspension during hand polishing of a composite sample

bonded or mechanically fixed to the substrate. The application of pressure helps lock the particles in the cloth and provides a fixed cutting surface. Furthermore, the high speed of the platen rotation, and the resultant centrifugal force, moderates the buildup of the alumina particles that are consistently applied to the cloth surface.

Summary of Rough Polishing

The following summarizes the rough polishing steps for both hand and automated techniques:

- *Method 1:* Hand (rough) polishing:
  a. 15 μm deagglomerated alumina suspension (12 g alumina powder to 1 L distilled water), silk cloth, counterdirection
  b. 5.0 μm deagglomerated alumina suspension (12 g alumina powder to 1 L distilled water), silk cloth, counterdirection. This step is not required!
  c. 0.3 μm deagglomerated alumina suspension (5 g alumina powder to 1 L distilled water), silk cloth, counterdirection
- *Method 2:* Automated (rough) polishing:
  a. 9 μm lapping film (Note: Use a high-quality lapping film; not all are the same), counterdirection
  b. 0.3 μm deagglomerated alumina suspension (5 g alumina powder to 1 L distilled water), silk cloth, counterdirection

The optimal conditions for all the parameters will rely on many factors, including the equipment available and specifications, the sample type and sizes, and mounting technique. However, after the 0.3 μm step, the surface should be nearly free of artifacts that can be seen at 100× magnification. As with all steps, it is necessary to thoroughly clean the samples between each polishing step so that there are no residual abrasive particles that can be transferred to the next step. If this occurs, scratches will be continuously found in the samples, because the larger particles can remain entrapped in cloths meant for use with the smaller-particle abrasives.

Final Polishing

The final polishing step transitions from rough polishing using napless cloths and high pressure to one of partial lapping with low speed and low pressure. A 0.05 μm deagglomerated alumina suspension (5 g alumina powder to 1 L distilled water) applied to a nonnap cloth or rubber pad is used for this final step. A premixed 0.05 μm deagglomerated alumina suspension (purchased in suspension) diluted with distilled water may also be used but must be diluted with distilled water (one part deagglomerated alumina suspension to 10 parts water). In the final polishing step, the lapping that may be performed is actually only partial lapping, because the abrasive particles are semifixed. As mentioned previously, one of the most useful materials for this final polishing step is a neoprene foam pad. This material is relatively soft, and, through the application of pressure, the particles can grab the surface and result in both semifixed and free-rolling particles. As with the previous polishing steps, the application of the suspension at one to two drops/second is usually adequate. It is necessary that the surface does not become dry in this step. A dry surface is especially a problem when using a neoprene pad, because the sample can grab the surface and damage the specimen and pad surface.

The critical parameters of fine polishing are very low vertical force, complementary rotation of the sample relative to the platen rotation, and a low platen speed. If the vertical force is too high, the fiber-resin interface will display cupping. In addition, if the platen speed is too high, the sample will become difficult to control if it is being hand polished, and the sample will want to stick to the platen. It should be noted that it is hard to control the sample when hand polishing in a counterdirection during this step. When hand or automated polishing is done, platen speeds less than 120 rpm with a complementary sample rotation give ¼ μm (or better) surfaces in 30 to 180 seconds. For hand polishing, the use of no wheel speed is also acceptable, with the sample rotated around the circumference of the pad. The platen may move in the direction the sample is being directed due to the pressure being applied on the wheel. Figure

3.17 shows how a polished surface of a composite material should look after this final step.

If available, a vibratory polisher can also be used for the final polishing step (Fig. 3.18). The vibratory polisher is slower than other automated polishing techniques and is easily contaminated but gives the best possible final surface. This process extends the polishing with lapping. Synthetic silk or Dacron (Invista, Inc.) cloth is recommended. Prestretched synthetic cloths (adhesive backing) on metallic plates work very well and are tough and durable. Silk can be used, but it is difficult to stretch over the platen surface. The same 0.05 μm deagglomerated alumina suspension mentioned previously is used with low applied pressure. Pressure can be applied to the specimens by placing a 1.3 kg (approximately 3 lb) weight on each of the samples.

The more traditional technique of final polishing with either a high- or low-nap cloth is not recommended. After as little as 20 seconds, rounding on the fiber-resin interface can be greater than 1 μm. This effect can be seen in Fig. 3.19. Unfortunately, this is still a widely used method and is found throughout the literature and often recommended by consumable material suppliers to new users.

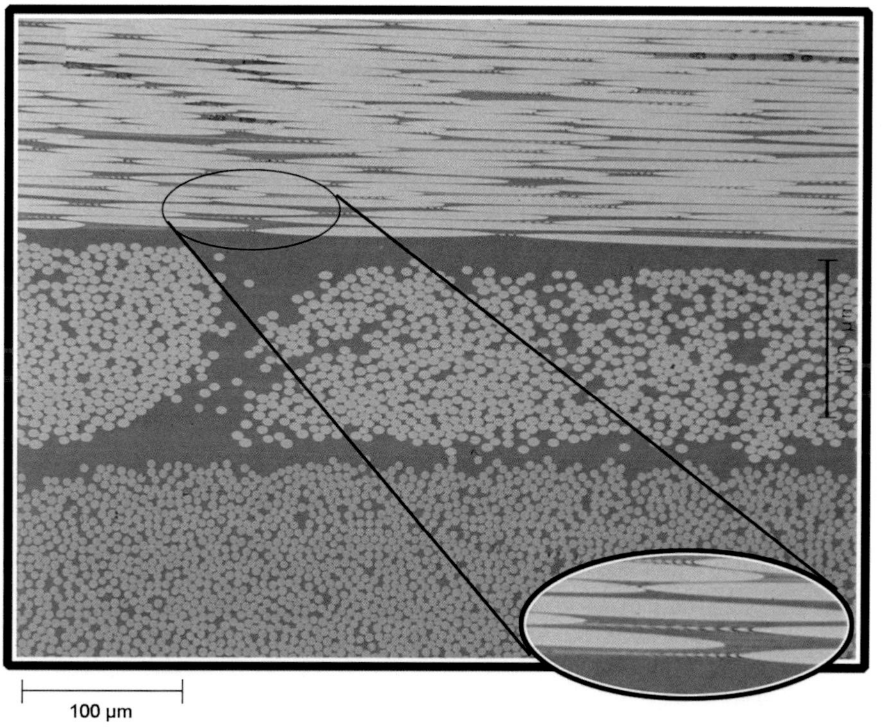

Fig. 3.17 Bright-field illumination (25× objective) of a composite specimen after final alumina polish. Note the interferometer bands on the longitudinal fibers. This is one way to check the uniformity of the polishing plane.

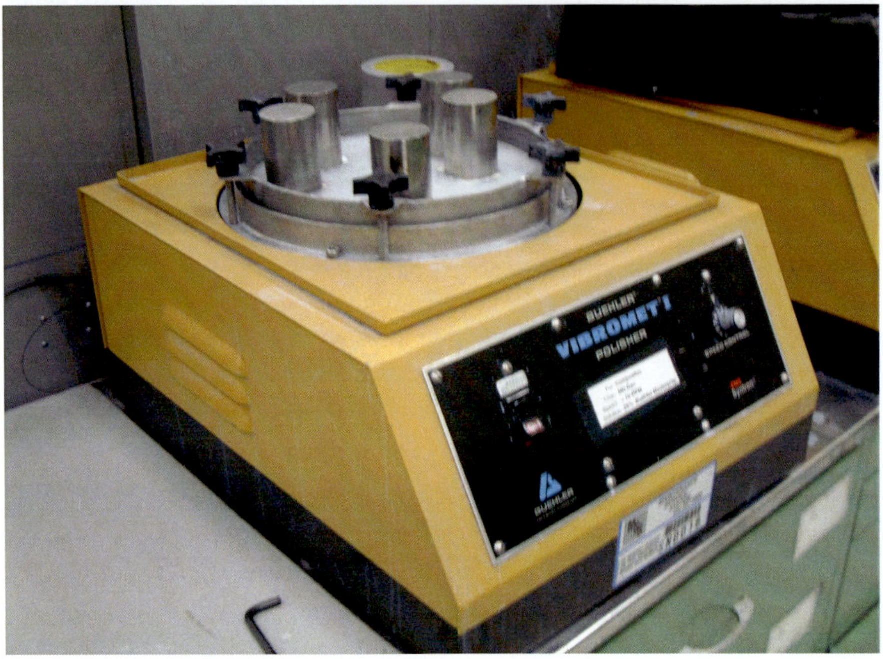

Fig. 3.18 Photograph of a vibratory polisher. This type of equipment provides the best final polish.

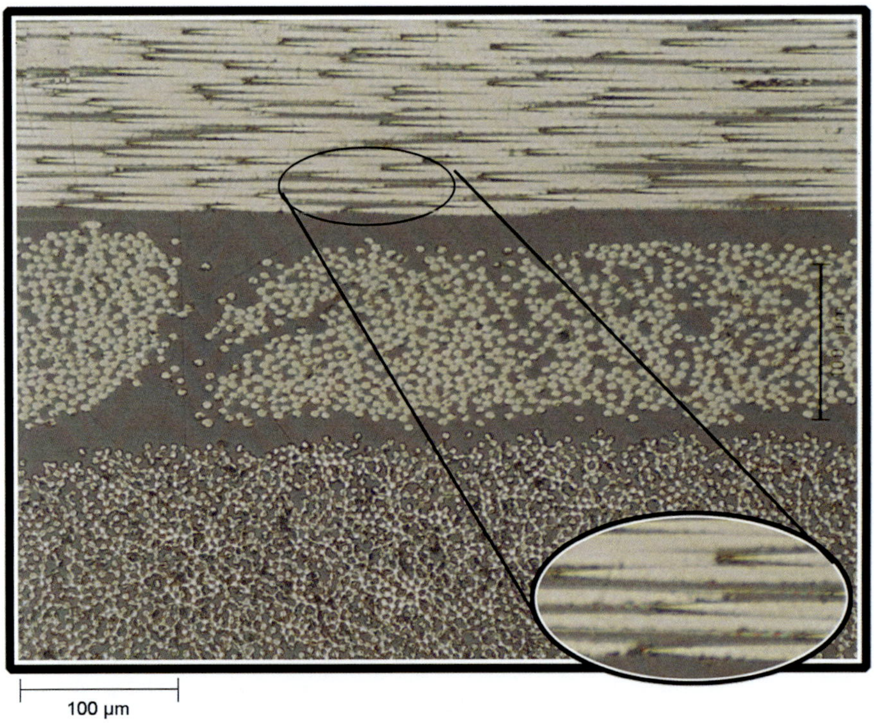

Fig. 3.19 Same specimen as in Fig. 3.17 but polished again with 6 and 1 μm diamond suspension on a nap cloth. Note the rounded fiber interface and the lack of interferometer bands on the longitudinal fibers.

Summary of Final Polishing

The following summarizes the final polishing steps for both hand and automated techniques:

- *Method 1:* Hand (final) polishing:
  a. 0.05 μm deagglomerated alumina suspension (5 g alumina powder to 1 L distilled water) or a commercial premixed 0.05 μm deagglomerated alumina suspension diluted with distilled water (one part deagglomerated alumina suspension to 10 parts distilled water by volume), neoprene cloth (pad), complementary direction
- *Method 2:* Automated (final) polishing:
  a. 0.05 μm deagglomerated alumina suspension (5 g alumina powder to 1 L distilled water) or a commercial premixed 0.05 μm deagglomerated alumina suspension diluted with distilled water (one part deagglomerated alumina suspension to 10 parts distilled water by volume), neoprene cloth (pad), complementary direction
  b. Vibratory polish with a synthetic silk or Dacron cloth

Common Polishing Artifacts

In preparing composite materials for microscopic analysis, there are many types of artifacts that can result from any of the steps, from rough grinding to final polishing. In the grinding and polishing steps, common artifacts that may be created include scratches, fiber pull-out, matrix smears, streaks, erosion of different phases, and fiber and sample edge rounding and relief. The most common form of damage that occurs in composite samples is surface scratches. Scratches that are difficult to remove with additional polishing may be imparted to the sample surface. This is usually a result of not cleaning the sample before proceeding to the next step, which not only can cause scratches in the present sample but also later samples processed using the same surface. Also, contaminants from other sources may have come in contact with the polishing cloths. In either case, the only remedy is to start with new cloths and make sure they are kept clean.

A variety of other processing sources can lead to damage. For example, another possible form of damage in composite materials is fiber pull-out. This can occur if the sample is not mounted correctly (for example, having unimpregnated, large-scale, damaged areas) and usually in combination with excessive sample pressure while grinding. Making sure the sample is completely impregnated will usually solve the problem. Scratches may also be found on the sample surface if fibers are removed from the sample during polishing. These harder fibers can remain in the polishing cloths and produce damage in following samples. The smearing of a composite matrix is usually not possible unless the matrix is soft and contains very few fibers or the fibers are made from a soft polymer. Thermoplastic matrices are more susceptible to smearing than thermosetting matrices. Even if the matrix is soft, the addition of a small amount of carbon fibers will result in enough support to the sample that smearing will not be an issue. Smearing usually is caused by insufficient coolant, holding the sample in one direction, or applying excessive pressure. The formation of streaks in a composite material usually is found with matrices that have hard inclusions or porosity. The streaks are caused by the samples being held in one orientation and are further noticed with high sample pressure. Erosion of different phases, fiber edge rounding, and relief are commonly observed with composites that have been polished using suspensions with abrasive particle concentrations that are too high, polishing cloths that are too soft or have nap, or overpolishing (such as excessive polishing time).

While the artifacts mentioned here comprise a majority of those observed in sample preparation, others may be found. If the techniques, materials, and processes for preparing composite materials that are described in this chapter are used without much deviation, the result should be high-quality polished specimens.

REFERENCES

1. B.S. Hayes and L.M. Gammon, Microscopy, *Composites,* Vol 21, *ASM Handbook,* ASM International, 2001, p 964–972
2. L.M. Gammon and B.S. Hayes, Microscopy of Composite Materials, *Structure—J. Materialogr.,* Vol 38, April 2001, p 16–18
3. L. Bjerregaard, K. Geels, B. Ottesen, and M. Ruckert, *Metalog Guide, Your Guide to the Perfect Materialographic Structure,* Struers A/S, Denmark, 1996

CHAPTER 4

Special Sample Preparation and Polishing

THE MOST COMMON methods and materials for preparing polymeric composites for microscopic analysis are described in previous chapters. These preparation techniques can be used for most fiber-reinforced composite materials. There are, however, a few composite materials that require special preparation techniques (Ref 1). These may consist of only special mounting materials or techniques, while other sample types may require completely different sample-preparation methods, from mounting to final polishing. Most of the special procedures that are required are due to differences in the material properties in the same sample. This may be a result of widely differing material properties within the composite material such as boron fiber/thermoset resin, or due to the combination of a fiber-reinforced composite with other classes of materials such as carbon fiber composite/titanium fastener. In either case, this makes sample preparation a challenge. Accordingly, the sample-preparation methods presented in this chapter should be used only when necessary.

Preparation of Titanium Honeycomb Composites

A material system that is used in some high-performance applications is a combination of titanium honeycomb sandwiched between polymer-matrix carbon fiber composite facesheets. The open and thin walls of the titanium honeycomb require a special mounting technique. This combination of materials cannot be prepared unmounted or by clamping alone. If this is attempted, the titanium foil will destroy the polishing cloth and cause significant damage to the sample. To make these types of specimens ready for

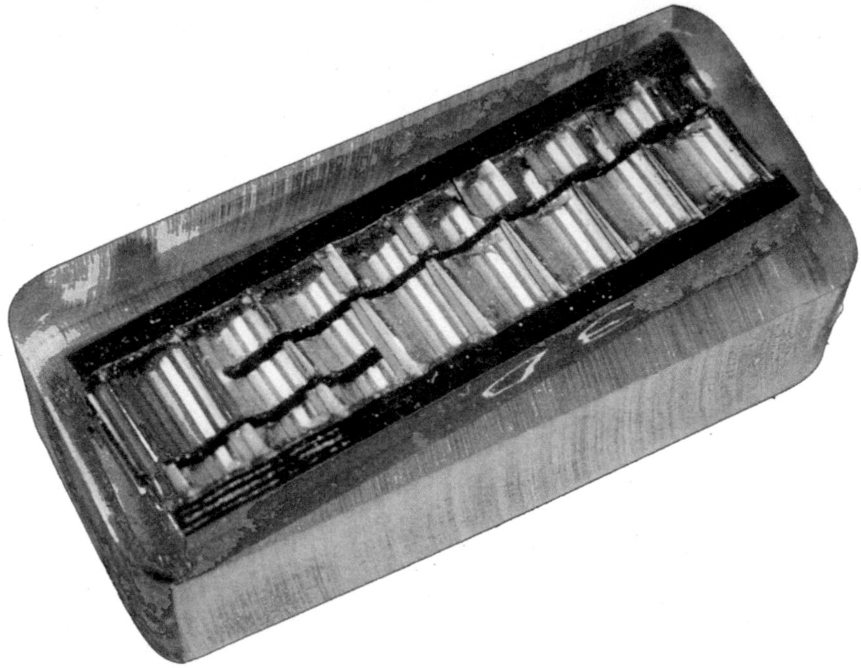

Fig. 4.1 Mounted titanium honeycomb specimen. Note the cuts introduced into the titanium honeycomb to facilitate resin impregnation.

grinding and final polishing, it is necessary that the honeycomb is impregnated with a mounting resin. This is accomplished using a thin abrasive blade and making cuts through the honeycomb core, but not far enough to separate the sample. An easy method to ensure the sample is not separated is to cut one-third of the distance from one composite facesheet, that is, over two-thirds the length of the sample, and make a second cut one-third the distance from the opposite facesheet and which exceeds the distance of the first cut. This will provide an opening in the honeycomb for impregnation with the mounting resin (Fig. 4.1). When the cuts have been made, the sample must be cleaned and dried before it is placed in the cavity mold. The sample can then be impregnated under vacuum with the mounting resin and cured with the application of pressure. The mounted specimen can then be ground and polished as described in Chapter 3, "Rough Grinding and Polishing," in this book. Preparation of other metallic honeycomb-cored composites will benefit from this technique as well as thin-walled ribbed composite structures.

Preparation of Boron Fiber Composites

Special preparation methods are required for composites that contain brittle and/or hard fibers, such as ceramic oxide, silicon carbide, or boron fibers, in a polymer matrix. All steps, from sectioning to final polishing,

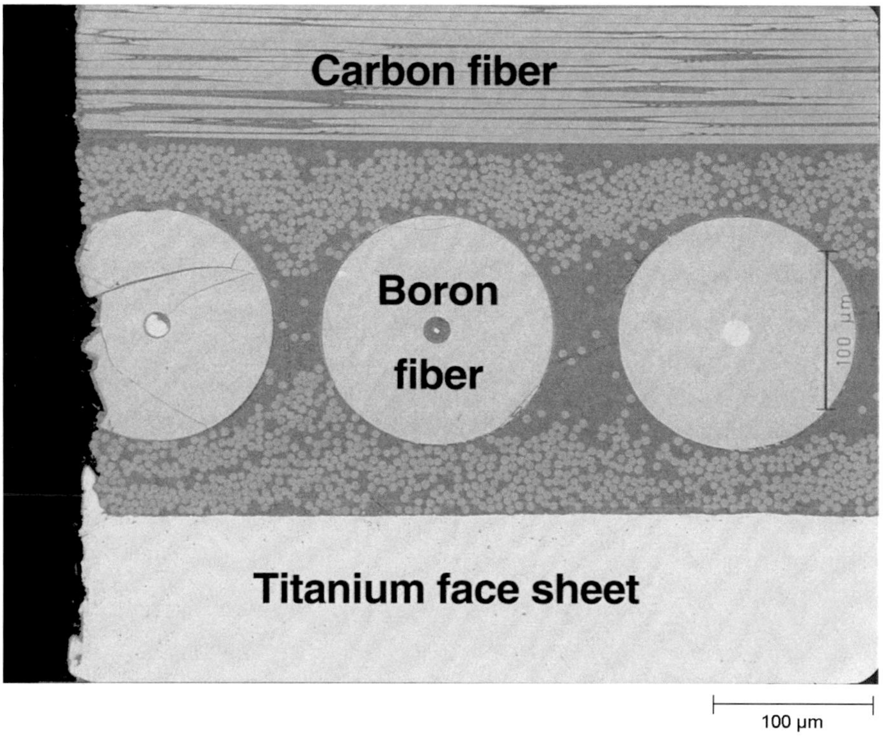

Fig. 4.2 Results of a diamond saw cut and the effect on the brittle boron fibers. The cracked fiber is easy to see, and scratches are evident in the micrograph. Bright-field illumination, 25× objective

are different from those described in Chapters 2 and 3 of this book. The sample-preparation technique for boron fiber composites highlights the effect of having widely different material properties contained in one composite material. Boron fibers are extremely brittle and the critical crack length is sufficiently short that any stress raiser or abrasion may result in the fiber shattering (Fig. 4.2). Given this, trying to cross section and polish boron fiber composites with conventional techniques can be frustrating as the boron particles break loose and gouge the polishing surface, frequently inflicting damage faster than it can be removed (Fig. 4.3). Although diamond polishing medium is more expensive than alumina, and usually not required or recommended for polymer-matrix fiber-reinforced composites, this is one case where it is very useful, if not necessary.

Mounting

The procedure for mounting boron fiber composites begins with preparing a 25.4 mm (1 in.) thick cylindrical blank compression mount using diallyl phthalate mounting compound. When the blank block has been made, two slots are cut into the block for insertion of the boron fiber composites. It is best to make the two parallel saw cuts in the blank such that

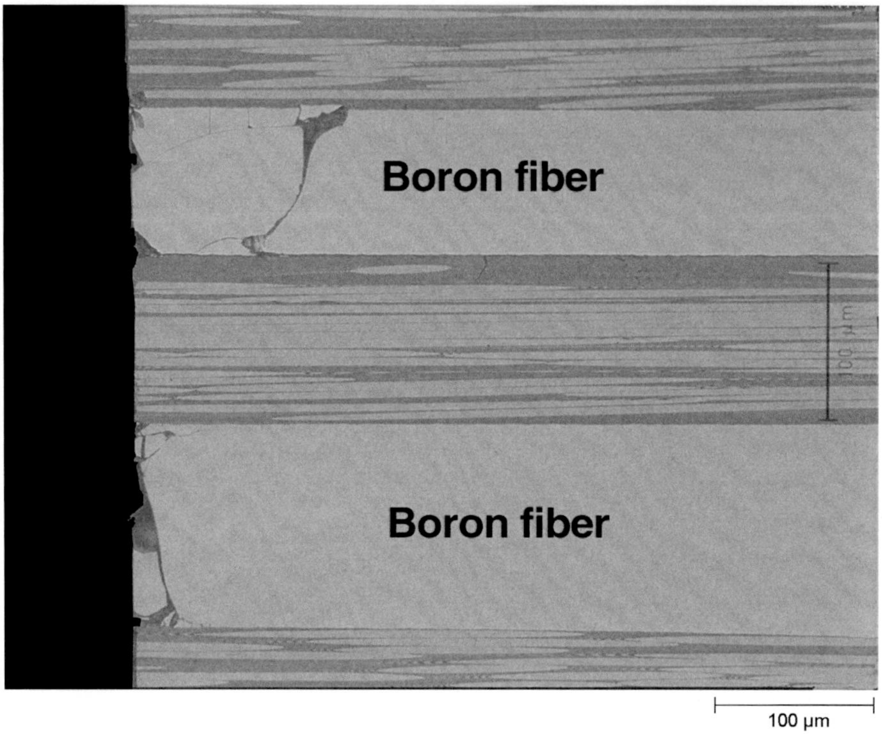

Fig. 4.3 Effect of a diamond saw cut on a boron fiber composite. Cracking can be seen to extend over 100 μm into these large brittle fibers. Bright-field illumination, 25× objective

the diameter of the blank is divided into three equal segments. Do not cut through the entire blank. Make the cuts as narrow as possible but wide enough to insert the boron fiber composites. Place the slotted diallyl phthalate blank into a rubber mold having an equal diameter, and insert the boron fiber composite specimens into the slots. Using a low-viscosity, low-shrinkage, room-temperature-cure epoxy, adhere the specimens in the slots (Fig. 4.4). It is best to perform this procedure in a vacuum chamber to remove entrapped air in the mounting resin, followed by curing under pressure, as described in Chapter 2, "Sample Preparation and Mounting." It is important that boron fiber composites are not mounted completely in a block of epoxy. Unlike diallyl phthalate, epoxy will load up on the diamond wafering blade in the following step and hinder cutting. Like epoxy, diallyl phthalate is a very hard material and offers excellent edge retention for the polishing procedure.

Sectioning and Polishing

Most of the problems associated with cutting and polishing boron fiber composites can be averted by combining sectioning and polishing into a single step. A thin diamond wafering blade is used to cut the mount through

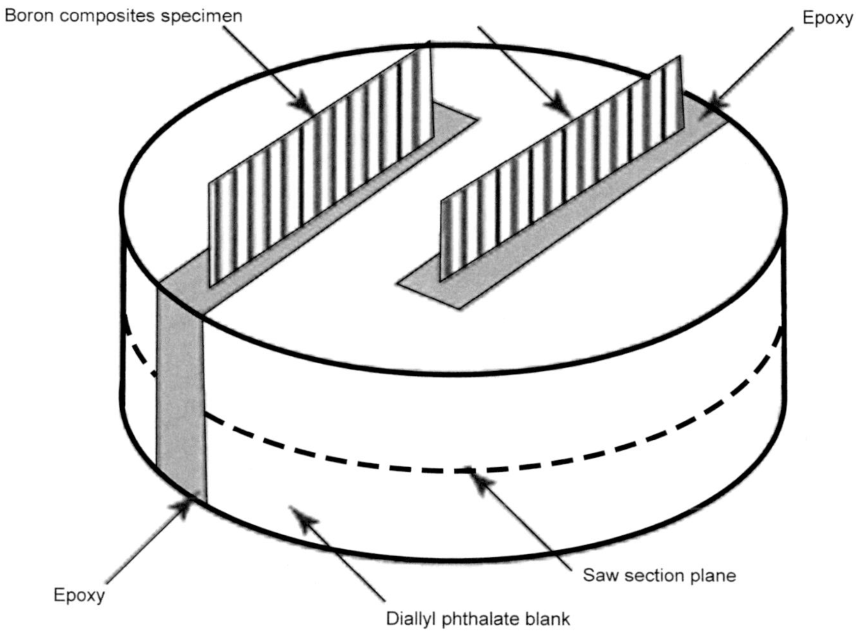

Fig. 4.4 Schematic showing the mounting of boron composite specimens in a diallyl phthalate blank

the midsection, as shown in Fig. 4.4. This process will bypass the steps that cause the majority of damage to the sample. After the sample has been sectioned with the diamond saw, it is ready to be polished. The diamond blade will provide a 320-grit equivalent roughness. Polishing is best performed using 1 μm polycrystalline diamond suspension. It is important not to use silicon carbide papers, because these will induce deep cracks in the specimen. Diamond polishing media are very effective for polishing materials with large differences in material hardness. Either a stainless steel mesh cloth or phenolic pad can be used in this procedure. Of these two polishing surfaces, the phenolic pad is best. However, both surface types will result in a slow polishing process. The vertical force that is applied during polishing should be high enough to lock the polishing compound into the polishing surface without overloading the equipment. A platen speed of 150 rpm should be used, with the head rotating counter to the platen at the same rate. The polishing surface must be kept wet with the solution of the diamond abrasive and the addition of ethanol. If the polishing plane is allowed to dry, the fibers can overheat and protrude from the surface of the sample. Destruction of the sample will follow. Ethanol has a high vapor pressure and therefore evaporates quickly, helping to cool the specimen-polishing plane. It is therefore imperative that the ethanol be applied at a rate that keeps the wheel continually wet. A small, automatic atomizer works well. This polishing step will take some time due to the extreme hardness of the fibers. Properly performed, this methodology will prepare specimens with no visible artifacts and with less than ¼ µm edge rounding. Up to this point in the preparation procedure, it is not recommended that hand polishing be used, because precise control of the sample is necessary, and deviations will cause the brittle fibers to fracture.

Because the fibers do not allow the diamond abrasive to penetrate very deeply into the sample, a final polish is not required. However, a neoprene pad with 0.05 µm deagglomerated alumina suspension and light pressure can clean up the last of the artifacts, if so desired (Fig. 4.5). This can be performed with automated polishing equipment or by hand, following the procedures described in Chapter 3, "Rough Grinding and Polishing."

The polishing steps for preparing boron fiber composites are summarized as follows:

1. Make a blank diallyl phthalate compression mount 25.4 mm (1 in.) thick, and cut two off-center slits on opposite sides with equal dimensions.
2. Bond the boron fiber composite specimens into the slits with epoxy mounting resin. Impregnation of the specimens in the slits under vacuum and the application of pressure during cure are recommended.

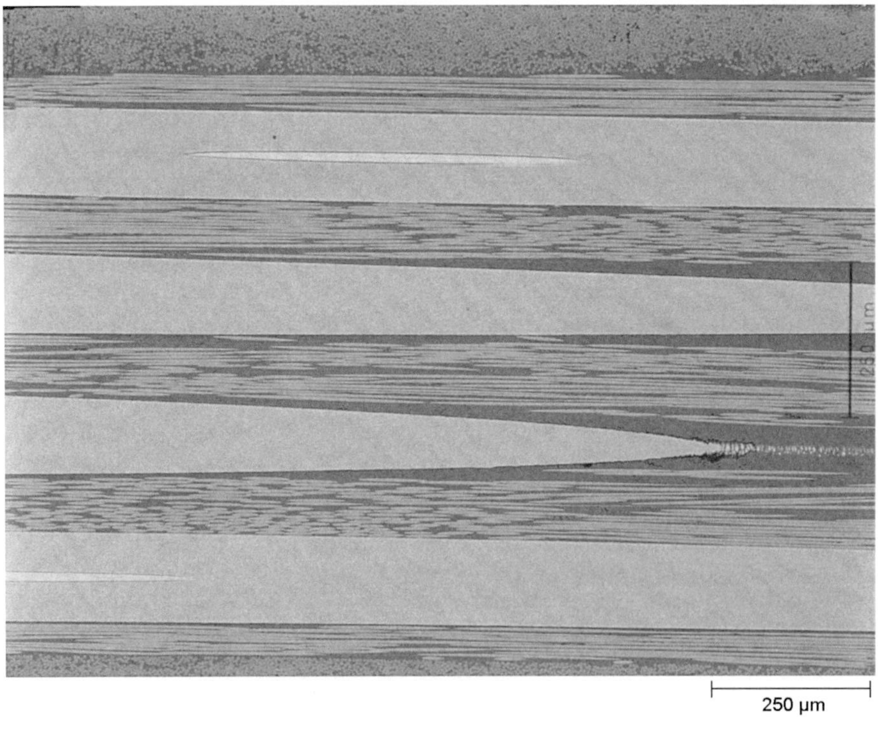

Fig. 4.5 Polished boron fiber composite cross section. Bright-field illumination, 10× objective

3. Cut the mount containing the bonded boron fiber composites through the midsection with a thin diamond wafering blade.
4. Polish the specimens using 1 μm polycrystalline diamond suspension. Also, apply ethanol to the phenolic pad or stainless steel mesh surface while polishing. Use a high vertical force with a platen and head speed of 150 rpm. Specimen movement should be opposite or in counterrotation to the direction of platen rotation.
5. Final polishing (if necessary) can be performed using 0.05 μm deagglomerated alumina suspension, as discussed in Chapter 3. Low pressure and platen speed

Preparation of Titanium/Polymeric Composite Hybrids

Hybrid composites may also require special preparation methods, depending on the difference in material properties of the two or more material types. With advances in structural materials and fastening technology, there is a need for the ability to perform microstructural analysis on titanium fastener/polymeric composite hybrid materials. Sample preparation and polishing of this type of hybrid material combination presents a challenge because the properties are quite different for each material. Traditional titanium preparation techniques will decrease the edge retention of the polymer composite near the interface of the titanium fastener. Conversely, if standard polymer composite preparation methods are used, the titanium microstructure will be destroyed. Given the challenges presented, a method was developed for sample preparation and polishing of titanium fastener/polymeric composite materials for microstructural analysis (Fig. 4.6). This preparation procedure is a compromise between methods used for titanium and composites. The subsequent processes use automated sample-preparation equipment.

Sectioning

As for all composite types, great care must be taken when sectioning hybrid materials so that artifacts are not created or the sample destroyed. There are two methods that work effectively for sectioning the titanium fastener/composite material. The first method is to cut the assembly just above the fastener with a standard abrasive cut-off saw or band saw, mount it in epoxy, and grind it down to just off the centerline of the fastener. With these types of saws, do not attempt to create the final section by cutting through the fastener. The heat introduced from the cutting process will damage the epoxy matrix and the titanium fastener. Cooling will not prevent damage from occurring. Cutting speeds of less than 1000 rpm are necessary to create as little damage to the assembly as possible. The heat damage introduced in the sample will be removed during the grinding pro-

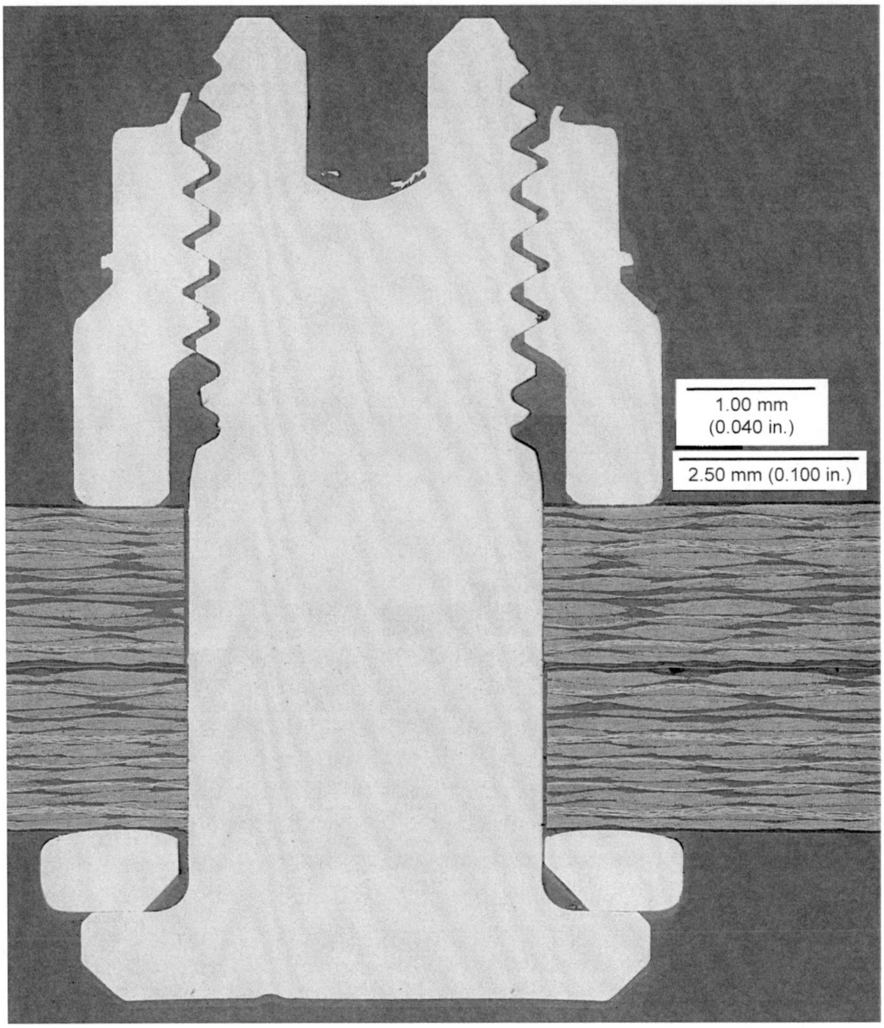

Fig. 4.6 Montage showing a polished titanium fastener/polymer composite assembly. Bright-field illumination, 65 mm macrophotograph

cess. If grinding proceeds past the centerline of the fastener, the fastener can fall out of the assembly. Care must be taken when handling the recently ground sections; the residual stresses at the fastener-composite interface may loosen the fastener, resulting in a distorted fastener-composite interface. Disadvantages of this method are that it is very time-consuming, and the possibility exists that damage introduced from the sectioning and grinding stages will not be removed during the subsequent polishing steps. The end result may be rounded interfaces between the titanium and composite materials and cold-worked titanium. The second sectioning method is accomplished with an abrasive wafering saw using 1000 rpm and applying a 400 to 1000 g (0.9 to 2.2 lb) load (Fig. 4.7). The specimen can be cut

Fig. 4.7 A carbon fiber composite with a titanium fastener is clamped in a fixture and aligned for sectioning. In this figure, the alignment is performed using a straight edge. This sample will be sectioned through the fastener, with the straight edge providing alignment of the edge of the saw blade. Coolant will be supplied during operation through the blue nozzle assembly.

near the final finish plane with very little damage. The section will take one to six hours to complete, depending on the specimen size and the type of blade. The blade must be constantly cooled, and it should be dressed often to reduce buildup of debris. This is the preferred method for sectioning if a wafering saw is available. High-speed saws must never be used because sparks will be created on contact with the titanium fastener, degrading and cracking the composite.

Mounting

After sectioning, the sample should be dried in a vacuum oven at approximately 40 to 50 °C (105 to 120 °F) before placement in the mold (Fig. 4.8a, b). An epoxy mounting resin should then be used to impregnate the interface gap between the composite and fastener while under vacuum. For best quality, the mounted sample should be cured under pressure. The vacuum and pressure can vary, depending on the equipment available. Any gaps at the interface will result in edge rounding. It is advantageous to dye the epoxy with Rhodamine B laser dye (Fig. 4.9). The dyed mounting epoxy under polarized light will appear differently than the matrix of the polymer composite. The procedure for adding the dye to the epoxy mounting resin is described in Chapter 2, "Sample Preparation and Mounting," in this book.

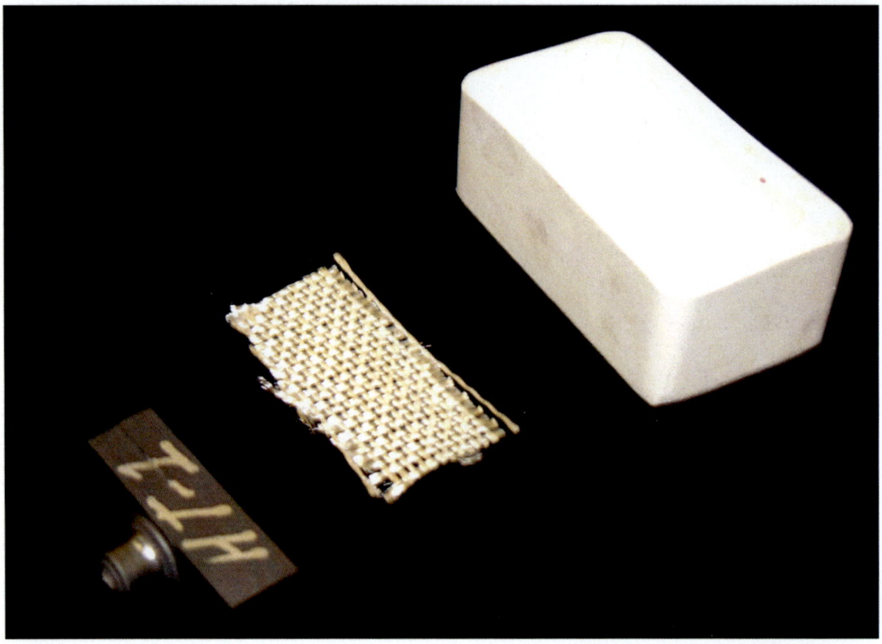

(a)

(b)

Fig. 4.8 Photographs showing the section of the titanium fastener/composite assembly prepared for mounting. (a) The cut titanium fastener/composite assembly is facing down, with a piece of glass fabric to be placed on the bottom of the mold to ensure impregnation of the face. (b) Mold containing the glass fabric and sample

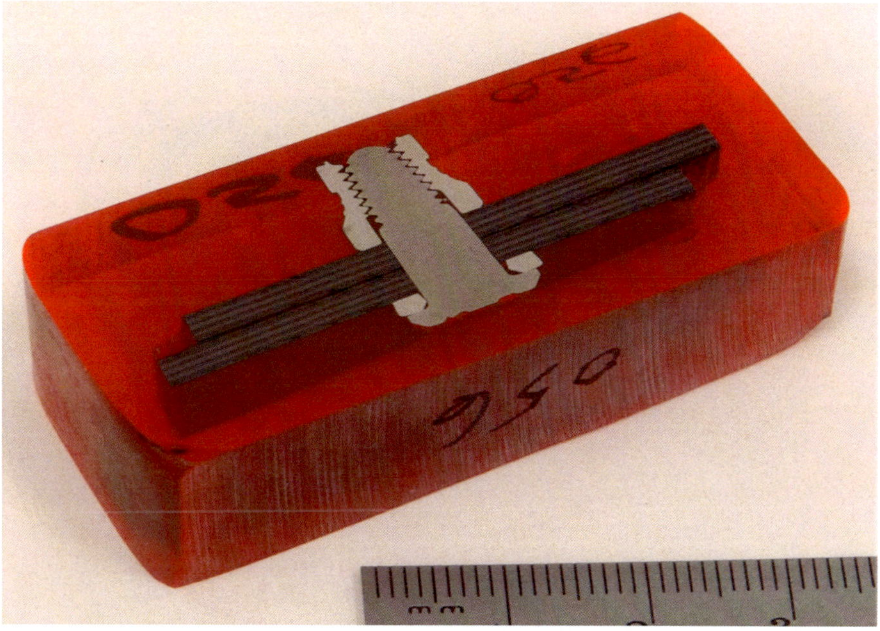

Fig. 4.9 Photograph of the titanium fastener/composite lap joint specimen mounted in Rhodamine-B-dyed epoxy resin. The mount is numbered on the sides for documentation.

Grinding

The sample location in the automated polishing head is different from what has been described for most composite materials, which was discussed in Chapter 2. Typically, the polishing head is loaded with the samples to be polished closest to the center, and the backing pieces of composites are placed on the outer edges of the sample holders. The best method for mounting the hybrid samples is to place them as far to the outside of the sample holder openings as possible. No backing pieces are required using this process. By placing the samples on the outside edges of the sample holder openings, the head can be programmed to travel partially off of the edge of the platen surface to allow the polishing fluid and removed sample material to be cleaned off. This is not usually performed for preparing composite materials, as described in Chapter 3, "Rough Grinding and Polishing."

The rough grinding process is best performed using silicon carbide papers. The successive reduction in grit size is the same as described in Chapter 3. While diamond-bonded disks can also be used, they are not recommended, because hybrid materials tend to damage the disks and therefore can be quite costly. The sample movement is complementary to the platen direction for preparing these hybrid materials, as compared to countermovement for most fiber-reinforced composites (Fig. 4.10). The platen and head speed are the same, 150 rpm, with 60 N (13.5 lbf) applied force.

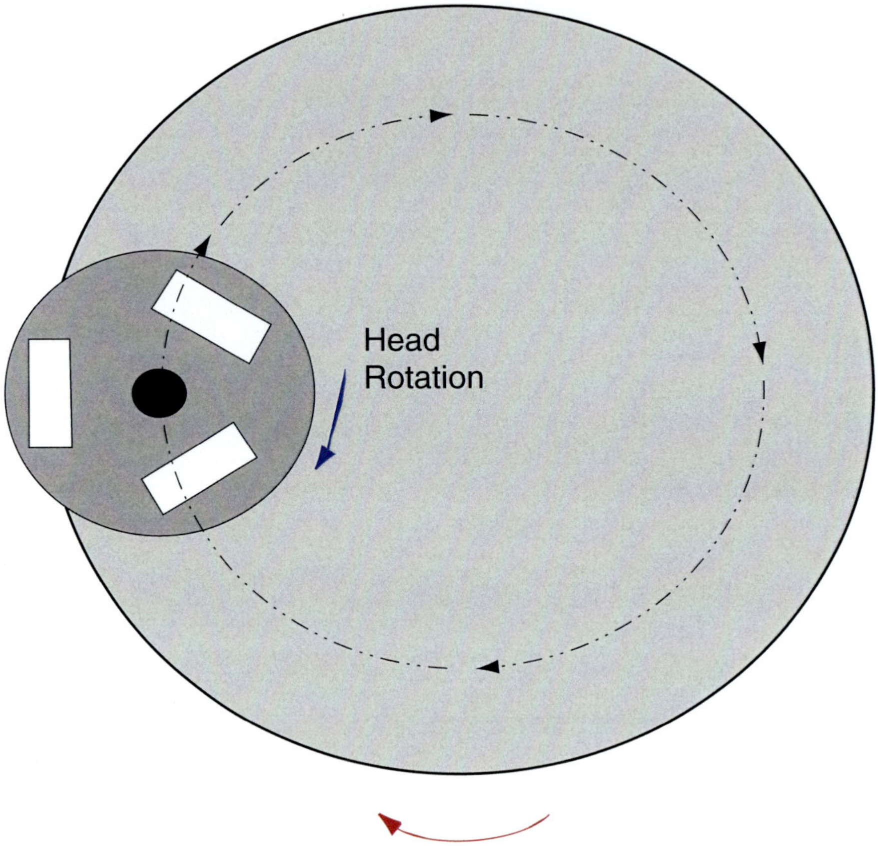

Fig. 4.10 Schematic showing the complementary rotation of the head and platen (polishing surface). The head is shown to travel partially off of the polishing surface to facilitate cleaning.

Polishing

After the last grinding step, there have been many different surfaces that have been used for the first step of the polishing process. These surfaces have included woven, nonnap cloths as well as diamond lapping films. The use of a 9 μm diamond lapping film provides good results but is quite costly due to the ease with which these substrates tear during polishing of hybrid materials. Although lapping films can be used, there is another option that provides better sample quality, lower process time, and less cost. It has been found that after the last grinding step, the next step of sample preparation is best performed with the use of a composite disk. The composite disk is used with a 9 μm polycrystalline diamond suspension (purchased premixed) and also a lubricant. A lubricant/coolant made from alcohol and glycol works very effectively with this process. The abrasive suspension and the lubricant are applied separately on the composite disk surface. The use of a composite (surface) disk provides partial lapping, because the diamond particles are not completely locked in place. The combination of using this type of disk and the process of the sample movement off of the edge of the disk results in effective cleaning of the polishing compound and removed material. These surfaces provide excellent edge retention and sample flatness. Again, this process is a compromise between metal and composite polishing. After this step, the polishing transitions to the use of 0.3 and 0.05 μm deagglomerated alumina suspension on woven nonnap cloths. The alumina concentrations are described in Chapter 3, "Rough Grinding and Polishing," in this book. The last polishing steps can be extended using a vibratory polisher.

In between each of the polishing steps, it is recommended that the titanium is etched so the composite can be more effectively polished. Titanium etching can be accomplished using diluted hydrofluoric acid (1 to 1.5 percent) or Kroll's reagent (92 mL distilled water + 6 mL nitric acid [68 to 70%] + 2 mL hydrofluoric acid [40%]). The titanium sample should be immersed in either solution for 15 seconds and then rinsed with distilled water. (Caution: Read the material safety data sheets for all materials to ensure proper utilization, and always use safe handling methods.) The last polishing step can include the addition of 50 vol% of 3% hydrogen peroxide solution into the 0.05 μm deagglomerated alumina suspension to help etch the titanium while polishing. This last process will take approximately one hour. The polishing surface that is recommended for the 0.3 μm deagglomerated alumina suspension is satin silk, while the polishing surface for the final 0.05 μm deagglomerated alumina suspension step is woven satin acetate. Although silk can be used for this last polishing step, satin acetate is recommended because it is harder than silk and provides a higher-quality final polish.

The applied pressure should be kept as high as possible without overloading the automated polishing equipment. A platen and head speed of 150 rpm, with sample movement complementary to the direction of the platen, is very effective for all polishing stages.

The following summarizes the polishing steps:

1. 9 μm polycrystalline diamond suspension (purchased premixed), composite disk, alcohol/glycol-based lubricant, complementary sample movement
2. 0.3 μm deagglomerated alumina suspension, satin silk, complementary sample movement
3. 0.05 μm deagglomerated alumina suspension (50 percent volume of three percent hydrogen peroxide), woven satin acetate cloth, complementary sample movement. A vibratory polisher can be used to further refine the polish, if desired.

To etch the titanium fastener in the assembly to show the microstructure, a few drops of an oxalic acid solution can be applied to the surface for approximately 15 seconds (Fig. 4.11).

Preparation of Uncured Prepreg Materials

All of the prior preparation techniques have focused on cured polymeric composites, but there is one uncured composite material that also commonly requires optical analysis. This material is thermosetting-resin-based

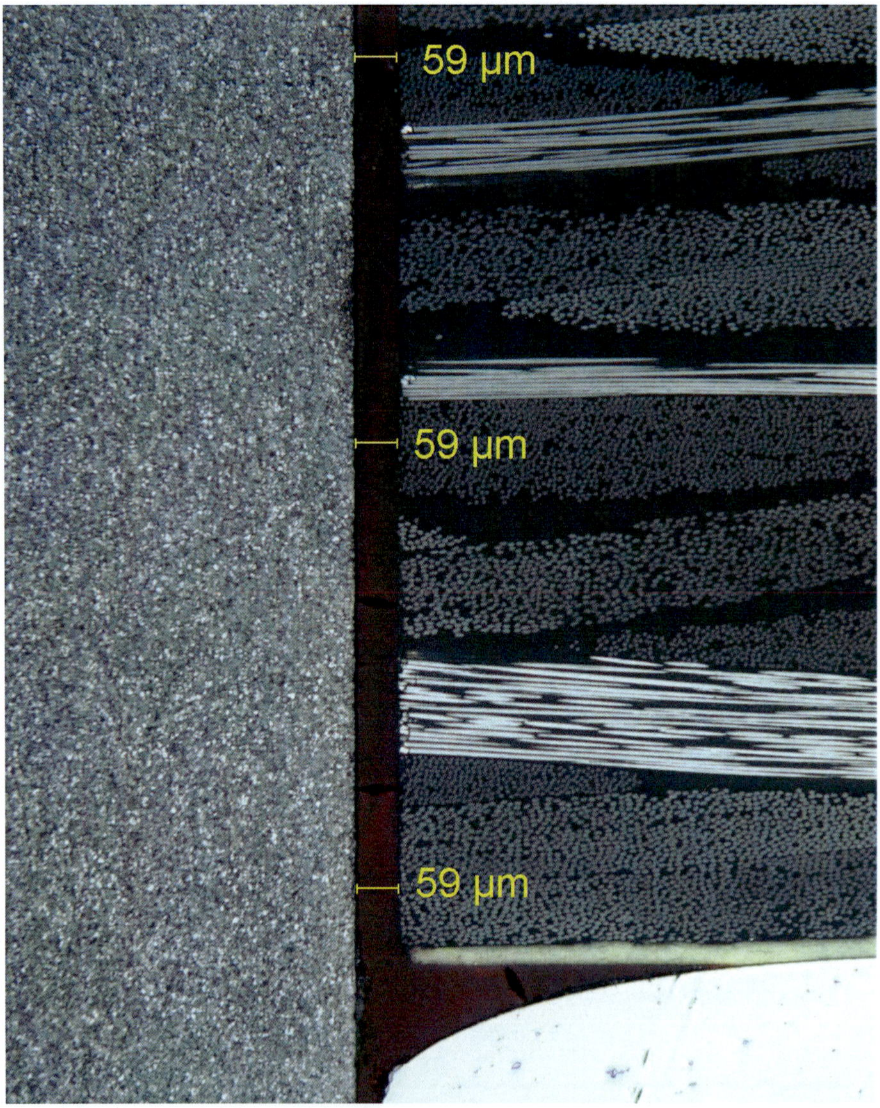

Fig. 4.11 Titanium fastener/polymer composite assembly etched with oxalic acid for 15 s. A gap of 59 µm can be seen between the composite and the etched titanium. Slightly uncrossed polarized light, 10× objective

prepregs, which are the precursors to many high-performance composite parts. These types of prepregs are made by preimpregnating continuous fibers or fabric with a controlled volume of uncured thermosetting resin. Optical analysis of uncured unidirectional or woven fabric prepreg provides information about the degree of impregnation, the resin and fiber distribution, and commingling of adjacent fiber tows (Ref 2, 3). This information can be used to understand and identify the effects of different prepreg processing parameters on the cured microstructure and their effects on the bulk factor. The resin and fiber distribution information also provides a better understanding of the prepreg tack, drape, and unidirectional prepreg transverse integrity (Ref 4, 5).

There are two methods that can be used for mounting prepreg materials. The type of mount is dictated by if the prepreg is to be prepared uncured (as received) or if the prepreg (matrix) will be staged (partially cured) before grinding and polishing. Each mounting method has its advantages and disadvantages in terms of artifacts, preparation, and analysis. The type of mounting method that is used determines the remaining preparation steps and techniques.

Mounting Uncured Prepreg Materials

If uncured prepreg (as received) is to be analyzed, usually only one or two pieces can be placed in the same mount. This is due to the lack of integrity of uncured prepreg. The preparation of a single piece of uncured prepreg involves bonding two plastic pieces on the outside of the sample, followed by grinding and polishing. This enables fast analysis that does not alter the resin impregnation of the prepreg sample. It is possible that two pieces of uncured prepreg can be bonded together, and then the two pieces can be bonded between pieces of plastic for support. Bonding of more than two pieces of uncured prepreg usually results in specimen failure during polishing.

To prepare a single piece of uncured prepreg for mounting, a similar procedure can be used as described in Chapter 2 for hand mounting composite samples. The first step is to cut two backing pieces of plastic, between which the uncured prepreg can be bonded. Thin pieces of thermoplastic or epoxy resin castings, approximately 6.5 mm (0.25 in.) thick, are optimal. The backing pieces must be sized so they will overlap the ends of the prepreg by at least a few millimeters to create a good bond with the adjacent backing piece. Backing pieces 50.8 mm (2 in.) long by 31.75 mm (1.25 in.) high have been found to work very well for this procedure. The uncured prepreg is cut with sharp scissors, keeping the release liner(s) on the prepreg surface. A sample 38.1 mm (1.5 in.) long by 19 to 25.4 mm (0.75 to 1.00 in.) high is recommended. Immediately after cutting the prepreg, "5 min epoxy" is mixed and applied to both backing pieces. The resin is allowed to react for a few minutes to increase the viscosity so this bonding resin does not impregnate the prepreg but still flows to provide the necessary adhesion. Before mounting, the release liner(s) must be removed from the prepreg, and then it can be placed between both pieces of the epoxy-coated backing plastic. A small binder clip can be used to hold the sample together while the bonded sample is curing. It is usually necessary to prestress the binder clips so the pressure is not too high on the prepreg sample. If the pressure is too high, it may cause the resin distribution in the prepreg to be altered. The bonded, uncured prepreg sample is set aside to allow the "5 min epoxy" to cure. When the epoxy is cured, the sample is ready for grinding and polishing.

Mounting of Staged (Partially Cured) Prepreg Materials

If the prepreg is staged, a stack of multiple pieces of staged pieces of prepreg can be mounted. The use of staged prepreg material provides the opportunity to prepare many sample pieces at one time or to provide a greater view along the length of a prepreg material. However, the staging process can take some time and can alter the resin distribution in the sample(s). If staging is performed at greater than room temperature, this reduces the viscosity of the matrix and can cause further resin impregnation into the prepreg. The temperature (and time) at which a prepreg material (matrix) is staged is the key parameter that dictates the amount of resin movement occurring in the prepreg. With some matrix materials, adequate staging at room temperature can be accomplished in as little as a few days, while others may take two months or more. Staging at elevated temperatures is faster and still allows the observation of the prepreg morphology, but the degree of resin impregnation may not be accurate.

Staging prepreg can be accomplished by either leaving the prepreg at ambient temperature for an extended time or in an oven at slightly elevated temperature. As mentioned previously, this depends on the information that is desired and the matrix chemistry. Mounting of the sample uses the same procedure as mentioned previously, where the unstaged (and will be staged as a complete mount) or staged prepreg plies are cut with sharp scissors and bonded together. The width of the overall sample, made from multiple plies of prepreg, that is in contact with the platen surface should be between 12.7 and 19 mm (0.5 and 0.75 in.) thick. The length of the prepreg should optimally be less than 50.8 mm (2 in.) and should be approximately 25.4 mm (1 in.) high for ease of holding. To make the sample, "5 min epoxy" is applied to one side of the prepreg surface and built up with additional cut prepreg plies, using adhesive between each ply, to the desired thickness. After the mount has been made, a small weight or large prestressed binder clip can be used to hold the sample together while curing. On the sides of the samples, there must be flat faces so the sample is not distorted and the polishing surface is planar. This can be accomplished using metal or composite backing pieces that are removed after the epoxy cures. A piece of the discarded release liner may be used in between the mount and the backing pieces to prevent adhesion. If the prepreg has not been previously staged, it can be put in a low-temperature oven for an appropriate amount of time, after the "5 min epoxy" resin cures.

An easy method to view the variation of prepreg quality across the width of the material is to cut the prepreg across the width in strips 25.4 to 50.8 mm (1 to 2 in.) wide. Then, apply "5 min epoxy" to one side of the uncured prepreg and keep folding it over, applying new adhesive to each side until it is folded up to a final length of approximately 50.8 mm (2 in.). One sample can be made from prepreg strips that are over 40 cm (15.75 in.) long, but this depends on the areal weight and prepreg thickness. The final thickness should be between 12.7 and 19 mm (0.5 and 0.75 in.). After folding, the adhesively bonded prepreg is placed between release films, hard backing pieces are placed on the outside, and the assembly is clamped together. After the "5 min epoxy" cures, the sample can be staged at ambient temperature or in an oven. When the degree of resin advancement is adequate to create a structural mount, the backing pieces can be removed and the mount polished.

Mounting of Prepreg Materials—General Comments

In mounting prepreg using either technique, the release paper, liner, or film on the prepreg must remain on until immediately before adhering the backing pieces or other pieces of prepreg. If there is only one release ply on the prepreg surface, this material should remain rolled up until the sample(s) is ready for bonding. The release liner keeps the resin held to the surface of the prepreg. When the liner is removed, the resin can wick into the more permeable fiber bed and make the initial impregnation from the prepregging process difficult to observe. With either method, the addition of the adhesive to the surface of the prepreg(s) prevents further matrix migration and helps the resin remain in its original location after mounting. Likewise, a piece of prepreg may be left out with the release liner off of both sides of the material for an extended period so that cold impregnation can be observed. For some resin-fiber systems, cold impregnation can occur rapidly. This can help in understanding why some prepreg materials change in tack or process differently (for example, prepreg table roll processing) after they are left with release liners off of the prepreg surface for periods of time. In most thermosetting prepreg systems, it is usually not the advancement of the resin (staging) that is occurring rapidly at room temperature but the cold impregnation of the resin into the fiber bed.

Grinding and Polishing of Unstaged and Staged Prepreg Materials

Grinding and polishing prepreg materials is usually performed only by hand. After preparing the staged prepreg samples, the grinding and polishing process is the same as described in Chapter 3, except that the last step using the 0.05 μm deagglomerated alumina suspension is usually omitted. The time required for each grinding and polishing step is usually found to be different for prepreg versus cured composite preparation, because the uncured matrix is not as hard as the fully cured materials. Figure 4.12 shows a micrograph of a polished stack of staged unidirectional carbon fiber prepreg. Unimpregnated regions are shown as black voided areas. Also shown is the thickness of the surface resin in contrast to the "5 min epoxy" adhesive between the prepreg plies.

A higher-quality final polish is usually found for staged prepreg samples compared to unstaged prepreg samples, due to the greater mount integrity and resin hardness. Furthermore, a disadvantage of polishing unstaged prepreg is that the specimens must be ground and polished with more care than a staged stack of prepreg. If the unstaged prepreg mount is held in one position for even a short period of time, some of the fibers may be pulled out and the surface smeared. It is also essential that a large quantity of water is used on the wheel for cooling. Water is the best coolant, because the majority of the resin matrices are hydrophobic.

The procedure for grinding and polishing unstaged prepreg is similar to the process described in Chapter 3, with a few exceptions. After the "5 min epoxy" has cured, the grinding step starts with 320-grit silicon carbide paper to remove the artifacts introduced by cutting the prepreg with scis-

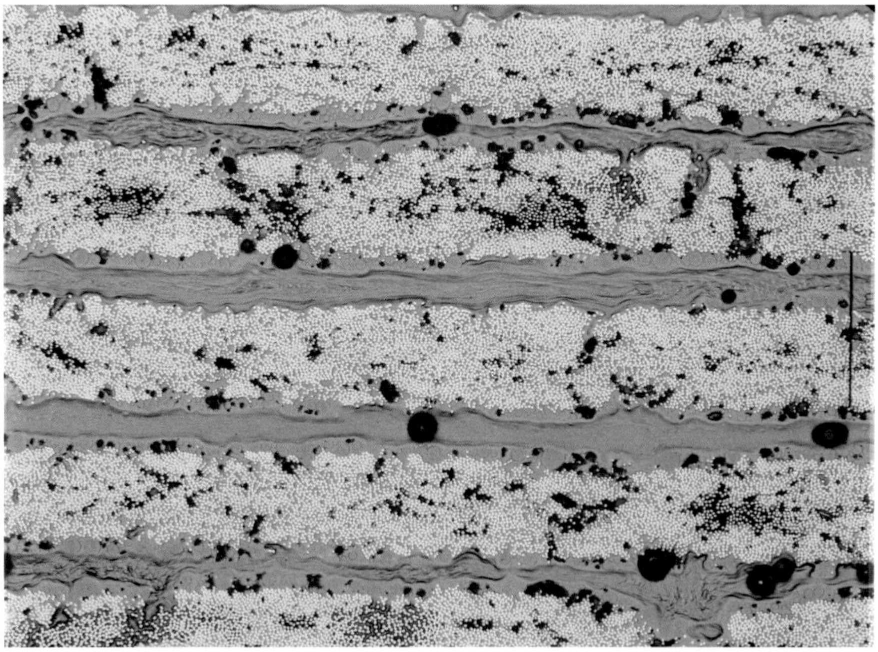

Fig. 4.12 Polished stack of prepreg that was bonded with "5 min epoxy" and then staged in an oven at 60 °C (140 °F) for 2 weeks. This micrograph is of 4½ plies of polished prepreg taken from the 20-ply mount. Bright-field illumination, 10× objective

sors. The use of 120-grit paper is not recommended as a first step because the larger abrasive size can pull out fibers and damage the specimen. A high platen speed (up to but not exceeding 1000 rpm) should be used, with moderate pressure applied to the sample. It is necessary that water is constantly applied on the platen surface during all stages of grinding and polishing. The sample must be moved continuously on the grinding/polishing surface, or the fibers will be oriented or pushed up into the mount. After the cut surface and any of the epoxy bonding resin are removed, the sample is subjected to the same procedure using 600-grit silicon carbide paper. These steps are usually very short and should be able to be performed in approximately a minute or two for each step. The sample should be picked up and reoriented multiple times during each stage so as not to give a false orientation to the fibers. After the grinding step, the first polishing step is performed using 15 μm deagglomerated alumina suspension (12 g alumina powder to 1 L distilled water) on a silk cloth. A high platen speed, up to 1000 rpm, is most effective, with moderate pressure. After approximately a minute of polishing, the final surface is prepared using 5.0 μm deagglomerated alumina suspension (12 g alumina powder to 1 L distilled water) on a new silk cloth, using the same process parameters. There is usually no reason to go beyond a 5 μm polish to be able to successfully obtain the information that is required. However, a final polish with 0.3 μm deagglomerated alumina suspension can be used, but it is not recommended because there is a greater chance to smear the surface and create artifacts. If necessary, this step should be very short, with the sample on the platen surface for no more than 10 s before picking it up and reorienting it on the platen surface. Only approximately 30 or 40 seconds in total should be required for this 0.3 μm polishing step.

After each of the aforementioned steps, it is necessary that the sample be cleaned with running water to remove all of the prior cut surface and grinding/polishing media. This cleaning should consist of using only water, no soap. Do not use sonication or a brush for cleaning these samples, because the sample can be altered. Figure 4.13a to c show the cross sections of uncured unidirectional carbon fiber prepregs that have been polished. These prepregs were made using different processing conditions. The gray regions within the sample are where the matrix resin is located, while the black regions are areas that were not impregnated by the matrix. In the grinding/polishing procedure, the unimpregnated areas of the prepreg usually absorb water. This water can create an artifact if not removed. The sample can be placed on a dry, lint-free cloth for a few minutes, and the water will wick out from the mounted prepreg material into the cloth.

After the final polishing step, the surfaces of the prepreg materials are not nearly as high a quality as that of cured matrices, because the resin is not a solid. These matrices vary in viscosity, viscoelasticity, and fillers, all of which affect the quality of the final polish. When working with a sample

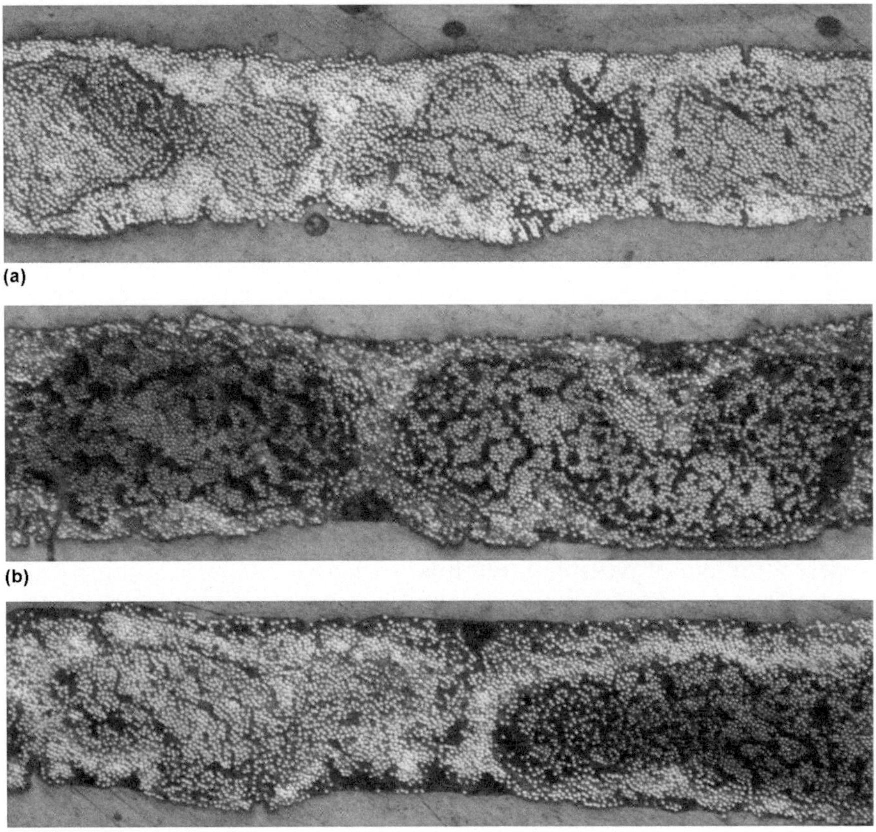

(a)

(b)

(c)

Fig. 4.13 Polished uncured prepreg materials that were made with different prepreg processing conditions. Bright-field illumination, 25× objective

of a staged prepreg composed of a soft matrix, it is difficult to prevent erosion of the matrix. In the preparation of an unstaged prepreg sample, the resin is so soft that the objective is not to smear it but to remove the resin from the place it is located. The result is not a true polished resin surface, because it is not a solid. However, the information that is required from the analysis is also different from cured composites and mainly involves resin impregnation, resin distribution, tow morphology, and intermingling of adjacent fiber tows.

REFERENCES

1. B.S. Hayes and L.M. Gammon, Microscopy, *Composites,* Vol 21, *ASM Handbook,* ASM International, 2001, p 964–972
2. C.J. Martin, J.W. Putnam, B.S. Hayes, J.C. Seferis, M.J. Turner, and G.E. Green, Effect of Impregnation Conditions on Prepreg Properties

and Honeycomb Core Crush, *Polym. Compos.,* Vol 18 (No. 1), 1997, p 90–99

3. B.S. Hayes, J.C. Seferis, and J.S. Chen, Development and Hot-Melt Impregnation of a Model Controlled Flow Prepreg System, *Polym. Compos.,* Vol 17 (No. 5), 1996, p 730–742
4. J.W. Putnam, B.S. Hayes, and J.C. Seferis, Prepreg Process-Structure-Property Analysis and Scale-Up for Manufacturing and Performance, *J. Adv. Mater.,* July 1996, p 47–57
5. E.C. Bruce, B.S. Hayes, J.C. Seferis, T. Pelton, and M. Wilhelm, Investigations of Unidirectional Transverse Prepreg Integrity in Relation to Processing Parameters, *Proceedings of the 44th International SAMPE Symposium and Exhibition,* May 24–27, 1999 (Long Beach, CA)

CHAPTER 5

Viewing the Specimen Using Reflected-Light Microscopy

WHEN VIEWING A POLISHED fiber-reinforced composite specimen, there are many factors to consider. The composite constituents (such as matrix and fiber type), the capability of the microscope, and the type of information that is desired will play an important role in the analysis. As with all analyses of composite materials using an optical microscope, the more that is known about the material or can be relayed to the microscopist, the easier the analysis. In most cases, the relevant detail may be resolved from 25 to 100× magnification using reflected-light bright-field illumination. However, there are many cases where the full features of an optical microscope may be necessary to obtain the required information. An understanding of the microscope is helpful to fully use its capabilities and therefore analyze composite materials. Due to the many excellent publications on this subject, only a brief overview of the illumination light paths and techniques is described in this chapter (Ref 1 to 7). Also, very helpful website references are available (Ref 8 to 11). The information in this chapter is designed so that a microscopy technique can be effectively selected for the specific analysis of a composite material (Ref 12).

Macrophotography and Analysis

Macrophotography is the merging of photo studio and optical microscopy techniques (Ref 13, 14). This combined technique has all the chal-

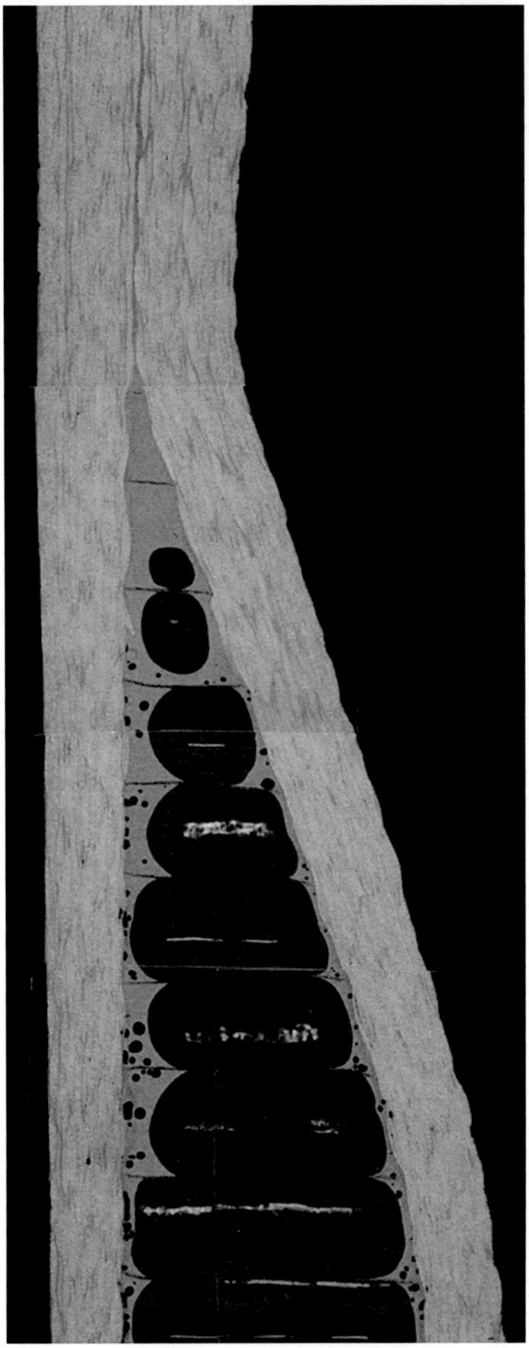

Fig. 5.1 Carbon fiber composite/honeycomb chamfer area. Bright-field illumination, 5× objective. 4 × 5 in. 14-picture (Polaroid) micrograph montage

lenges and methods of both. There is a tendency to overlook the macro aspect of the analysis and go straight to the cross-sectioning or electron-microscopic portions of the analysis. All forms of microscopy are complementary to each other. Photo studio and macroimaging work are at the heart of any research project or investigation. This can be the most challenging and difficult aspect of any microscopic project. Once the specimen is dissected, it is often impossible to go back and capture the as-received image. As the specimen is broken down or dissected, it is impossible or, at best, very difficult to reassemble it to document the section locations. Macrophotography uses all the techniques of both microscopy and the photo studio. In any microscopy work, there is no substitute for a thorough understanding of lighting techniques.

The capture of a bright-field macroimage (1 to 30×) on polished materialographic cross section is a significant challenge. On a microscope, only the light filament needs to be aligned, because the rest of the optics are preset. Successful use of the macroscope requires that each operating parameter be customized to the task at hand. The light source, filament aperture, distance, intensity, half-mirror, and alignment of the specimen flat to the film plane are all critical. It is difficult (if not impossible) to produce an instrument that will give a high-quality macroimage with preset conditions. With the correct components and careful alignment, a high-quality image can be obtained (Fig. 5.1).

Microscope Alignment

Aligning your microscope with Kohler illumination is essential in capturing the fine detail of your specimen. The first step in aligning the microscope is ensuring that the specimen is oriented normal to the light path (that is, 90 degrees to the light path). The importance of this step becomes apparent, because the resolution limit is a function of the alignment as well as the numerical aperture. Likewise, the integrity of the light path must be maintained.

The light source, internal mirrors and lines, apertures, and cleanness of the whole system play an important role. Although there are rare samples that have multiple focal planes and thus benefit from an off-centered aperture, these are the exception, not the rule. The back focal plane of the microscope may be observed using an internal Bertrand lens or a phase-focusing telescope so the rays of light can be properly aligned. In addition, the aperture opening must be optimized for each objective. If using a light meter, fully open the aperture and note the time for exposure. Dim the light until the time has doubled; the aperture is now correctly set for optimal resolution. Note that as the objectives are changed, the aperture setting will need to be adjusted. In general, the larger the magnification, the smaller the aperture.

Bright-Field Illumination

Reflected-light (such as incident light, episcopic, or epi) microscopy is the most common technique for examination of composite materials, but other illumination and contrast methods are often required due to the lack of reflectivity of some materials. Different composites present different types of resoluble features, depending on the reflectivity of the individual components present. The carbon fibers tend to be very reflective (show a high degree of contrast) and are easily resolved in polymeric resins. By comparison, glass fiber composites tend to absorb light equally across the spectrum, making it difficult to distinguish the individual components.

Reflected-light bright-field illumination is used in most cases for viewing fiber-reinforced polymeric composites. When using bright-field illumination, it is essential that the specimen be oriented 90 degrees to the light path. The light is reflected from a half-mirror through the lens, off the sample, and back through the lens to the eye. Therefore, the flat surfaces are bright using this illumination technique, but any voids, microcracks, or indentations are darker because the light is reflected off the sample at an angle. Bright-field illumination is particularly useful for imaging samples for ply counts, fiber-orientation verification, resin-to-fiber ratio determination, void studies, and most microcrack investigations (Fig. 5.2). Gener-

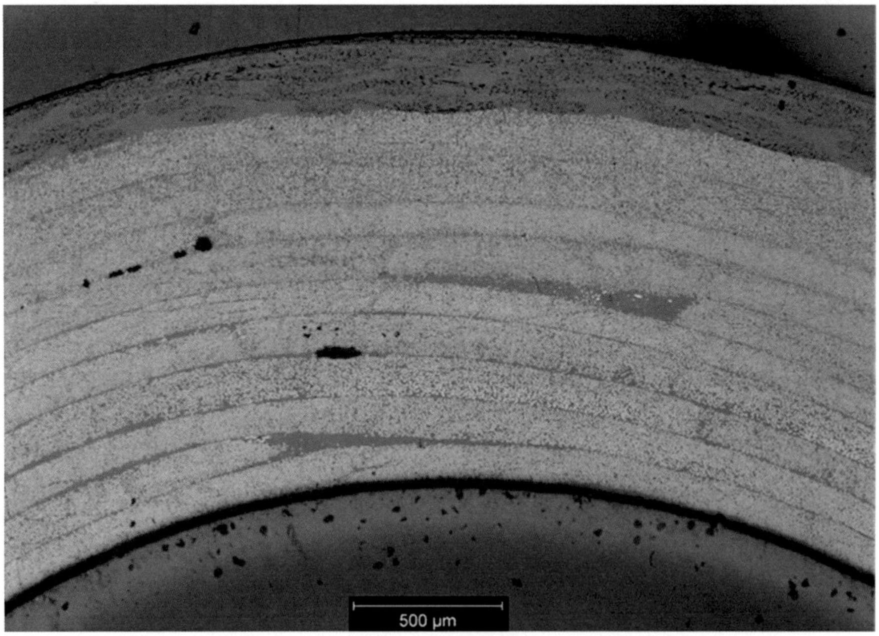

Fig. 5.2 Composite cross section showing many of the different facets that are usually investigated using reflected-light bright-field illumination. Shown in the cross section are voids (dark areas), ply terminations (i.e., ply drops), carbon fiber plies having different thicknesses, different prepreg material combinations (glass fabric prepreg and carbon fiber prepreg), and the number of plies. Bright-field illumination, 5× objective

ally, it is advisable to observe all polished specimens with bright-field illumination before continuing with other illumination methods.

Void analysis of composite specimens can be performed using low-magnification bright-field illumination of the cross section. Usually, 25 to 50× magnification is sufficient for documentation of voids, but magnification up to 100× is sometimes required. It is important to realize that this is the total magnification observed and not just the objective magnification. An example of a carbon fiber-reinforced polymeric composite material containing voids is shown in Fig. 5.3. The voids are shown as dark (black) irregular regions on the polished composite surface. Macro bright-field illumination (less than 30× magnification) can be difficult to achieve. Most microscopes are not set up for this range, and those that are have difficulty achieving the quality of a dedicated macroscope. The lowest magnification of the microscope's capability and a montage of multiple micrographs are often required if a macroscope is not available for documenting voids.

Although voids in composites are most often viewed normal to the *xz* or *yz* plane, the analysis of voids normal to the *xy* plane can offer a more complete understanding of the void morphology. Likewise, more information about microcracks in composites and fracture areas may also be possible from viewing the features in more than one dimension. Techniques

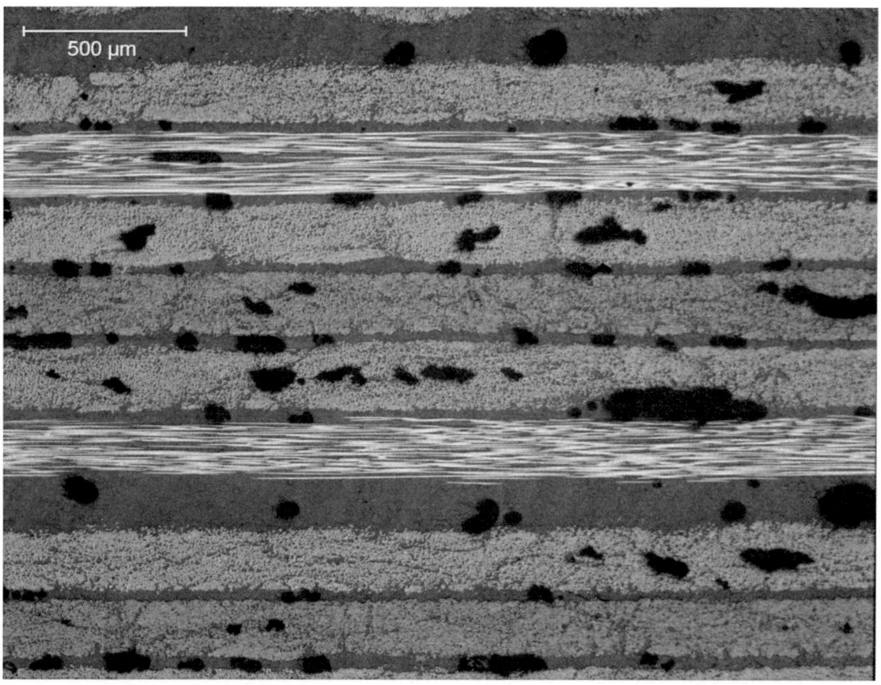

Fig. 5.3 Composite part made from unidirectional prepreg showing a large quantity of voids in the cured structure. Bright-field illumination, 5× objective

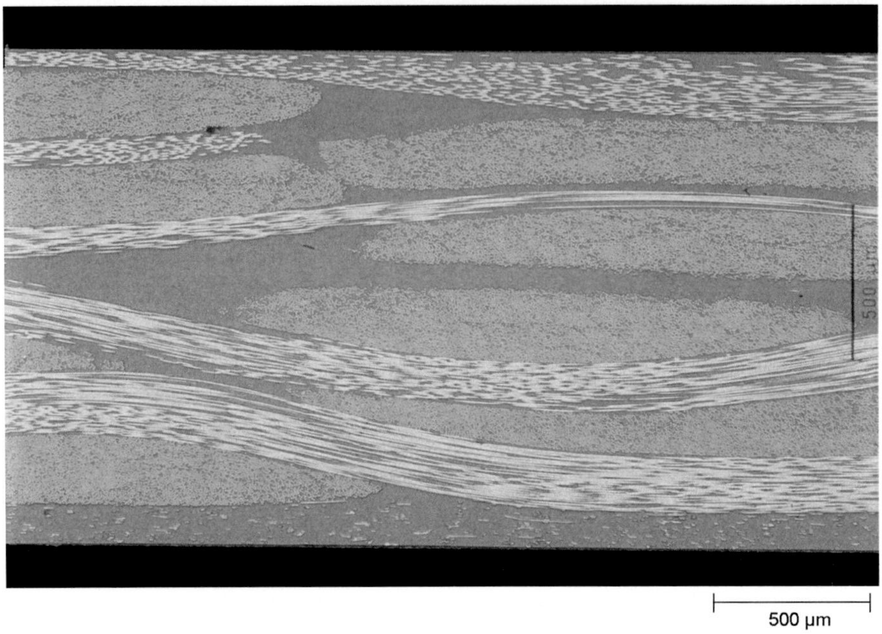

Fig. 5.4 Cross section of a 3k-70 plain weave carbon fabric composite. Also shown is a surfacing film on the bottom of the composite. Bright-field illumination, 5× objective

for analysis of voids and microcracks in composite materials are described in Chapters 8 and 9, respectively, of this book.

Ply counts can best be performed on carbon fiber fabric composites using low-magnification (25 to 100×) bright-field illumination of the cross section. Since the 90-degree tows can hide in the node zone, it is important to use a large field of view because of the structure of the weave (Fig. 5.4). A montage of a large area or cross section can provide even more information and may be necessary for documentation of the com-posite.

The determination of ply angles and the number of prepreg plies or layers of reinforcement can best be accomplished by sectioning the composite at a 45-degree angle to the plane view, followed by mounting and polishing. Cross sectioning the composite on an angle through the thickness elongates the fibers for easier angle determination. This is the easiest method to identify the ply angles using only bright-field illumination (Fig. 5.5). More details on this method are discussed in Chapter 7, "Composite Structure Analysis," in this book.

Dark-Field Illumination

Dark-field illumination (epi-dark field) is used to bring out subsurface features such as microcracks. With dark-field illumination, the light path is blocked from the objective center and is directed to the specimen at an

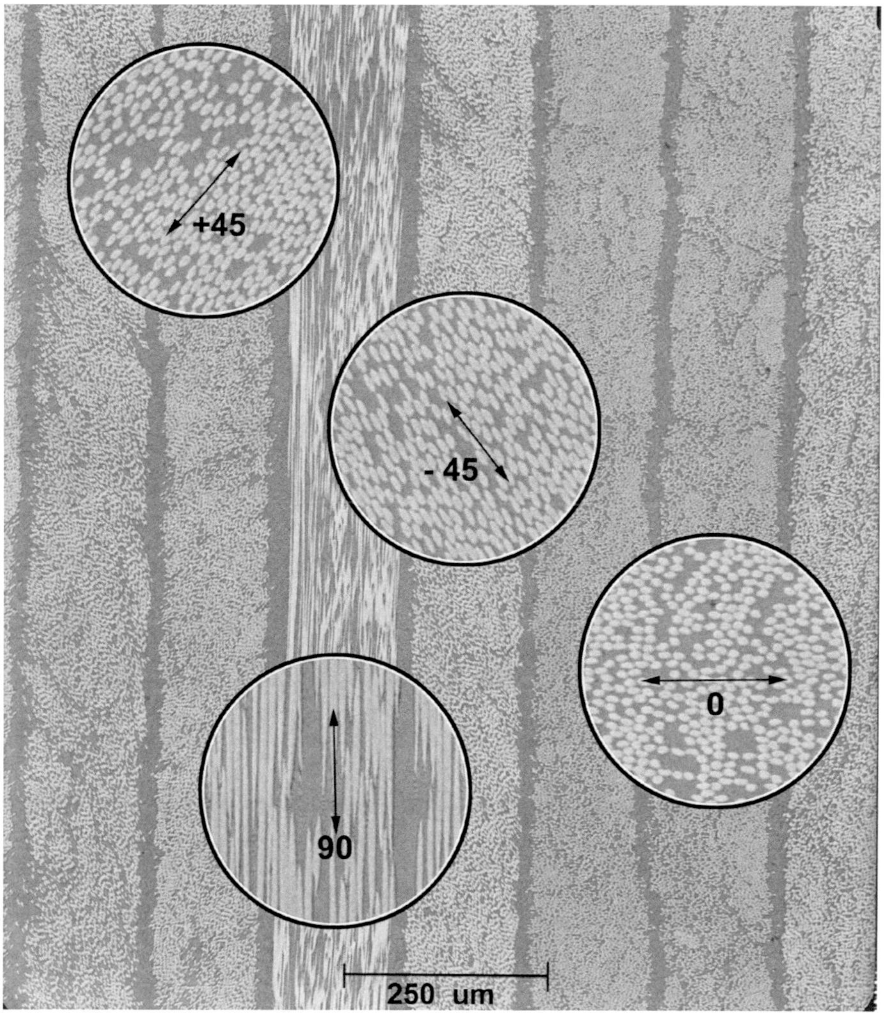

Fig. 5.5 Bright-field illumination of a unidirectional carbon fiber composite showing the ply angles. Bright-field illumination, 10× objective (insets 25× objective)

oblique angle. This eliminates the reflected image in favor of the image formed as light passes through the sample subsurface. Therefore, a dark background is shown for planar surfaces; voids, edges, microcracks, and indentations are shown brighter with dark-field illumination. Microcracks in fiber-reinforced polymeric composites are easily resolved using dark-field illumination. Similarly, samples with low surface contrast, such as glass fiber composites, are often visualized using dark-field illumination. However, there may be cases where subsurface features remain indistinct. The use of penetration dyes is often effective in highlighting these indistinct features (Fig. 5.6). This technique is discussed further in the sections describing dyes, etchants, and stains for composite materials.

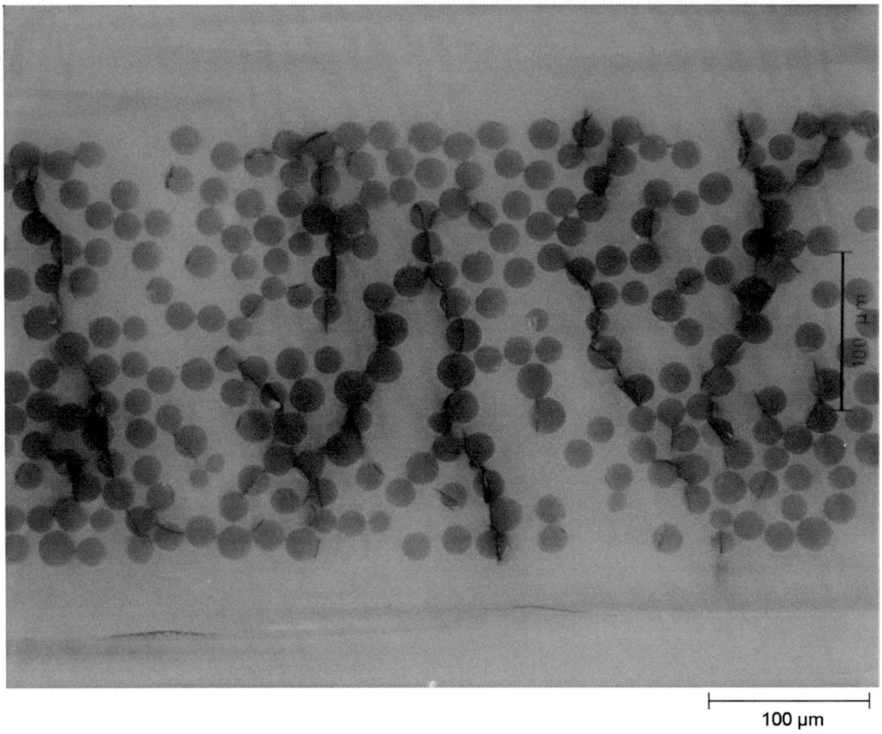

Fig. 5.6 Microcracked nylon fiber composite dyed with Magnaflux Spotcheck SKL-H (Magnaflux Corp.) penetrant. The microcracks appear red. This technique works well for most translucent fibers. Dark-field illumination, 10× objective

Polarized-Light Microscopy

Polarized light (epi-polarized light) is similar to dark-field illumination, except there is a polarizing element at the light source and an analyzer-polarizer between the sample and the eye. These additional elements allow the visualization of contrast created as light is both reflected and polarized by the sample surface. One advantage of using polarized light is best exemplified with fiber-orientation studies of composites made from unidirectional carbon fiber prepreg. The carbon fibers oriented normal to the view (zero-degree fibers) are dark (blacked out) using polarized light, while the transverse fibers (90-degree fibers) depolarize the light and thus are seen as bright features. With a 540 nm wave shift, the zero-degree fibers will appear as first-order magenta; the 90-degree fibers will appear white. While these two orientations will appear in this manner with any type of illumination, the predominant advantage of polarized light is the ability to discern other orientations (for example, ±45 degrees). Using polarized light, the ±45-degree fibers appear pink. As the fiber angle becomes closer to zero degrees, the color will turn progressively to first-order ma-

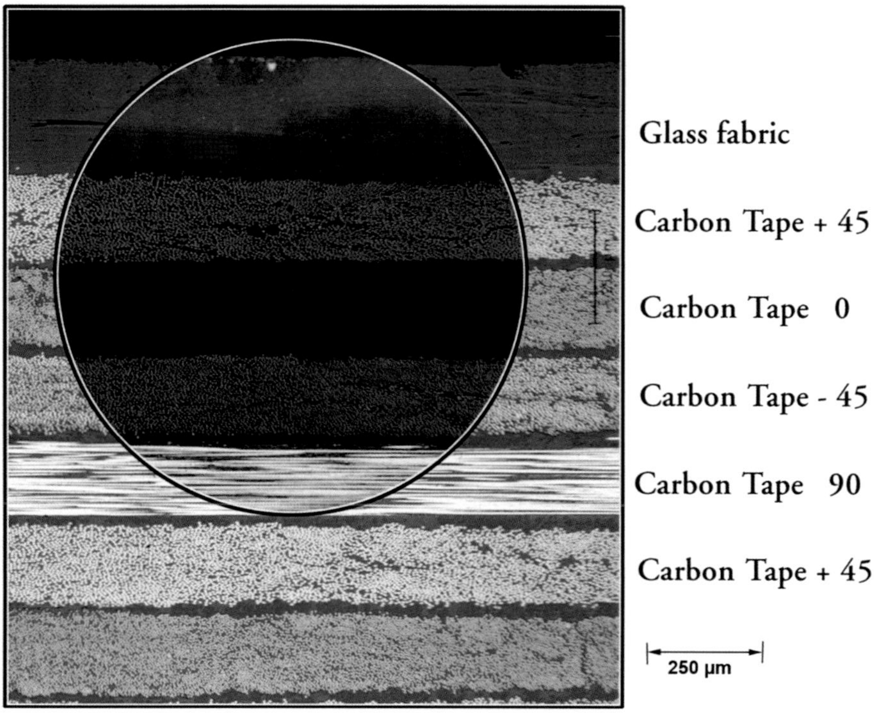

Fig. 5.7 Cross section of a glass fabric/unidirectional carbon fiber composite part showing a bright-field illumination background and a polarized-light center inset. Note the lack of contrast of the glass fabric when viewed using bright-field illumination as compared to the carbon fibers. 10× objective

genta, while fibers that are closer to 90 degrees will fade, eventually becoming white (Fig. 5.7).

Epi-circular polarized light describes the condition where the polarizer and analyzer-polarizer are crossed at 90 degrees to each other in conjunction with a ¼ wave plate (130 nm) in the light path that is tilted at a 15-degree angle. This wave plate is mounted on the low-magnification objective and is aligned 45 degrees off the polarizer. This type of circular polarizer is effective only with low-magnification objectives. For some composite surfaces, the prevalence of nonreflective surfaces scatters the light, leaving the image with a fogged appearance or a center "hot spot." Circular polarized light removes both the fogging and the hot spot from the image.

Interference and Contrast Microscopy

It is often found that polymeric matrices in composite materials do not display much contrast when viewed using the aforementioned reflected-light techniques. As a consequence, dyes that bind to various phases or etchants for dissimilar material phases are often used to enhance the contrast. While these methods can be effectively used for many composite materials and matrices, significant time may be required to determine a functional dye or etchant material, and it is possible that a useful dye or etchant may not be found. Before expending the time and energy on researching dyes and etchants, it is best to observe the sample with one of many different types of illumination techniques that have been developed to improve the contrast in specimens. While these techniques are most commonly used with transmitted light, some of the techniques can be very effectively used with reflected light. A technique that can be used with etch-relief surfaces is reflected-light (episcopic) differential interference contrast (DIC) such as Nomarski. This type of contrast technique distinguishes height differences and discontinuities on composite surfaces and results in a partial three-dimensional topography (Fig. 5.8). With this technique, it is possible that softer phases within composite matrices may become visible due to slightly greater removal during the polishing procedure. In reflected-light DIC, one Nomarski prism is used that both separates and recombines the polarized light. The two separated light beams are focused at the objective, reflected from the sample surface, and recombined in the Nomarski prism. The image contrast is due to the optical path gradi-

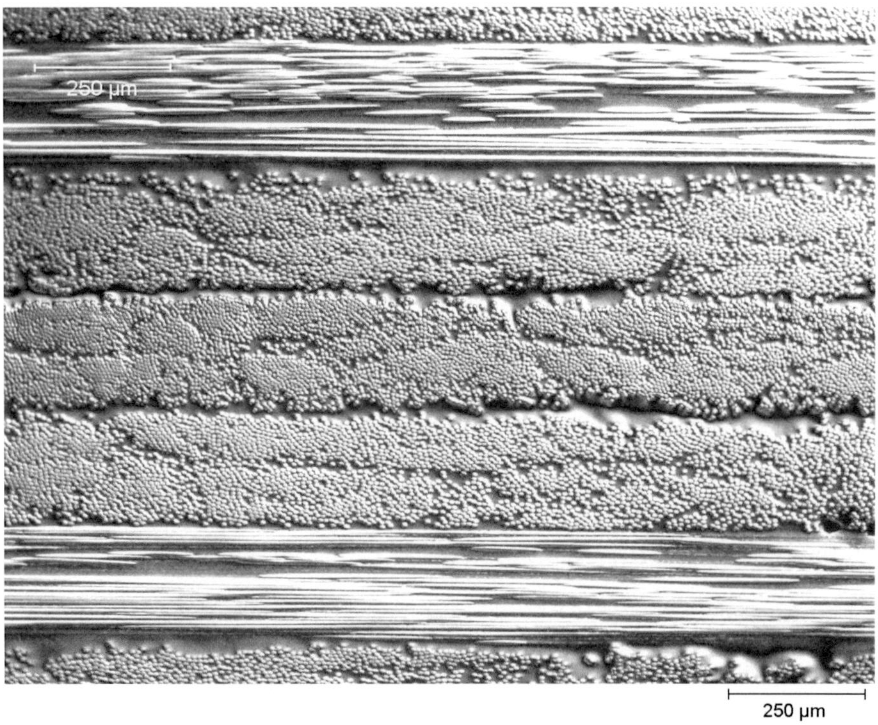

Fig. 5.8 Cross section of a polished interlayer-toughened composite that was lightly etched showing height differences on the sample surface using reflected-light differential interference contrast. 10× objective

ent across the specimen surface and is therefore differential, with steeper gradients generating more contrast (Ref 8). Accordingly, for many composite materials that have been properly polished, the reflected-light DIC image will be similar to an image obtained using reflected-light bright-field illumination.

Another technique that is worth mentioning but is rarely used for enhancing the contrast of composite specimens is reflected (incident)-light phase contrast. One reason this is not often used is that there are very few microscopes that have this capability. The more common technique, reflected-light DIC described previously, can be used for resolving similar features. In reflected-light phase-contrast microscopy, phase differences arise from relief on the specimen surfaces.

Fluorescence Microscopy

Epi-fluorescence microscopy uses an ultraviolet light source (mercury or xenon) for illumination. The light is transmitted though an excitation filter that blocks all but a narrow bandwidth of the light spectrum. The narrow bandwidth of light strikes the sample surface and is mostly reflected. However, some of the light is absorbed and reemitted, causing a lengthening of selected light bands. The image is passed through a dichroic mirror that filters some of the spectrum, with the remaining bands passing through a barrier filter. The barrier filter removes all light except that which was emitted (fluoresced) from the sample. It is important to note that with fluorescence microscopy, you do not observe detail but see only the fluorescing illumination. The selection of an appropriate light source and filters is critical. All microscope manufacturers offer a wide range of filter combinations for this purpose. One of the most useful combinations is the following: a 390 to 440 nm excitation filter, a 460 nm dichroic mirror, and a 475 nm barrier filter. Although most polymeric resins will naturally fluoresce in the 390 to 440 nm range, some do not. If a longer-wavelength excitation filter (450 to 490 nm) is used, only the fluorescing material, such as Rhodamine B dye, will be observed, with very little to no polymer fluorescence. The ability to see the polymer fluorescence is essential in most cases to provide contrast. Otherwise, only the fluorescing dye will be observed, with the remaining background appearing black, showing none of the composite features.

In some multiphase resins, it is possible to selectively dye certain phases with a laser dye for increased contrast (Fig. 5.9a, b). However, the most common application of fluorescence is to observe microcracks that remain obscure with any other type of microscopy. It is common that thermosetting carbon fiber composites and thermoplastic-matrix composite materials have microcracks that cannot be observed with bright- and dark-field illumination or polarized light (Fig. 5.10a). These microcracks can be easily identified with commercially available laser dye solutions that are

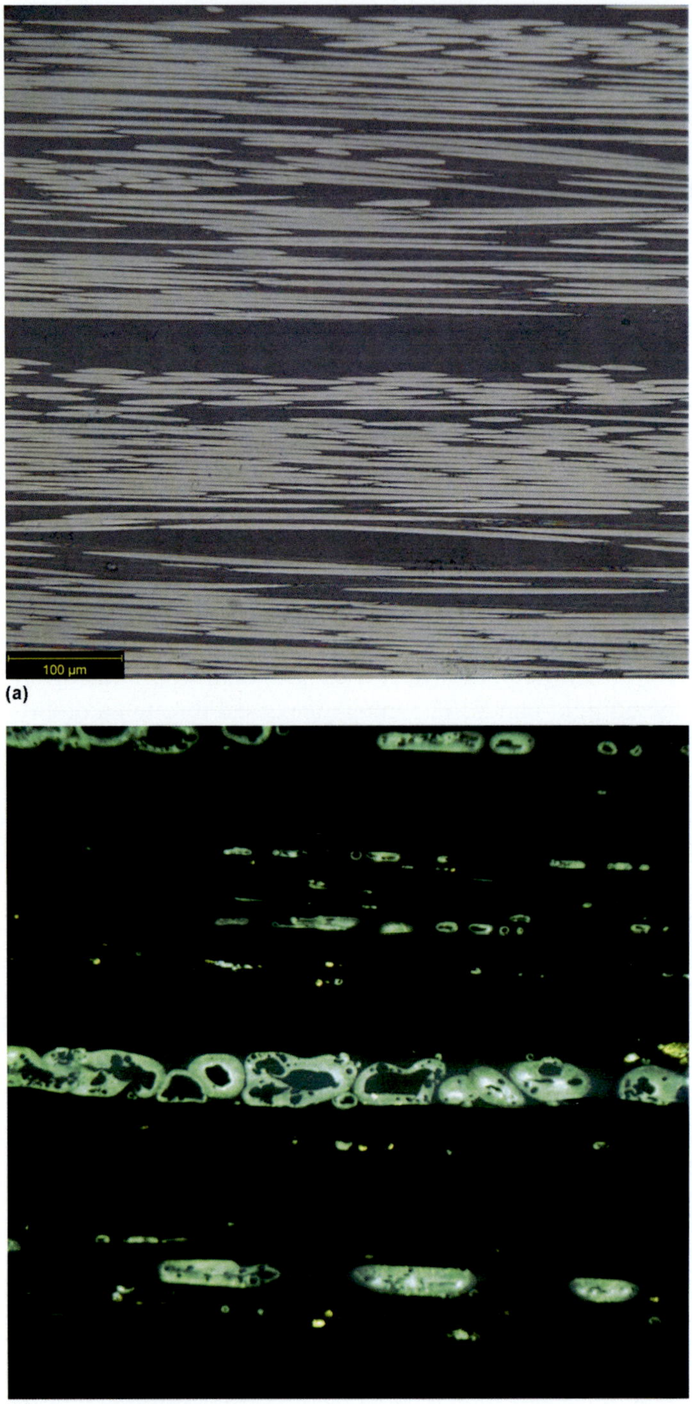

Fig. 5.9 Cross sections of an interlayer-toughened composite material. (a) Bright-field illumination, 25× objective. (b) Same view but after the addition of a solvent-based laser dye (Magnaflux Zyglo, Magnaflux Corp.) to the sample surface. The laser dye is preferentially absorbed into the rubber particle phase. Epi-fluorescence, 390–440 nm excitation, 25× objective

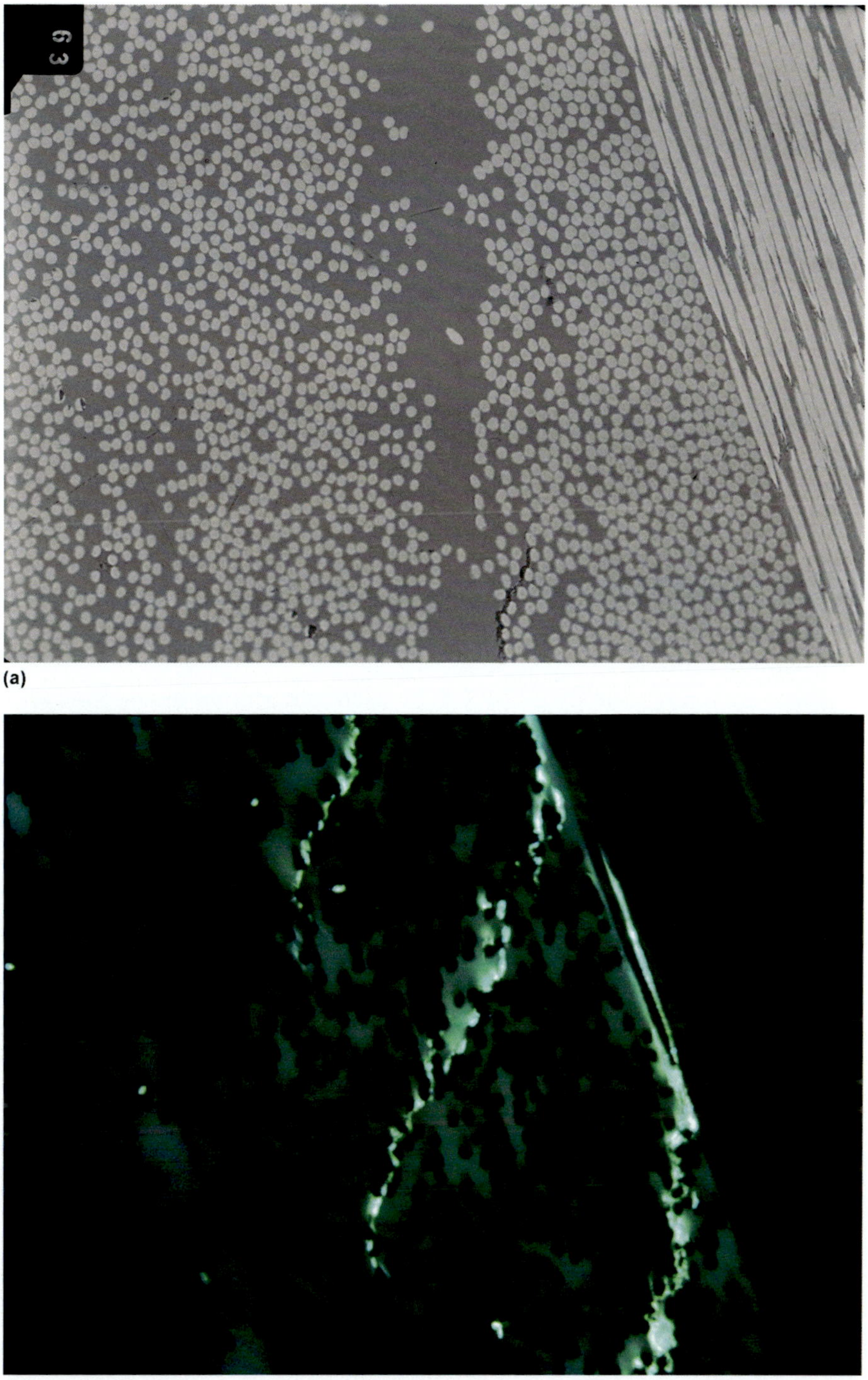

(a)

(b)

Fig. 5.10 Microcracks in a composite material that are difficult to observe using epi-bright-field illumination. (a) Bright-field illumination, 25× objective. (b) Same location viewed after applying a fluorescent penetrant dye (Magnaflux Zyglo) to the surface and back-polishing. Epi-fluorescence, 390–440 nm excitation, 25× objective

wicked into the microcracks (Fig. 5.10b). Another comparison of the advantages of fluorescence is shown in Fig. 5.11(a and b). This is a sample of a fiber-reinforced composite part that shows the surfacing film, primer, and paint layers. In this polished sample, the larger microcracks can be observed without the use of fluorescence, but using the fluorescence technique provides greater detail concerning the extent of the microcracks.

The use of laser dyes for enhancing microcracks is a simple process. It is usually best to vacuum dry the polished section between approximately 40 and 50 °C (105 and 120 °F) before applying a laser dye penetrant. The laser dye is wiped or brushed onto the dried polished surface and allowed to dry. The excess laser dye is then wiped off the surface with a solvent-dampened cloth. Acetone is the recommended solvent because acetone usually does not readily attack the polymer matrix, and it evaporates rapidly. Drying the sample in a warm oven at approximately 40 to 50 °C (105 to 120 °F) and also under vacuum will speed the process. Often, the sample can be viewed as is, but, in most cases, it is best to back-polish the mount before viewing. This is performed by subjecting the specimen to the final polishing step for a very short time—less than 30 seconds. This removes any excess laser dye that remained on the surface after the solvent-wiping process.

There are many laser dyes that can be used to highlight the microcracks in fiber-reinforced polymeric composite materials. An excellent laser dye for this application is Magnaflux Zyglo ZKL-H (Magnaflux Corp.), but other similar products can be used. The advantage of this type of laser dye is that the dye is in a solvent that will dry on the surface and can be back-polished. If the dye remains wet or is water based, it cannot be back-polished. In this case, the dye can still be used by wiping off the surface excess, but it may be more of an art to obtain the necessary detail and contrast.

Penetration Dyes

There are many types of features in fiber-reinforced polymeric composite materials that can benefit from increased contrast. In Chapter 2, "Sample Preparation and Mounting," the technique of adding a fluorescing dye to the mounting resin was described. This works very well if the sample contains damaged areas in which the mounting resin can impregnate (Fig. 5.12a, b). However, some features, such as microcracks and small voids, may not be able to be impregnated by the mounting resin. This is not only due to the higher viscosity of the mounting resin, but also the location of the feature may not allow impregnation by the mounting resin. In this case, a dye applied after the final polishing process can provide the necessary contrast (Fig. 5.13). Many of the dyes that can be used to highlight these features are solvated, and some may require even further dissolution with

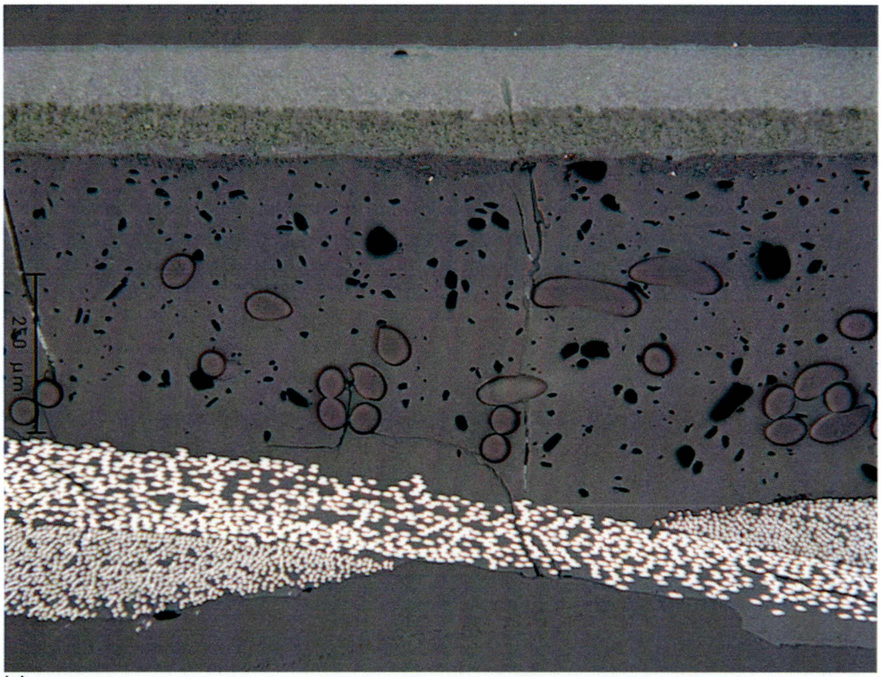

(a)

(b)

Fig. 5.11 Composite part containing microcracks that extend to the surfacing film, primer, and paint layers. (a) Slightly uncrossed polarized light was used to contrast the paint layer (10× objective). (b) A fluorescent penetration dye (Magnaflux Zyglo) was applied on the surface of the specimen to enhance the contrast of the microcracks. Epi-fluorescence, 390–440 nm excitation, 10× objective

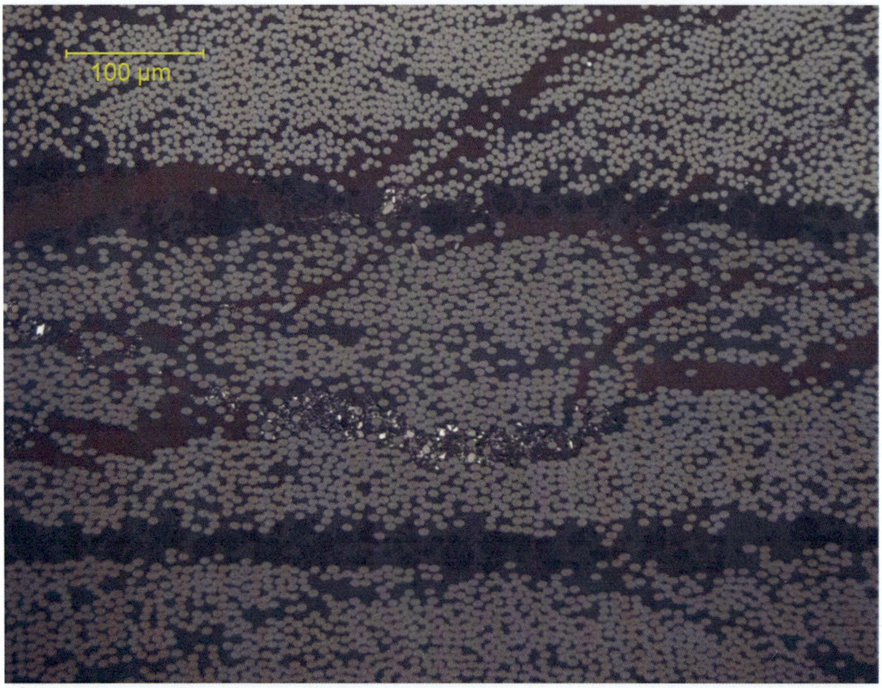

(a)

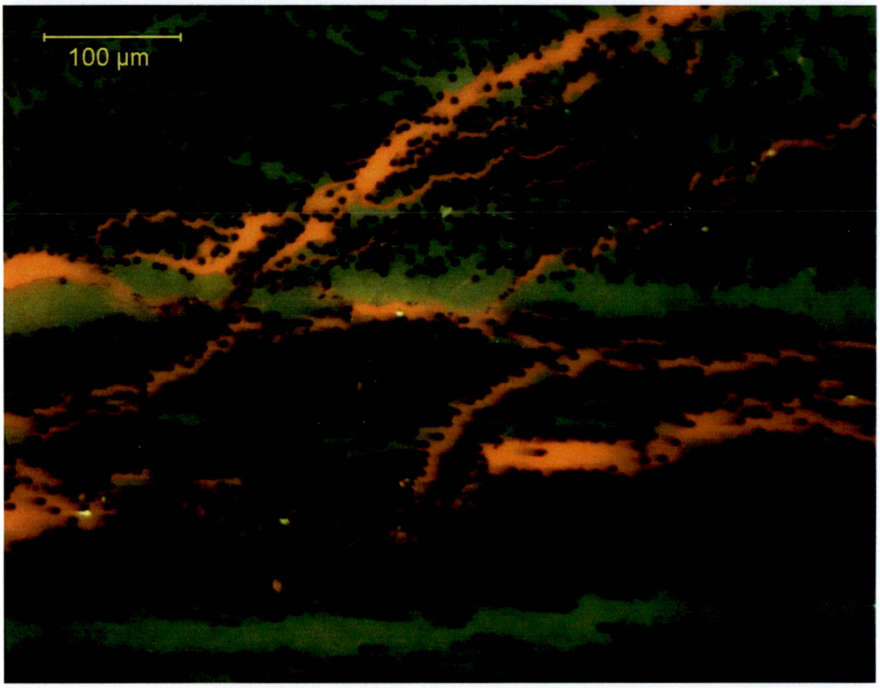

(b)

Fig. 5.12 Composite part that was impact damaged. The composite sample was impregnated with epoxy casting resin that was dyed with Rhodamine B to reveal the details of the microcracks. (a) Slightly uncrossed polarized light, 25× objective. (b) Epi-fluorescence, 390–440 nm excitation, 25× objective

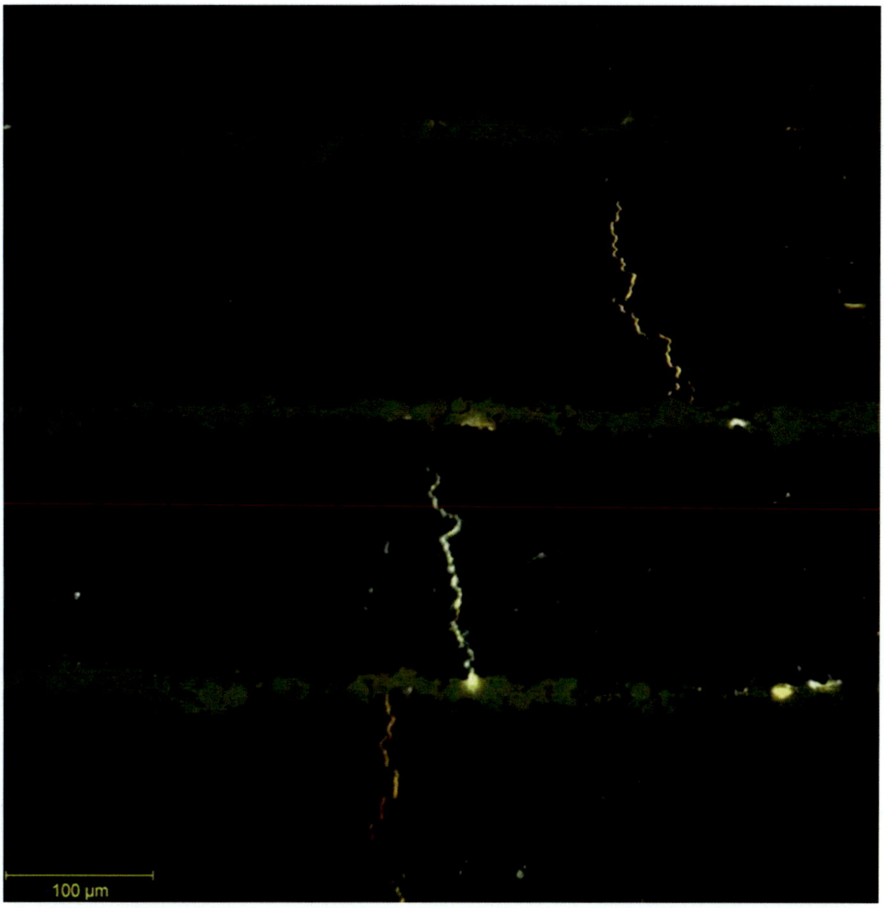

Fig. 5.13 Polished composite where Rhodamine B was added to the mounting resin and a laser penetration dye (Magnaflux Zyglo) was added after polishing. The microcracks impregnated by the Rhodamine-B-dyed mounting resin appear red, and the microcracks impregnated by the (Magnaflux Zyglo) penetration dye appear yellow. Epi-fluorescence, 390–440 nm excitation, 25× objective

solvents. As described previously, the use of laser dyes may be a good option if a microscope is available that incorporates this capability. There are other cases where the use of laser dyes is not necessary, and color contrast without fluorescing provides the required detail.

Because of the translucent nature of polymers, there is a wide range of techniques available for viewing polymeric composite materials. If bright-field illumination is the only method used, much of the data can be overlooked. Enhanced contrast is required to study defects having ultrahigh aspect ratios (length but virtually no width), such as microcracks. Polymer composites made with translucent fibers, such as glass, Kevlar (E.I. du Pont de Nemours and Company), polyester, and polyamide, can be examined using colored dyes (Fig. 5.14). Without the dye, the features can be so

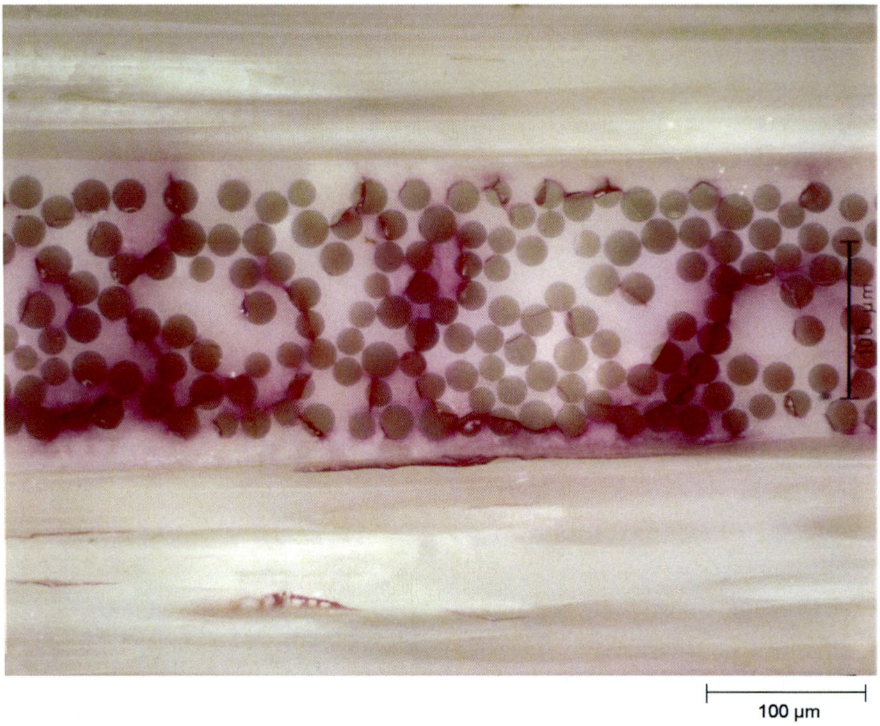

Fig. 5.14 Thermoplastic fiber-reinforced composite with the microcracks dyed using Magnaflux Spotcheck SKL-H. Dark-field illumination, 25× objective

Table 5.1 Penetrant contrast dyes for composite materials(a)

Red or black permanent ink felt-tip pen: Translucent fiber composites
DYKEM Steel Red (Illinois Tool Works, Inc.) layout fluid: Translucent fiber composites
Magnaflux Spotcheck SKL-H (Magnaflux Corp.): Translucent fiber composites
Magnaflux Zyglo ZKL-H (Magnaflux Corp.): Nontranslucent fiber composites
Magnaflux Zyglo (Magnaflux Corp.) penetrants: Nontranslucent fiber composites

(a) Read the material safety data sheets for each of these materials. The dyes are toxic by ingestion and also can be absorbed through the skin.

subtle that they go undetected. When viewed with dark-field illumination or polarized light, the dyed features can be easily observed.

A list of commonly used penetrant dyes and the function they perform is shown in Table 5.1. The polished mount should be vacuum dried at approximately 40 to 50 °C (105 to 120 °F) before application of all dyes. Longer drying times may be required, depending on how deep the moisture from the polishing process has wicked into the features. After the drying process, the dye is added to the sample and allowed to absorb into the cracks or voids. The surface is usually then repolished, using the final polishing step for less than 30 seconds.

One of the easiest ways to dye microcracks in composites that have translucent fibers is by using a red permanent-ink felt-tip pen. The pen is rubbed several times over the surface of the polished mount. Adequate pressure and rubbing in multiple directions is required. After this is performed, the mount is repolished, using the final polishing step to remove the excess permanent ink from the surface. The contrast of the microcracks and the depth that the dye penetrated is easily observed using this technique.

For polymeric composites with translucent fibers, colored dyes work very well for providing contrast, but composites that do not have translucent fibers, such as carbon, boron, and ceramic, require a different approach. With composites containing these other fiber types, the use of colored dyes will not provide any more contrast enhancement for features such as microcracks and voids than using only bright-field illumination. It is therefore best to use fluorescent penetrant dyes and epi-fluorescence to view the features, as described previously.

Stains and Dyes for Polymeric Material Dispersed Phases

For the majority of specimens, the microstructure is readily visible after polishing, although some features may need to be etched, dyed, or stained to be fully visible. The features of a multiphase polymer matrix, such as a thermoset containing a second-phase thermoplastic or elastomer, are usually not resolved using bright-field illumination. Although some phases may be observable with dark-field techniques, most multiphase polymer matrices require additional sample preparation after polishing. One option is the development of an ultrathin section from the material of interest, as described in Chapter 6, "Thin-Section Preparation and Transmitted-Light Microscopy," but other methods, such as etching, staining, and dyeing, are less time-consuming and work well for elastomers and thermoplastics in thermosetting matrices. The type of elastomer or thermoplastic in the thermoset matrix will dictate if it is better to etch, stain, or dye the phase. This may or may not be known and can require multiple trials to obtain the necessary contrast. In general, stains provide contrast by reacting with the polymer phase or microstructure, while dyes preferentially impregnate the phase but do not react.

Stains have been used to provide contrast in multiphase-matrix systems for many years. Table 5.2 shows some frequently used stains for high-

Table 5.2 Stains and dyes for polymer phases(a)

Osmium tetroxide: Rubber (double bond)
Ruthenium tetroxide: Rubber (double bond), epoxy resin (oxidizes aromatic rings), polyethylene
Rhodamine B: Rubbers and thermoplastics (1 g Rhodamine B + 25 mL methanol + 75 mL methylene chloride); soak for 30 s

(a) Read the material safety data sheets for each of these materials. The dyes are toxic by inhalation and can be absorbed through the skin.

lighting dispersed phases in thermosetting multiphase composite materials. A common stain that has been used for identification of butadiene-acrylonitrile rubber toughening domains in thermosetting resins is osmium tetroxide (Ref 15). This material is usually dissolved 2.5 percent in tert-butylphenol. Osmium tetroxide reacts with the double bonds of the butadiene and stains the phase black (viewed using epi-bright-field illumination). A stronger oxidizing agent, ruthenium tetroxide, can be used in place of osmium tetroxide and will stain samples much faster. This material is also stated to oxidize aromatic rings and will even stain bisphenol-A-based epoxies (Ref 16). Care must be exercised, because it is possible that the phases can be overstained with either of these stains, resulting in problems identifying phase boundaries. In the case of using ruthenium tetroxide, the bulk matrix may also be stained, leading to incomplete contrast. Both of these materials are highly toxic, and therefore, extreme caution should be used when working with them.

In contrast, thermoplastic and elastomer phases can be dyed using a laser dye. An example of a dyed multiphase epoxy-matrix carbon fiber composite, containing both a dispersed thermoplastic and elastomer phase, is shown in Fig. 5.15. This specimen was dyed with a Rhodamine B solution and viewed at 390 to 440 nm using fluorescence microscopy. The composition of this dye is shown in Table 5.2. This dye is used to illustrate the different phases in a polymer matrix. In this case, the Rhodamine B dye is absorbed by the elastomer and thermoplastic phases at different wavelengths. This provides contrast between the two phases and contrast within the thermoset matrix. In Fig. 5.15, the yellow phase is a thermoplastic polymer, the red phase is an elastomer particle, and the light-blue area is the bulk thermoset matrix. This type of dye technique can be used with many different types of elastomer- and thermoplastic-modified thermosets to provide contrast between the phases.

Etches for Polymeric Matrices

The selective etching of materials with a solvent or acid that is capable of dissolving or eroding a dispersed phase is one of the easiest and best techniques if you know the composition of the phase. For example, if the thermoset is toughened with polyetherimide (PEI), which creates a dispersed phase, a good solvent to etch or dissolve the PEI phase is methylene chloride (Ref 17). After the sample is polished, it can be placed face down in the solvent for 2 to 24 hours, depending on the level of etch required. The sample surface may need to be subjected to the last polishing step to clean up the surface after solvent etching. Small glass crystallization dishes work very well for the solvent and samples and should be covered to prevent evaporation of the solvent. All of these materials are toxic and must be handled accordingly. The time that the sample spends in contact with the solvent depends on many factors, including the molecular weight of

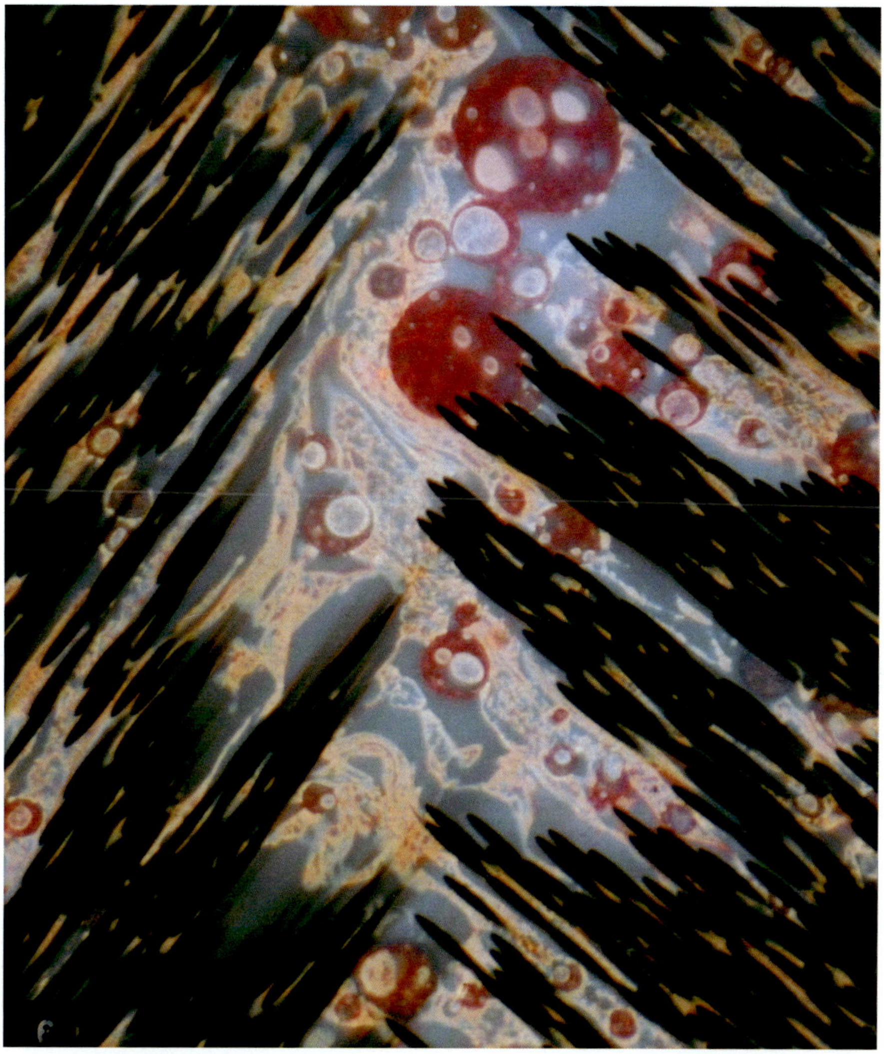

Fig. 5.15 Image of a composite cross section dyed with Rhodamine B in solution. Viewed using epi-fluorescence, 390–440 nm excitation, 25× objective. The use of the dye was necessary to distinguish the multiple phases within the matrix.

the polymer in the dispersed phase, the solubility parameter, the degree of phase separation, the concentration of thermosetting material in the dispersed phase, and the desired degree of contrast. In many cases, only a few minutes may be required to see the contrast, and more time may completely remove the dispersed phase. This can provide more information as to the structure of the phase and the interphase with the continuous phase. As with the use of dyes, if the dispersed phase is not known, many trials may be required to find an effective solvent. Care must be exercised, because it is also possible to overetch the sample, which can lead to loss of

Table 5.3 Etchants for multiphase polymeric-matrix materials(a)

Methylene chloride: Thermoplastic etch (excellent for polyetherimide phases), 2 to 24 h
Methyl ethyl ketone: Thermoplastic etch (polysulfone, polyether sulfone phases), 24 to 72 h
CrO_3 (12.5 g), H_2O (50 mL), HNO_3 (50 mL): General thermoplastic and rubber etch, 10 to 120 s
Formic acid: General thermoplastic and rubber etch, 10 to 120 s
Tetrahydrofuran: Rubber-phase etch, 16 to 48 h
British Petroleum thermoplastic etch: The etch lifetime is approximately 1 h.
1. Dissolve 0.4 g fine-ground potassium permanganate in 4 mL water at 30 °C (86 °F).
2. Add 10 mL ortho-phosphoric acid and then 10 mL sulfuric acid while stirring.
3. Etch for 2 min.
4. Wash samples in 30% hydrogen peroxide solution, then with water.

(a) Read the material safety data sheets for each of these materials. Use good health and safety practices.

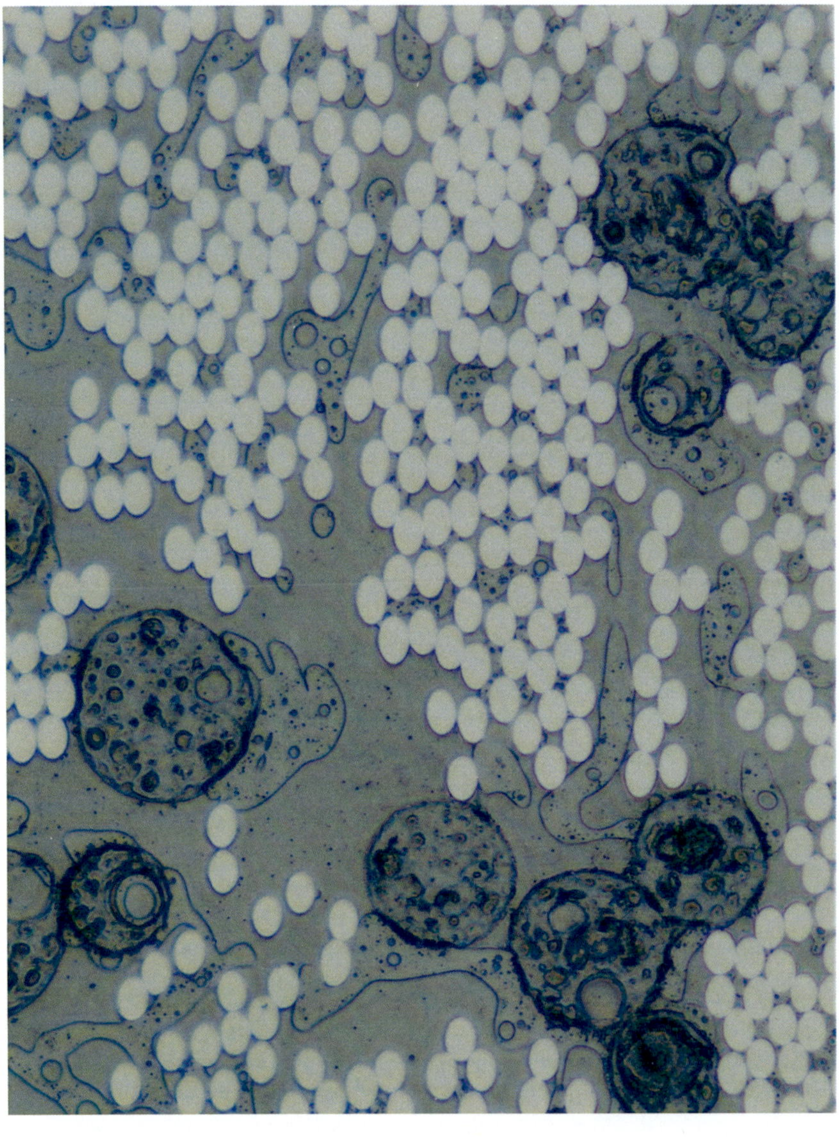

Fig. 5.16 Composite material having a dispersed phase that was acid etched for 30 s using the CrO_3/HNO_3 etch described in Table 5.3. Reflected-light phase contrast, 50× objective

contrast. Some of the common materials that are used to etch various polymers and the approximate process times are shown in Table 5.3. If the indicated polymers are not present as a dispersed phase in the thermoset, the solvents may not have a significant effect on the microstructure or surface. These materials represent only a few of many etches that can be used.

Solvents can be used to remove the dispersed phase but also may be used to swell the phase or cause a thermoplastic area to craze. This technique can provide exceptional contrast. Tetrahydrofuran can be used to swell or dissolve rubber phases in thermoset matrices. This solvent swells cross-linked rubber phases, such as those made from core-shell rubber particles, and can selectively dissolve rubber phases made of common reactive liquid polymers, such as carboxyl-terminated butadiene-acrylonitrile rubber, or similar materials. The use of this solvent commonly requires 16 to 48 hours to etch the rubber.

As with solvents, acids and acid mixtures can be used effectively to etch various phases in thermosetting matrices (Fig. 5.16, 5.17). It is also possible to overetch a sample if acids are used. A few of the acid etches that can be used for both thermosetting and thermoplastic phases are shown in Table 5.3. Caution must be used with these etches.

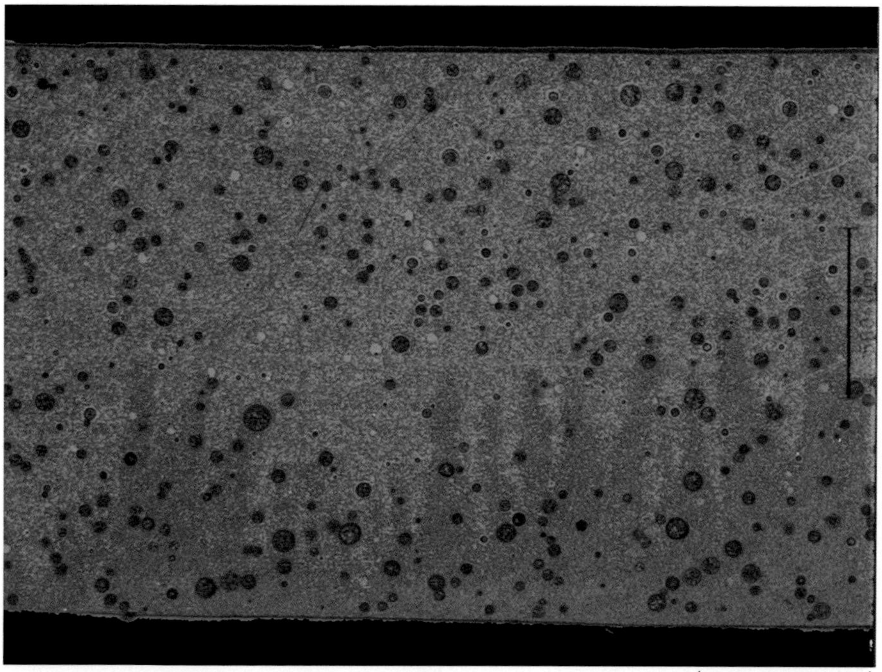

Fig. 5.17 Etched (CrO_3/HNO_3, Table 5.3) composite specimen after being subjected to a varied solvent-sensitivity stress test. Darker striations show the stress effects in the material after testing. Slightly uncrossed polarized light, 25× objective

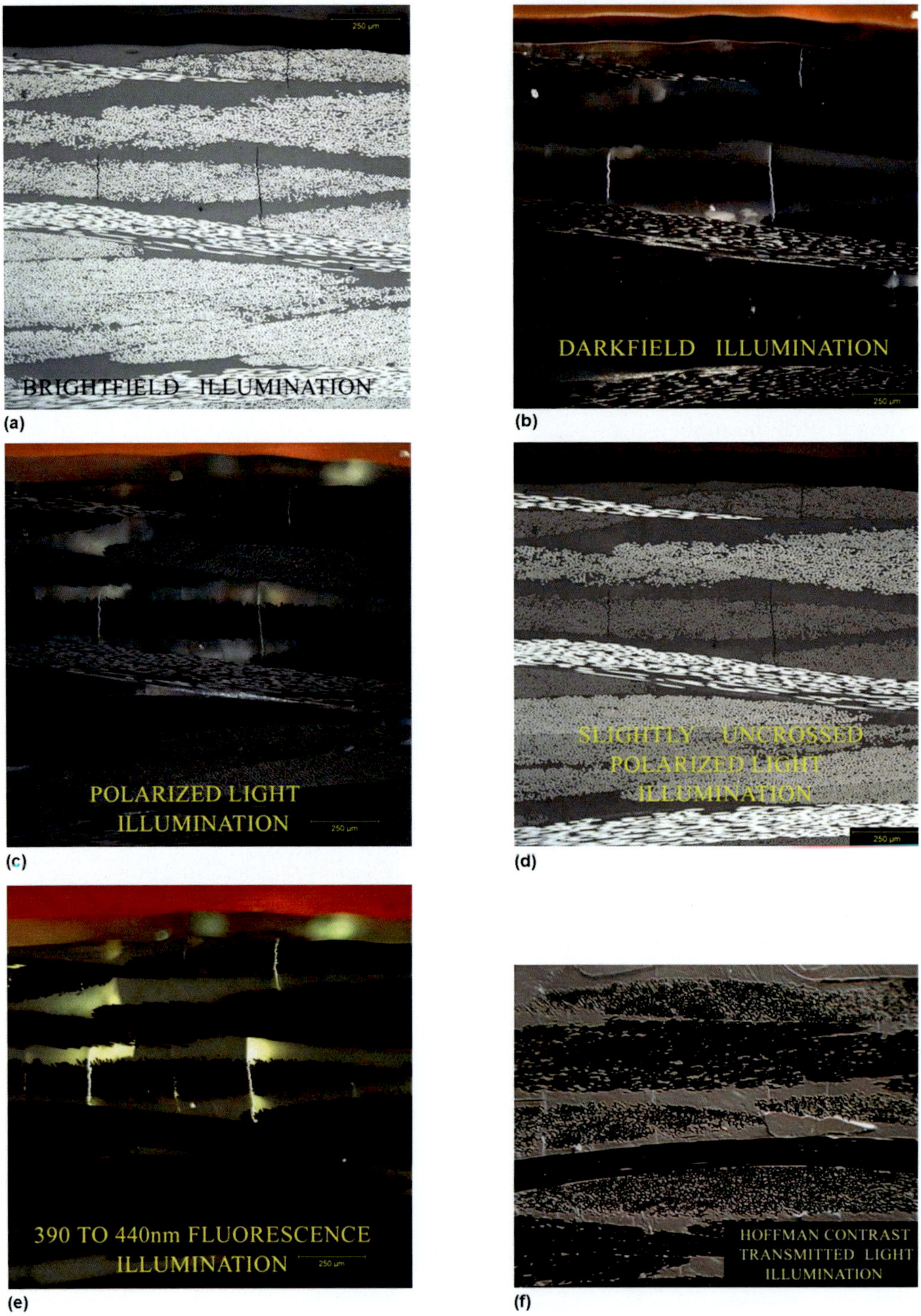

Fig. 5.18 Micrographs of a composite cross section showing the differences in contrast methods. The composite morphology and microcracks appear significantly different using these epi-illumination modes. One transmitted-light method is shown for reference. (a) Bright-field illumination, 25× objective. (b) Dark-field illumination, 25× objective. (c) Polarized light, 25× objective. (d) Slightly uncrossed polarized light, 25× objective. (e) Epi-fluorescence, 390–440 nm, 25× objective. (f) Transmitted light, Hoffman modulation contrast, 20× objective

If the composite matrix is made from a thermoplastic matrix with dispersed phase inclusions, the use of etches can be very difficult, because it is possible that the entire matrix is dissolved. Therefore, the selected etchant and the exposure time are critical. It is also possible to etch the amorphous regions from the crystalline regions in thermoplastic-matrix composite materials by using solvents. The solvent will attack the amorphous regions faster than the crystalline region. In this case, overetching is possible if the matrix is exposed to the solvent for too long a time period.

Summary

The analysis of composite materials using optical microscopy is a process that can be made easy and efficient with only a few contrast methods and preparation techniques. However, a complete "toolbox" of preparation materials, such as dyes, etches, stains, and the available microscopy contrast methods will enable a more efficient and comprehensive sample analysis. Figure 5.18 is a micrograph comparison summary of the contrast methods described in this chapter and the methods that will most likely find wide use for analysis of fiber-reinforced polymeric composites. All six illuminations are provided for the same sample, with the five (epi) illuminations taken from the same location. The thin-section procedure and transmitted-light analysis are described in Chapter 6, "Thin-Section Preparation and Transmitted-Light Microscopy," in this book. As with most things, greater experience in preparing and viewing samples leads to an even greater understanding of the material and the analysis.

REFERENCES

1. M. Abromowitz, Contrast Methods in Microscopy: Transmitted Light, *Basics and Beyond,* Olympus Corp., Lake Success, NY, 1987
2. F.D. Bloss, *An Introduction to the Methods of Optical Crystallography,* Saunders College Publishers, Philadelphia, PA, 1989
3. P.F. Kerr, *Optical Mineralogy,* McGraw-Hill Co., New York, NY, 1977
4. C.W. Mason, *Handbook of Chemical Microscopy,* 4th ed., John Wiley & Sons, New York, NY, 1983
5. W.C. McCrone and J.G. Delly, *The Particle Atlas, Principles and Techniques,* 2nd ed., Ann Arbor Science Publishers, Ann Arbor, MI, 1973
6. L.C. Sawyer and D.T. Grubb, *Polymer Microscopy,* 2nd ed., Chapman & Hall, New York, NY, 1996
7. D. Shelley, *Optical Microscopy,* 2nd ed., Elsevier, New York, NY, 1985
8. "MicroscopyU, The Source for Microscopy Education," Nikon Instruments, Inc., http://www.microscopyu.com/ (accessed July 7, 2010)

9. "Molecular Expressions, Images from the Microscope," The Florida State University, http://micro.magnet.fsu.edu/index.html (accessed July 7, 2010)
10. "Microscopy Primer," Olympus America Inc., http://www.olympusmicro.com/primer/index.html (accessed July 7, 2010)
11. "Microscopy from the Very Beginning," Carl Zeiss MicroImaging GmbH, http://www.zeiss.de/C1256B5E0047FF3F?Open (accessed July 7, 2010)
12. B.S. Hayes and L.M. Gammon, Microscopy, *Composites,* Vol 21, *ASM Handbook,* ASM International, 2001, p 964–972
13. L.M. Gammon, Macro Techniques and Digital Microscopy, *Microsc. Microanal.,* Vol 12 (Suppl. 02), 2006, p 1682–1683
14. L.M. Gammon, Macro Imaging Technique and How It Merges into the Digital World, *Microsc. Microanal.,* Vol 11 (Suppl. 02), 2005, p 2102–2103
15. J.H. Klug and J.C. Seferis, Phase Separation Influence on the Performance of CTBN-Toughened Epoxy Adhesives, *Polym. Eng. Sci.,* Vol 39 (No. 10), 1999, p 1837–1848
16. R.W. Smith, The Staining of Polymers, *Microsc. Microanal.,* Vol 8 (Suppl. 02), 2002, p 190–191
17. B.S. Hayes and J.C. Seferis, Variable Temperature Cure Polyetherimide Epoxy-Based Prepreg Systems, *Polym. Eng. Sci.,* Vol 38 (No. 2), 1998, p 357–370

Optical Microscopy of Fiber-Reinforced Composites
B.S. Hayes and L.M. Gammon

CHAPTER **6**

Thin-Section Preparation and Transmitted-Light Microscopy

THE TECHNIQUE that is used to prepare fiber-reinforced composite specimens for optical analysis dictates the information that can be obtained and the level of detail. While most common analyses on fiber-reinforced composites are performed using reflected-light optical microscopy, a sample-preparation technique that enables transmitted-light analysis is extremely valuable (Ref 1). At the ultrathin level, even opaque materials become optically transparent. This preparation is needed to resolve and analyze microstructural features at the theoretical limit for optical microscopy. Thin sections allow the use of several types of transmitted-light microscopy contrast methods on materials normally considered opaque. Transmitted-light methods reveal more details of the morphology of fiber-reinforced polymeric composites than are observable using any other available microscopy techniques (Ref 2).

Ultrathin-section development of composite materials for transmitted-light optical analysis is time- and labor-intensive, and it typically can only be justified when the required data cannot be obtained by one of the many reflected-light methods. Therefore, ultrathin sections are typically developed and used as a research tool rather than a production tool. Investigations related to problem solving, failure analysis, and advanced material development are common areas where this technique is invaluable. An ideal thin section is one where the only residual stress and strain that should be observed is that from the original specimen rather than introduced by the preparation technique (Ref 3, 4). Polymeric materials cannot be microtomed (which induces stress and strain as artifacts) to obtain the high resolution or detail that can be achieved with a polished ultrathin section. The following sections describe the various aspects relating to the selection and preparation of ultrathin-section specimens of fiber-reinforced polymeric composites for examination by transmitted-light microscopy techniques. Examples of composite ultrathin sections that are analyzed using transmitted-light microscopy contrast methods are shown throughout and at the end of this chapter.

Procedure and Selection of the Rough Section

The procedure for the preparation of the rough section depends on the nature of the starting bulk material from which the specimen is to be extracted and the type of information sought. For example, the primary interest may be in examining the resin-rich interlayers between prepreg composite plies. In this case, a planar specimen should be prepared by cutting the laminate through a horizontal plane at a slightly oblique angle to the fiber plane. The outcome of this type of mount is shown in Fig. 6.1. This procedure usually results in only one face or sample in a mount, due to the width of the sample. In contrast, multiple samples can also be mounted on edge, as shown in Chapter 2, "Sample Preparation and Mounting," in this book, and viewed normal to the *xz* or *yz* plane. If the features that are desired to be analyzed are noticeable on the surface of a composite, such as a fracture area, this area is ideally centered in the section that is taken. This provides assurance in the following preparation procedures, because the edges of the specimens may be removed or damaged during the ultrathin-section development. If the features to be investigated are not evident on the surface of a part, sections should be taken throughout the part and analyzed using reflected-light illumination to find the areas of interest. When the location is determined, the reflected-light sample can be further prepared as an ultrathin section. This adds little additional work to the section to be examined, because the same polished first face is required prior to developing the ultrathin section. Also, it is possible that the information that is desired can be determined with only reflected light, and further preparation may not be required.

When preparing ultrathin sections, the orientation of the section relative to the fiber alignment is a major consideration for fiber-reinforced composites. The angle at which the section is taken is important when quantifying anomalies such as voids and microcracks. If extensive microcracking of the resin matrix has occurred, thin sections perpendicular to the fibers would be extremely fragile, and the specimen may tend to break apart. In this situation, a section parallel to the longitudinal fibers will be beneficial, because the fibers provide support to hold the cracked matrix together.

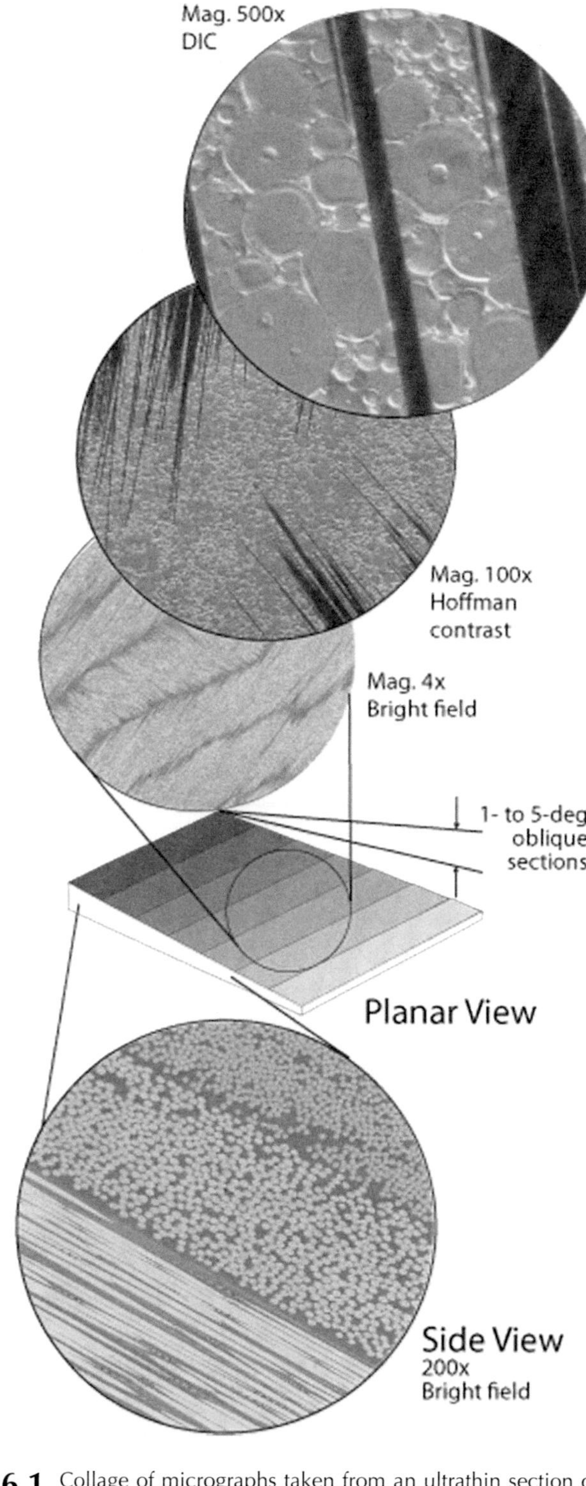

Fig. 6.1 Collage of micrographs taken from an ultrathin section of an interlayer-modified carbon fiber composite. The morphology of the interlayer area is shown with the use of transmitted-light Hoffman modulation contrast and differential interference contrast (DIC). Reflected-light bright-field micrographs are shown of the planar and side view of the composite for comparison.

The development of ultrathin sections and transmitted-light analysis is useful or necessary for some of the following features that occur in fiber-reinforced polymeric composites:

- A single, small area, such as the fracture origin in a composite laminate. The feature can be a small zone that requires several sections for a broader sampling.
- Dispersed particles, polymer phases, and short fibers embedded in a matrix or adhesive. A common feature of interest is a toughening phase or particle either dispersed throughout the matrix or in the interlayer region.
- Large, complex structures with an array of different materials. Frequently, there are zones with intermingled phases (for example, honeycomb sandwich composites/prepreg matrices, honeycomb core adhesives and dip resins, film adhesives, primers, and paints) between components that must be documented.
- Surfaces where thermal oxidation, ultraviolet light degradation, or contamination has occurred
- Areas that have been subjected to lightning strikes

Preparation of the Rough Section for Preliminary Mounting

The samples for thin-section development are best mounted in a reusable silicone or ethylene propylene diene rubber mold that is 57 by 25 by 25 mm (2.25 by 1 by 1 in.). A hard epoxy casting resin should be used for making the mount. The procedure is described and illustrated in Chapter 2, "Sample Preparation and Mounting," in this book. Backing pieces are necessary on the outside edges of the sample(s) of interest. As with all mounts, it is best if the backing pieces are of similar hardness, but this is one case where that may not be possible. The provision of a dark border or edge is advantageous during the ultrathin-section development. This allows visual observation of the thin-section development process and is an indicator for uniform sample removal (Fig. 6.2). If, for example, glass fiber composite backing pieces were used on the edges of a glass fiber composite sample, there would be no way to observe the sample thickness, and it would therefore be difficult to visually determine when the sample is thin enough to move to the next step in the process. In all cases, carbon fiber composites are recommended for the backing pieces.

The specimen length and backing pieces should extend the length of the mold. This will ensure that the material to be prepared for the final thin section extends the entire length of the glass slide face that is mounted to the first polished face. In addition to the ability to visually monitor the sample thickness during preparation, the edges of the sample are protected by the backing pieces. Therefore, artifacts are not created in the actual

Fig. 6.2 Photograph of an ultrathin section made from a glass fabric composite material. In this sample, the backing pieces are made from carbon fiber composite to provide contrast with the glass fabric composite samples in the center of the section (which cannot be seen due to the ultrathin dimension!).

specimen during the preparation process. Also, by having a longer specimen, the area of interest can be more effectively located as close to the center of the mount as possible. The removal of material, in the polishing steps, is better controlled from the center of the sample than at the edges. Therefore, smaller mold sizes should not be used, even though the final ultrathin section will be the same size as a standard petrographic slide, 46 by 27 mm (1.8 by 1.1 in.). The mounted samples must be essentially void-free, with excellent adhesion between the sample and backing pieces. If not, during subsequent sample preparation, distortion or complete sample destruction may occur.

Grinding and Polishing the Primary-Mount First Surface

The same procedures described in Chapter 3, "Rough Grinding and Polishing," are used to grind and polish the samples. This operation can be performed either by using automated grinding/polishing equipment or by hand grinding/polishing. The "first surface" that is polished will become the surface bonded to a glass slide. It is essential that there are no artifacts created in the polished first face, or it will affect the optical light path on the final section. Flatness of the surface is also critical, not only on the section plane but also on the structure interface of different components (for example, carbon fiber/matrix interface). After the polishing is complete, the sample surface can be viewed using reflected-light optical microscopy. If the quality of the surface is not excellent, it can be subjected to longer polishing or reworked using the initial polishing steps.

Mounting the First Surface on a Glass Slide

Mounting the first polished surface to a standard petrographic slide, 46 by 27 mm (1.8 by 1.1 in.), is the next step in the process. The epoxy resin that is used for this mounting must be water white and transparent in addition to providing excellent adhesion to the glass slide and polished substrate. The selected resin must be low viscosity (that is, less than 300 cP mixed viscosity) to provide a very thin bondline. The lack of a homogeneous adhesive bond between the sample and the glass slide will decrease the quality of the thin section. Trapped air and/or contaminants can cause the sample to separate from the slide during the final polish, resulting in the destruction of the thin section. The following steps should be used to mount the sample onto a glass slide:

1. Clean a standard 46 by 27 mm (1.8 by 1.1 in.) petrographic glass slide with an appropriate solvent.
2. Clean and dry the polished sample surface.
3. Apply the adhesive resin to the polished first face using a vacuum impregnation apparatus. This ensures that any cavities opened by the polishing are filled and sealed. Alternatively, apply vacuum to the mixed adhesive resin to remove the entrapped air, and then manually apply it to the polished first face. After the adhesive resin is applied on the polished specimen surface, it can again be placed under vacuum to remove any air entrained from the application process.
4. Place the glass slide onto the polished first face/adhesive.
5. Clamp the sample to the slide, and allow it to completely cure. The easiest way to apply pressure to the sample and glass slide is to place an approximately 1 kg (2 lb) weight on top of the sample. Do not apply heat to cure this assembly. Heat can cause large residual stresses at the sample/glass slide interface.

Preparing the Second Surface (Top Surface)

The preparation of the second surface is time-consuming and can only be accomplished by hand. This process relies heavily on "feel" and becomes more natural as the microscopist produces multiple samples. As additional samples are produced, more information is learned relative to when each step is complete, how the sample is controlled in each step, and what causes damage to or artifacts in the specimens. Initially, many samples can be destroyed in the final steps in the preparation of the second face. This is common, but, with more experience, samples lost due to damage typically are less than one percent. Like the first surface, the second surface must be free of artifacts. Any artifact, especially on the top surface, can present a problem. When using transmitted-light microscopy, the residual stress and strain associated with the scratches will affect the optical path and interfere with the microscopic analysis.

The second surface is prepared by initially cutting and shaping the mount. This is done so that it can be trimmed down using a vacuum chuck and also so it will fit inside of a sacrificial hand vise for further preparation. The overall procedure consists of grinding and polishing the mounted sample from a 1 mm (0.04 in.) rough-cut section to a nominal 5 to 0.5 μm thick sample. The second (polished) surface must be parallel to the first (bonded) surface. The following sequence of steps is recommended for the preparation of the second surface after mounting the first surface to the glass slide.

Step 1: Trimming the Rough Sample

The first step in the second-face preparation is to trim the flashing from the mounted sample around the glass slide so it will fit into a vacuum chuck (Fig. 6.3). The excess flashing must be trimmed so the mount matches the dimensions of the petrographic slide. This also prepares the sample so it can be placed in a sacrificial hand vise, followed later by a standard petrographic hand vise.

Step 2: Trimming the Sample

The trimming of the sample is best performed using a vacuum chuck to provide a precision wafer cut (Fig. 6.4). A final sample thickness of 1 mm (0.04 in.) is usually targeted for this first cut. If a wafering saw with a diamond blade is not available, an abrasive saw can be used. In this case, the sample should be cut to between 2 and 3 mm (0.08 and 0.1 in.) thickness, because more damage may occur to the sample, and it may not be as planar. The trimming procedure using a wafering saw is as follows:

1. Place the primary mount with the bonded glass slide in a vacuum chuck.
2. Clamp the vacuum chuck and primary-mount assembly for preparation of the 1 mm (0.04 in.) cut.
3. Use a wafering saw to cut off a nominal 1 mm (0.04 in.) thick section, not including the bonded glass slide (Fig. 6.3). The manufacturer's recommendation for speed and pressure should be followed, because these parameters will vary with the resins used and the materials that

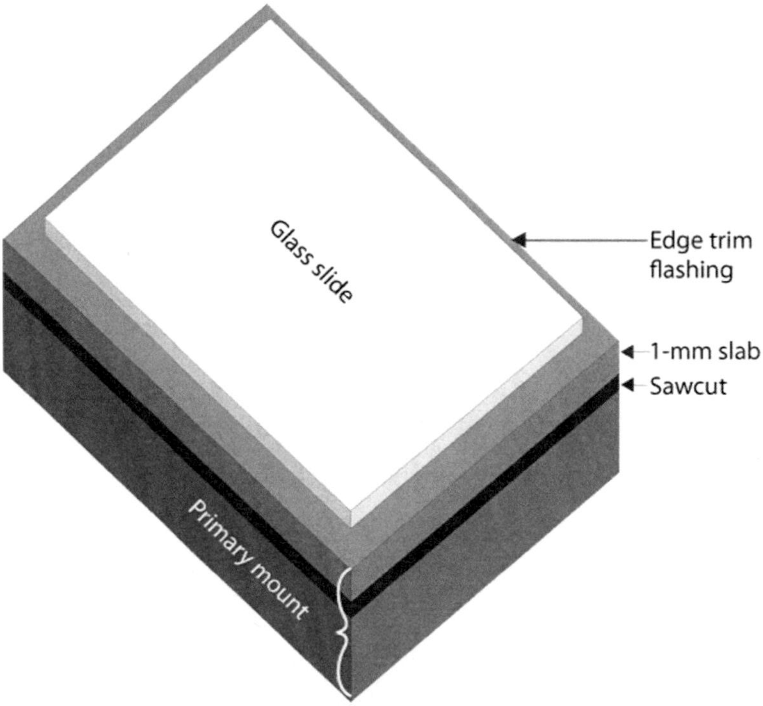

Fig. 6.3 Sketch of the primary mount and flashing that must be trimmed to fit into a vacuum chuck and hand vise

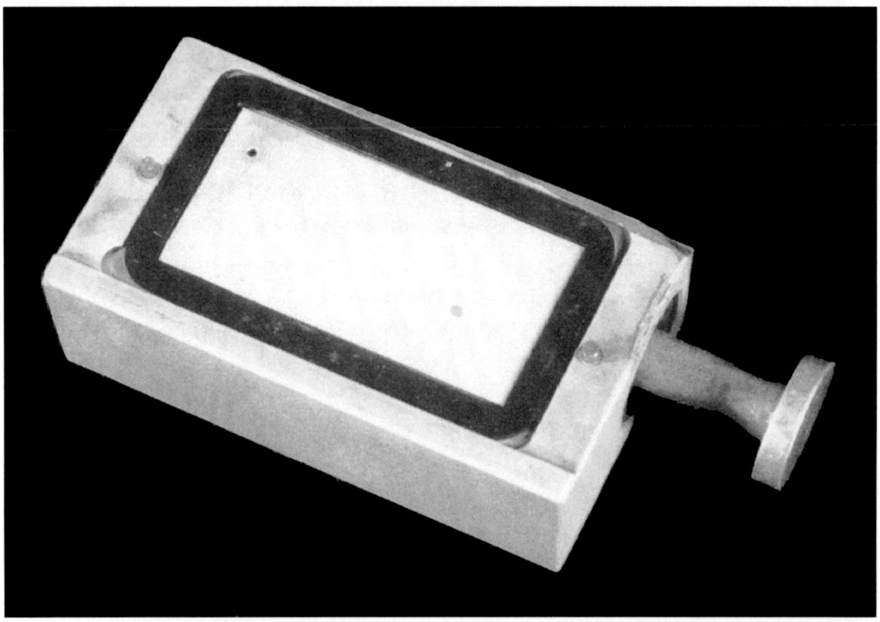

Fig. 6.4 Vacuum chuck used for holding the sample while cutting a 1 mm (0.04 in.) thick section from the primary mount. The glass slide that is bonded to the polished first face of the mount is placed against the vacuum chuck and held secure while sectioning.

are cut. The remaining rough-cut specimen may be reprocessed from the primary mount to obtain a series of thin sections.

Some polymer resins may load up the blade and prevent it from cutting. When cutting the 1 mm (0.04 in.) section, it is best to use a new blade and to change or dress it every few samples or when it first begins to load up with resin. The blades can also deflect as a result of material buildup, resulting in nonplanar cuts. If a wafering saw is not available, sample removal can also be done by grinding a thicker specimen, using 120-grit SiC paper, to a final thickness of 1 mm (0.04 in.). This is often time-consuming, depending on the original specimen thickness.

Step 3: Mounting the Wafer in the Vise for Grinding the Second Surface

Coarse and fine grinding is performed with the aid of a sacrificial hand vise (Fig. 6.5). These are custom made for the purpose of ultrathin-section development. The purpose of this type of vise is to increase the surface area that is in contact with the grinding/polishing surface and to provide a holder for the sample. This provides excellent control of the sample and planar surfaces after each step. The sacrificial hand vise is made from an

Fig. 6.5 Photograph of a sacrificial hand vise used for rough and fine grinding the second face of the sample

aluminum cylinder 25 mm (1 in.) high by 60 mm (2.375 in.) in diameter and has a 46 by 27 mm (1.8 by 1.1 in.) movable (and removable) rectangular piston located in the center. A threaded hole is located on one side of the aluminum block for a set screw that is used to provide the piston stop. The rectangular piston is best made from a material harder than the aluminum vise body, such as stainless steel. The harder piston is not damaged from the pressure of the set screw, and the surface is more scratch resistant. It is advantageous to make the hand vise inset and rectangular piston with slightly rounded corners, so the sample edges are not as apt to catch on the polishing cloths. Hence, the glass slide/sample corners of the 1 mm (0.04 in.) thick rough section also require rounding to fit into the hand vise. This is best done by grinding the corners with 120-grit SiC paper until the sample fits in the cavity. The following procedure is used for mounting the rough section in the sacrificial hand vise:

1. The surface of the piston and the surface of the hand vise must be synchronized on the same plane for each use, which is accomplished by grinding them to a common plane. This is usually accomplished by grinding on 120-grit SiC paper, followed by grinding on 320-grit SiC paper. After the piston and the hand vise have been "planed," the piston should be reset to 375 μm, plus the thickness of the glass slide, into the cylinder. The easiest way to set the standoff distance is to use a blank slide and 375 μm of shim material (Fig. 6.6).
2. Before mounting the specimen in the sacrificial hand vise, it is necessary that the edges of the specimen are rounded to fit into the vise.
3. At this point, a few drops of glycerin are placed on the recessed piston, and the glass-bonded 1 mm (0.04 in.) rough section is inserted into the cylinder. The sacrificial hand vise with the glass-bonded 1 mm (0.04 in.) rough section is now ready for the grinding process (Fig. 6.7).

Step 4: Grinding the Second Surface

The coarse and fine grinding is accomplished by using the sacrificial hand vise and can only be performed manually (that is, by hand grinding and polishing). The grinding parameters are described in Chapter 3, "Rough Grinding and Polishing," in this book. In all of these steps, the pressure applied is high enough for fast sample removal while also keeping the sample under control and the surface planar. The following procedure is recommended for this task:

1. Grind the material protruding from the vise surface with 120-grit SiC paper until the excess sample is removed and it is flush to the aluminum vise surface. Do not use coarse paper, because damage and deep scratches may result. Note: Use reflection off-surface to identify high spots where additional grinding is required (Fig. 6.8).

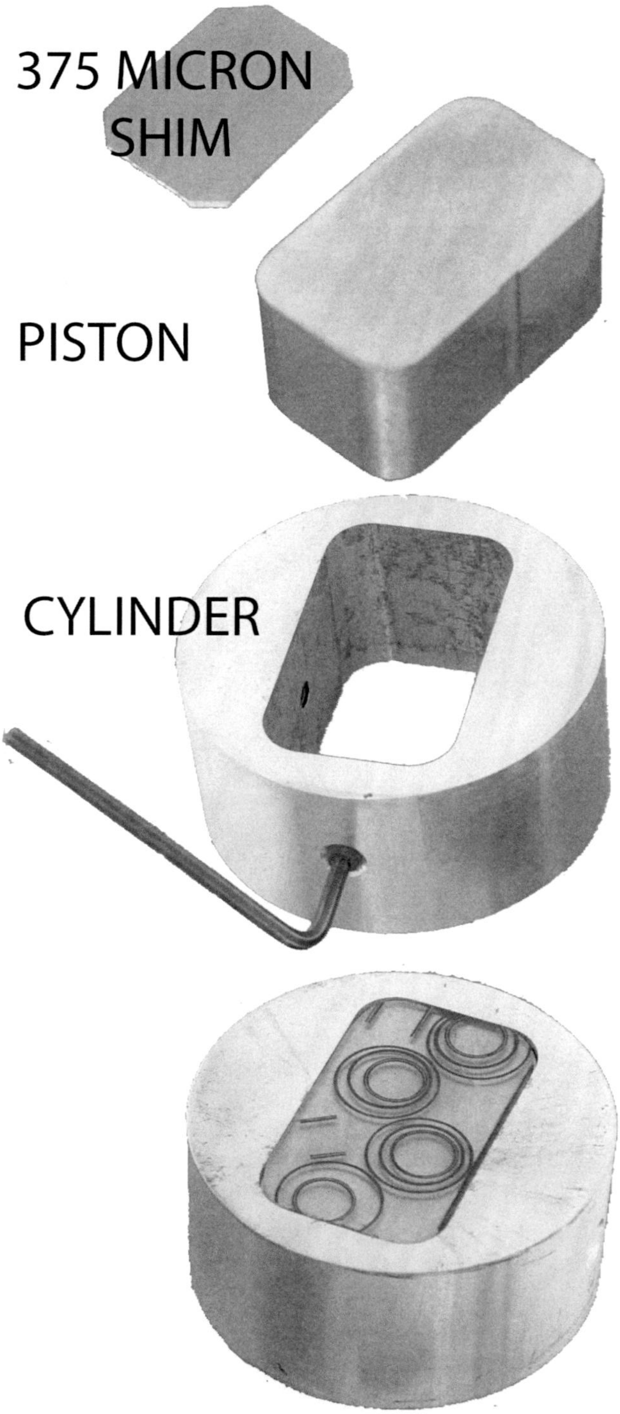

Fig. 6.6 Sacrificial hand vise showing the piston, shim, and mounted specimen

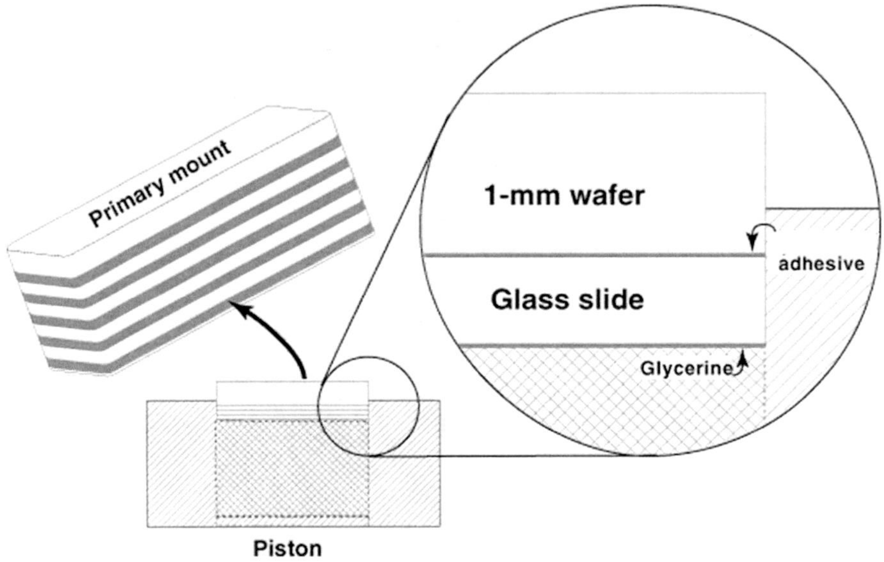

Fig. 6.7 Schematic of the sacrificial hand vise assembly showing specimen placement and the use of glycerin for holding the specimen in the cavity

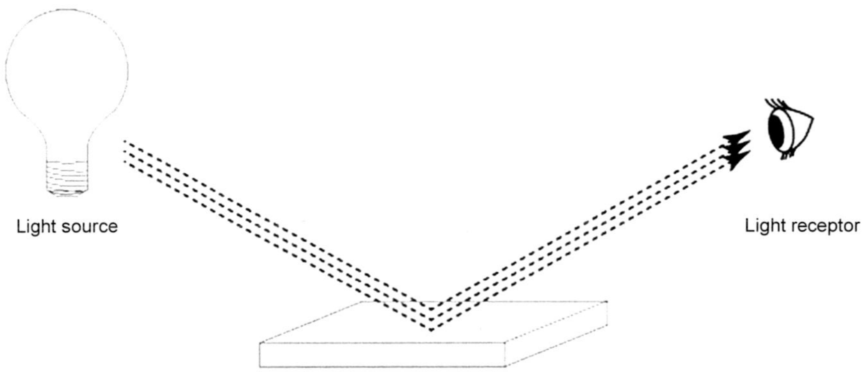

Fig. 6.8 Use of light reflection as a tool for checking the sample plane

2. Continue the grinding process in the vise with 320-grit SiC paper down to a level of roughly 50 μm above the first surface. This involves grinding the aluminum block of the vise as well. The exact thickness of 50 μm may be hard to initially determine. The specimen can be removed and measured with the aid of a micrometer if necessary. However, if the piston is pushed up to remove the sample, it must again be placed on a hard flat surface to ensure that it is mounted planar again.
3. Finish the grinding process in the vise using 600-grit SiC paper. The sample thickness should be approximately 25 μm. This completes the grinding process, and the polishing process can now begin. The sacrificial hand vise will not be used for the polishing steps.

Step 5: Polishing the Second Surface

The polishing of the ultrathin section requires the specimen to be mounted in a petrographic hand vise (Fig. 6.9). The purpose of the hand vise is to serve as a handle to hold the slide while polishing. A hand vise can be purchased commercially, or it can be made. A 22 mm (0.875 in.) high by 60 mm (2.375 in.) diameter stainless steel cylinder works very well with a 1 mm (0.04 in.) recessed 46 by 27 mm (1.8 by 1.1 in.) rectangle machined out of the center. In the center of the machined rectangle, a hole extending through the vise is necessary so that an object can be used to push the sample out of the cavity. A few drops of glycerin should be applied to the cavity to hold the specimen secure while polishing. The specimen can be removed at any time for microscopic evaluation and reinstalled in this vise for further preparation. The polishing is performed by applying pressure on the hand vise using the same procedure as described in Chapter 3, "Rough Grinding and Polishing," in this book. When polishing using the 15 μm deagglomerated alumina suspension, the polishing step should be stopped when the thin section is within 1 to 3 μm of the final thickness. Shaping is done by center-weighting or off-center-weighting to control the final plane of the ultrathin section so it is parallel to the first surface or tapered slightly to create a wedge (Fig. 6.10). This is the last opportunity to do any major shaping of this plane.

Fig. 6.9 Mounted specimen in a hand vise ready for polishing

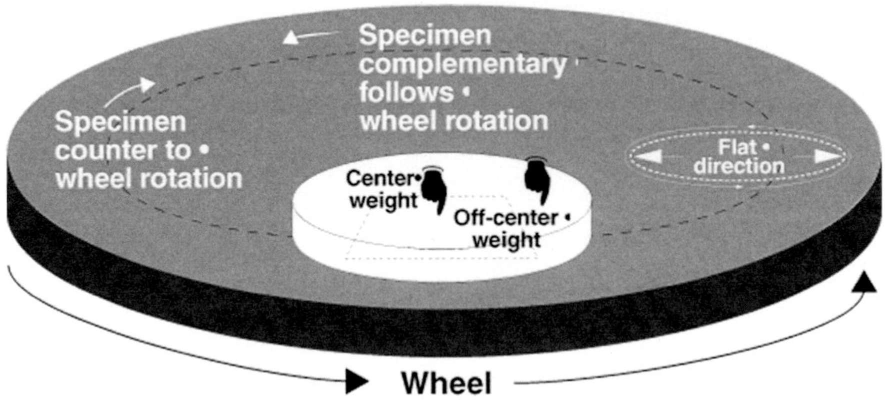

Fig. 6.10 Polishing wheel showing specimen weighting for feathering the sample plane

The second polishing step uses 0.3 μm deagglomerated alumina suspension on a silk cloth, as described in Chapter 3. This step removes the last 1 to 2 μm. This step will achieve the final plane and leave the surface at a point where it can be examined without further processing. It is important to maintain center-weighting. Overpolishing can cause the sample to disbond from the glass slide. This step will test the effectiveness of the adhesive resin. If necessary, a final polishing step can be performed by hand or by vibratory polisher using diluted 0.05 μm deagglomerated alumina suspension.

Summary for Ultrathin-Section Sample Preparation

It is important to pay attention to all aspects of polishing (for example, center-weighting, hand pressure, and cutting rate). Even the sound of the abrasive paper and the polishing cloth as the specimen is prepared can help guide the microscopist in maintaining the desired plane. While proceeding through the steps involved in preparing a specimen, a good microscopist will continuously check the specimen by viewing the light reflecting off the surface. This technique saves time and is an excellent tool. Knowing when to stop and when to continue to the next step is critical and is usually done by feel and experience. Each step leaves damage of decreasing magnitude. When too much material is removed at any given step, the sample, in most cases, is either ruined or distorted. If not enough of the sample is ground away at any given step, it becomes difficult, and sometimes impossible, to reach ultrathin dimensions. Feathering out is defined as the process of creating a slight wedge that is spread across the entire sample thickness, varying from 0 to 5 μm. The advantage of feathering is that it allows the entire thickness range to be scanned using the thickness effect on the optical contrast, refractive indices, and resolution. This allows the microscopist to obtain the maximum information and photographic detail.

The biggest advantage of this is that different optical contrast modes can be used for different portions or thicknesses of the sample. By carefully controlling the weighting of the sample during polishing, the specimen can be tuned to any shape or thickness needed. It is possible to obtain a near-parallel surface; however, most polymeric composite materials have an optimal thickness that corresponds to differences in contrast. Therefore, tapering the surface of the specimens may offer more information while also providing easier identification of sample removal.

Transmitted-Light Microscopy

Transmitted-light microscopy methods use several contrast-enhancement methods that are not commonly used in metallurgical reflective microscopy. These methods include:

- Bright field
- Polarized light
- Phase contrast
- Differential interference contrast (DIC), also known as Nomarski
- Modulation contrast, also known as Hoffman modulation contrast
- Epi-illumination
- Optical staining

The various methods of transmitted optical microscopy are used to enhance or selectively detect changes in light intensity or color. The human eye and photodetectors are sensitive to differences in light intensity or wave amplitude and changes in color, indicating changes in the frequency of light. An example of several contrast methods applied to one material was shown in Fig. 6.1. At low magnification, bright field shows the composite with good contrast. Hoffman modulation contrast at 100× and DIC at 200× are used to bring out the sample features at higher magnification.

Color and/or light-intensity effects shown in the image are related to the rate of changes in refractive index and the thickness of the specimen, or both. Optical path differences are sometimes referred to as optical thickness or optical color staining (Ref 5). These microscopy methods are based on the fundamentals of optics and optical crystallography and polymer microscopy, which are explained in detail in Ref 5 to 12. Key terms are briefly defined here:

- Microscopy contrast or relief is low when the image of a particle is nearly invisible, and high when features are readily resolved or a grain has distinct outline:

$$\%\ \text{contrast} = \frac{(\text{Intensity BACKGROUND} - \text{Intensity SPECIMEN})}{\text{Intensity BACKGROUND}} \times 100 \quad \text{(Eq 1)}$$

- Refractive index ("n" or RI) is defined by Snell's law as the ratio of speed of light in vacuum to the speed of light in medium. The "n" of liquids and solids is greater than one, because light travels slower through these than through vacuum or air. The RI of a material is determined by the chemical composition and crystallographic properties of the material. The Gladstone-Dale equation states that a material must have a change in composition (chemical bonding) or in density (crystal volume) to change RI (Ref 13). Changes in RI must be considered in context. Changes may indicate compositional changes, crystallographic changes, or intermingling of phases, and these may cause changes in physical properties.
- Birefringence is the numerical difference, $|n_2 - n_1|$, between the maximum and minimum RI of a crystal, fiber, or film. A birefringent material has crystallinity or strain. An amorphous material has one RI or a variable RI, indicating variable composition.
- Retardation (r) is the product of the feature thickness and the birefringence. This is written in units of nanometers and is seen as a distinctive color in the Newtonian interference series:

$$r = t\,|n_2 - n_1| \qquad \text{(Eq 2)}$$

- By measuring the ultrathin-section or particle thickness, the retardation or observed change in color can be measured to determine the change in RI (Ref 7).

Optimization of Microscope Conditions

A number of steps can be taken to maximize the resolution and contrast detected. These steps can be applied for all transmitted-light microscopy. The commonly accepted theoretical resolution limit for a given microscope objective can only be approached by careful attention to these details:

- Align the light path to obtain back-focal plane illumination, with the light source imaged in the plane of the objective, frequently referred to as Kohler illumination. This is required for optimal focus and control of light to prevent light scatter (Ref 5 to 12):

$$\text{Resolution limit} = \frac{k\lambda}{NA} \qquad \text{(Eq 3)}$$

  where k = 0.5 or 0.61, NA is the numerical aperture = $n_0 \sin \alpha$, λ is the wavelength of light used, and n_0 is the RI of the objective space.
- Use high-NA condensers and objectives. The consideration of the quality of both components is essential, because the effective NA is ½(NA objective + NA condenser). This step optimizes the resolution.
- Use RI oils selected to increase relief and contrast of the specimen.

This requires a basic knowledge of the RI of the composite or polymer components in order to select appropriate oils. A small scraping of polymer or other material is immersed in a series of oils to determine the RI by comparison of relief (Ref 5, 7). A difference of 0.10 to 0.20 between specimen RI and mounting oil is recommended for contrast enhancement. A range of RI oils, commonly known as Cargille RI oil sets, can be readily obtained from microscopy supply catalogs.

- Use glass cover slips 0.16 to 0.18 mm (0.006 to 0.007 in.) thick (No. 1 or 1.5) over the RI oils. In general, microscope objectives for transmitted light are designed to be used with this thickness of cover slips. Thicker glass, 0.22 mm (0.009 in.) (No. 2), can be used for less critical microscopy than examination of ultrathin sections.
- Control the thickness of the thin section or polymer film. A specimen such as a polyamide with very fine crystal features (i.e., for example, less than 15 μm) and high birefringence requires ultrathin sections for optimal resolution and contrast. The presence of very fine-grained residual structures from liquid crystallinity also requires ultrathin sections. The maximum specimen thickness, T_{max}, can be calculated through the use of Eq 4 for a uniaxial spherulitic structure for optimal resolution (Ref 6):

$$T_{max} = \frac{\lambda\varepsilon\omega}{NA\,|\,\varepsilon^2 - \omega^2\,|} \quad \text{(Eq 4)}$$

where λ is the illumination in nanometers, ε is the RI of the major axis of the structure, and ω is the RI of the minor axis of the structure.

- Maximize contrast by use of a green filter for the best effect when phase-contrast microscopy is used. A single wavelength of light has less scatter than white light, plus black-and-white film sensitivity is maximum for green light.

Microscopy of the Section

After producing ultrathin sections and optimizing the microscope conditions, the specimen is ready to be analyzed by microscopy techniques. Standard texts on microscopy should be consulted to obtain and interpret the images recorded by micrography (Ref 5 to 12). Changes in heat and the type of stress the material has seen in use or during testing may be detected in the ultrathin section (Ref 3, 4).

Examples of Ultrathin-Section Analyses

The following are examples of ultrathin-section analyses:

- A thermoset-matrix composite was examined to determine differences in morphology produced with heat-ramp rates of 2.78 and 0.56 °C/min

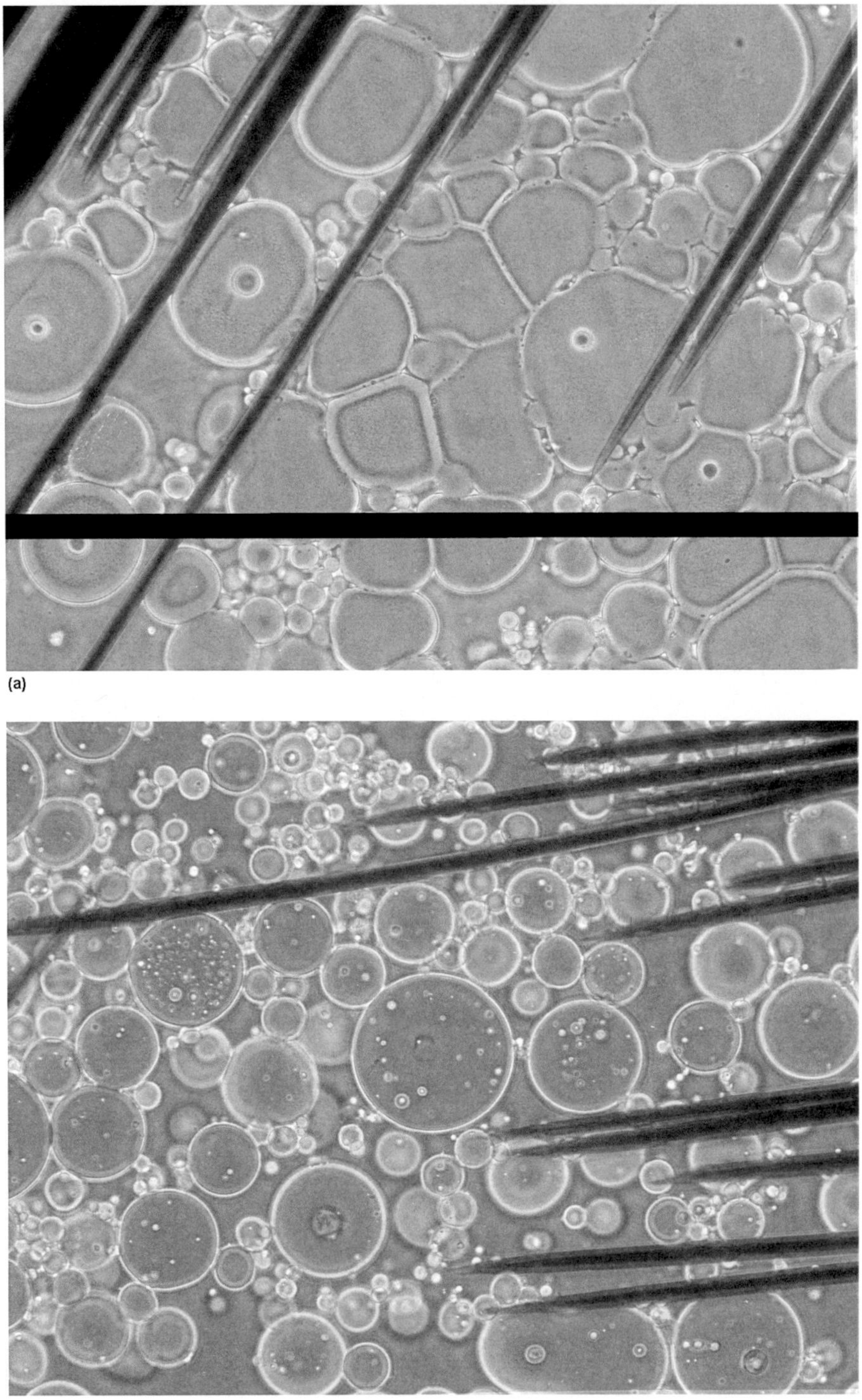

(a)

(b)

Fig. 6.11 Micrographs of a thermoset-matrix carbon fiber composite material comparing the use of two different ramp rates in the cure cycle. (a) 2.8 °C/min (5 °F/min). Transmitted light, phase contrast, 20× objective. (b) 0.56 °C/min (1 °F/min). Transmitted light, phase contrast, 20× objective

(5 and 1 °F/min) during the cure cycle. Transmitted-light phase-contrast microscopy of the ultrathin sections of these materials revealed significant differences in products, as shown in Fig. 6.11(a and b), respectively. These very-low-contrast sections show the development of smaller inclusions as a second phase at 0.56 °C/min (1 °F/min) ramp rate. This material had lower physical properties when compared with the composite that was ramped at 2.78 °C/min (5 °F/min). Other imaging methods could not illustrate this second phase.

- Transmitted-light DIC was used to characterize a glass-fiber-filled engineering thermoplastic to determine if voids were present. The section (Fig. 6.12) showed significant porosity due to air entrapment during the injection molding cycle. The fibers are randomly oriented, with many fibers perpendicular or parallel to the section plane. Probable cause of the observed fiber orientation is fiber movement during air bubble formation. The typical fiber orientation would be predominantly in the injection flow direction.
- Transmitted polarized light was used to study a nominal 1.5 μm section of a high-temperature thermoplastic composite material to determine if crystals were present (Fig. 6.13). These are seen as spherulites but can be resolved at 1000 times to have liquid crystalline polymer character. Careful examination can determine the presence of true three-dimensional polymer spherulites with definite edge boundaries

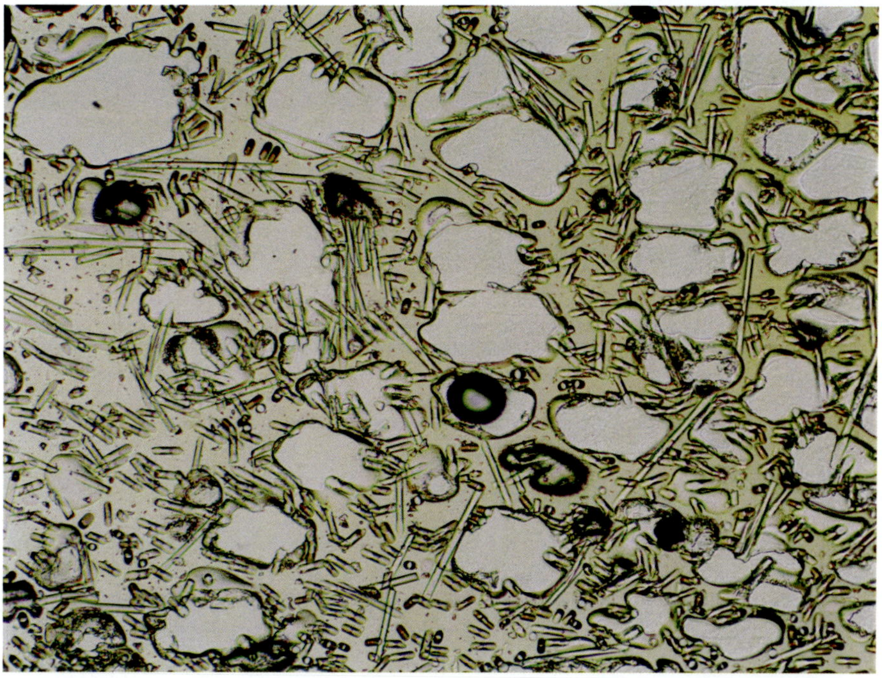

Fig. 6.12 Voids in a glass-fiber-filled engineering thermoplastic matrix. Transmitted light, differential interference contrast, 40× objective

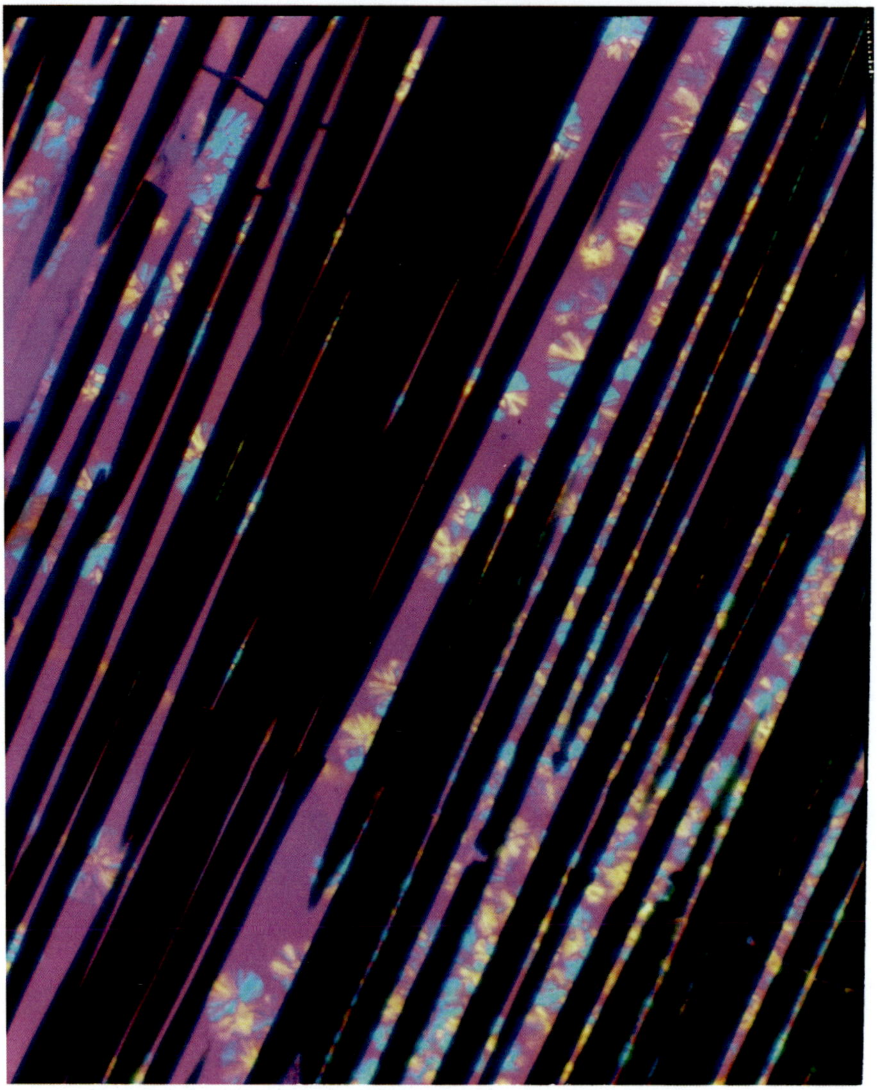

Fig. 6.13 Crystals formed in a high-temperature thermoplastic-matrix composite. Transmitted polarized light, 40× objective

compared with contrast changes due to stress set up by an inclusion that does not have distinct edges (Ref 6).

Conclusions

The production of ultrathin sections greatly expands the resolution and contrast available from numerous methods of optical microscopy. No chemical etchants are needed to bring out the contrast or phase boundaries with these methods. The images presented are representative of the materials examined. The microscopist must invest a great deal of time and effort in the study of polymer microscopy and optical crystallography theory. The comparison of sections from normal and abnormal materials is essential. Discussions to define the heat or stress environment of the material being examined are also essential to interpretations of ultrathin sections of composites and engineered materials.

REFERENCES

1. B.S. Hayes and L.M. Gammon, Microscopy, *Composites,* Vol 21, *ASM Handbook,* ASM International, 2001, p 964–972
2. L.M. Gammon and D.J. Ray, Preparation of Ultra-Thin Sections for Fiber-Reinforced Polymer Composites and Similar Materials, *Proceedings of the 30th Annual IMS Convention,* July 1997 (Seattle, WA)
3. N.T. Saenz, Ultrathinning Section Techniques for the Characterization of Brittle Materials, *Microstruct. Sci.,* Vol 18, 1991, p 147–159
4. L.M. Gammon, Optical Techniques for Microstructural Characterization of Fiber-Reinforced Polymers, *Microstruct. Sci.,* Vol 19, 1992, p 653–657
5. M. Abromowitz, *Contrast Methods in Microscopy: Transmitted Light, Basics and Beyond,* Olympus Corp., Lake Success, NY, 1987
6. C. Viney, *Transmitted Polarized Light Microscopy,* McCrone Research Institute, Chicago, IL, 1990
7. F.D. Bloss, *An Introduction to the Methods of Optical Crystallography,* Saunders College Publishers, Philadelphia, PA, 1989
8. P.F. Kerr, *Optical Mineralogy,* McGraw-Hill Co, New York, NY, 1977
9. C.W. Mason, *Handbook of Chemical Microscopy,* 4th ed., John Wiley & Sons, New York, NY, 1983
10. W.C. McCrone and J.G. Delly, *The Particle Atlas, Principles and Techniques,* 2nd ed., Ann Arbor Science Publishers, Ann Arbor, MI, 1973
11. L.C. Sawyer and D.T. Grubb, *Polymer Microscopy,* 2nd ed., Chapman & Hall, New York, NY, 1996
12. D. Shelley, *Optical Microscopy,* 2nd ed., Elsevier, New York, NY, 1985
13. *Microscopic Determination of Nonopaque Minerals,* U.S. Government Printing Office, 1952

CHAPTER 7

Composite Structure Analysis

THE ANALYSIS OF THE STRUCTURE of a composite material is essential for understanding how the part will perform in service. Often, the mechanical properties of composite materials are assumed to be only a function of the materials and lay-up that are used in manufacturing the part. Although this can be the case, and is most often used for modeling purposes, the manufacturing process/cure cycle adds another variable that dictates the final performance. While through-thickness uniformity is usually assumed, fiber volume variations, void content, ply orientation variability, foreign objects, and other factors can degrade the composite performance. Some of these variables can be controlled through tight quality-control standards, while others are a function of the design and manufacturing of the composite part (Ref 1). The use of optical microscopy in the development phase of the composite product can provide the insight necessary for optimizing the structure and therefore the performance.

The analysis of the composite structure is typically performed using bright-field illumination and low magnification. In some analyses, the use of slightly uncrossed polarized light can provide additional contrast. Depending on the information that is desired, the analysis may be quite easy. In many cases, the lay-up is simple, and only the number of prepreg plies or reinforcement type is required to be determined. Usually, the number of plies is quite easy to determine, but composites having high fiber volumes or parts that are filament wound can be a challenge, and the composite may need to be sectioned at a greater angle to increase the interlayer area. This can sometimes make the determination of the number of plies easier and also help in the identification of the fiber angle.

Ply Terminations and Splices

In some composite designs, the lay-up includes ply terminations and splices that need special attention as to how they are integrated within the composite structure. It is common for these designs to have ply thickness variations within the composite part and other defects in the areas around the ply terminations. A low-magnification montage of a carbon fiber composite cross section having prepreg ply terminations within the part is shown in Fig. 7.1. In this figure, the number of prepreg plies and ply terminations can easily be observed, as well as how the adjacent plies accommodate the ply-drops. Where both ply terminations are found, the area is free of voids, and the transition is smooth. It can be seen that the prepreg plies that are of the same orientation are constrained by the ply terminations. Ply separations or terminations in the center of composite laminates are also found as part of the design or as a result of manufacturing problems. In either situation, the part requires analysis to observe how the adjacent plies are affected. Figure 7.2 shows an example of how prepreg ply terminations can affect the adjacent fiber plies in a composite material. In the vicinity of the ply terminations, no voids are observed, and the adjacent plies are found to fill into the open space. In other designs, ply splices may be required, as shown in Fig. 7.3. A smooth transition of the splices is shown in this composite material.

Composite parts often have defects that are induced by the manufacturing process. Common defects include voids in the composite structure (Chapter 8, "Void Analysis of Composite Materials," in this book), ply thickness variations, and ply or fiber wrinkling. With any defect, the overall performance of the part is degraded. A composite cross section showing wrinkles in the plies is shown in Fig. 7.4. The wrinkle started on the surface and disturbed the plies almost to the center of the composite lay-up.

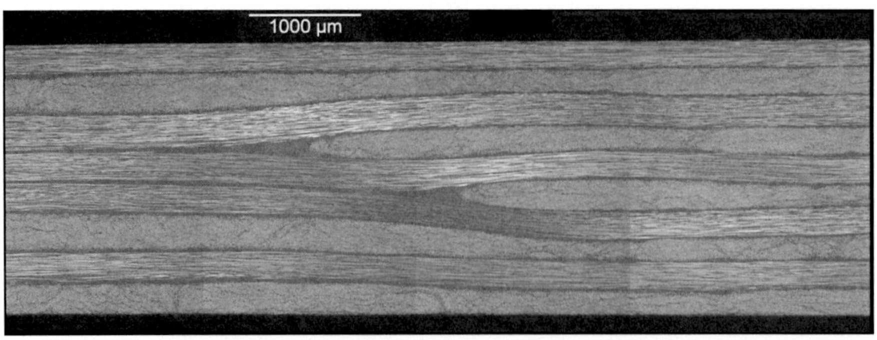

Fig. 7.1 Montage of micrographs taken of a cross section of a composite made from unidirectional prepreg that shows the termination of two prepreg plies near the center of the part. Bright-field illumination, 5× objective

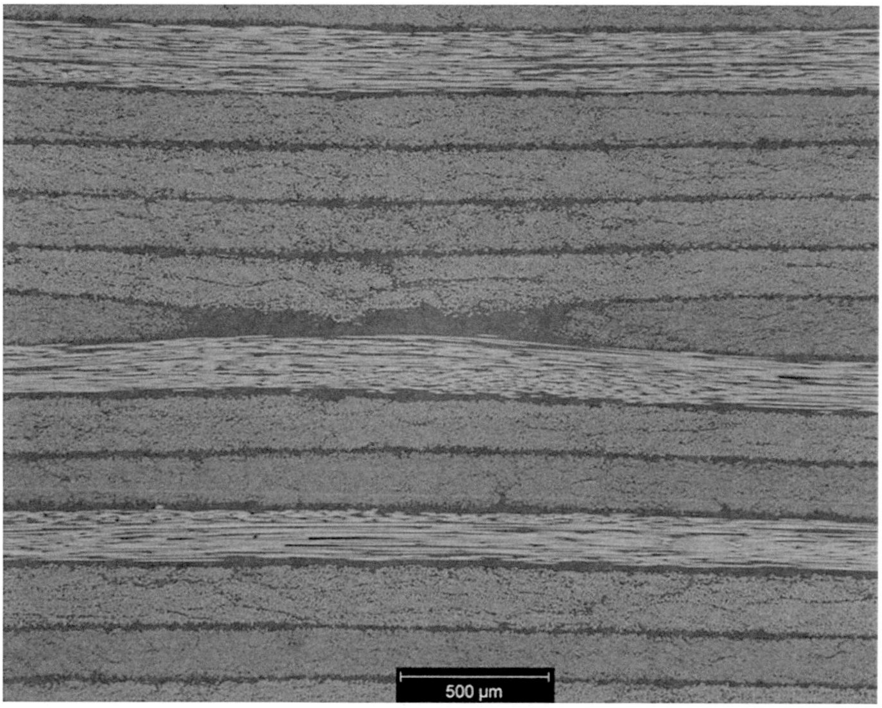

Fig. 7.2 Cross section of a composite material made with unidirectional carbon fiber prepreg that shows a ply separation in the part. Bright-field illumination, 5× objective

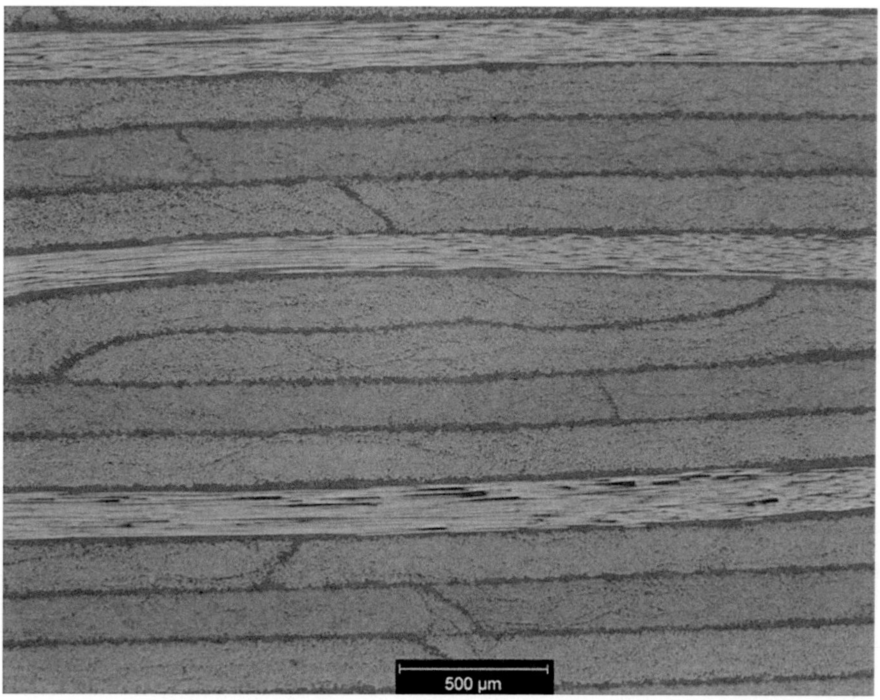

Fig. 7.3 Ply splices shown in a carbon fiber composite cross section. Bright-field illumination, 5× objective

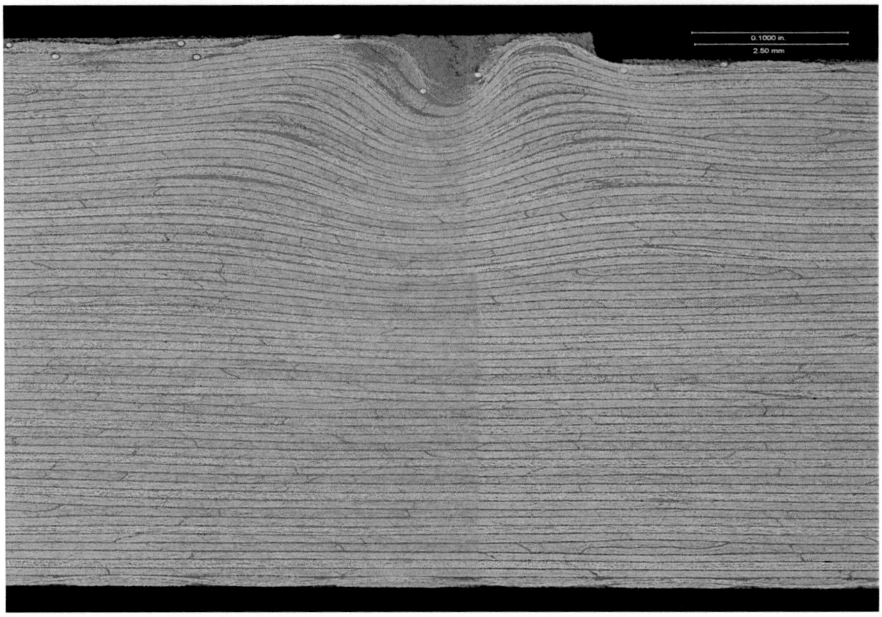

Fig. 7.4 Montage of ply wrinkling in a composite material developed during manufacturing. Bright-field illumination, 5× objective

Multiple Material Combinations

Often, a composite part consists of a single prepreg material or fibrous preform, but in some applications, various material combinations are used to achieve the desired properties. In many cases, the selection of the fiber constituents is based only on optimizing the modulus and strength of the part. Figure 7.5 shows a part made with two different unidirectional carbon fiber prepreg materials having different thicknesses (in this case, areal weights). It can be seen that there are eight thinner prepreg plies in the interior (alternating angle orientation) and three thicker plies of the same orientation with a partial ply that was milled off on the top surface. In this application, various prepreg plies having different thicknesses and material properties were used to optimize the mechanical performance.

In some composite parts, the addition of material combinations may be used for improving characteristics such as damping, damage tolerance, wear and abrasion performance, or environmental resistance. The addition of woven fabric or different fiber materials may be combined to achieve the desired performance. Figure 7.6 shows a cross section of a composite part made with woven glass fabric prepreg and two different unidirectional carbon fiber prepreg materials. Using this combination of materials, the required balance between durability, damping, and stiffness was achieved.

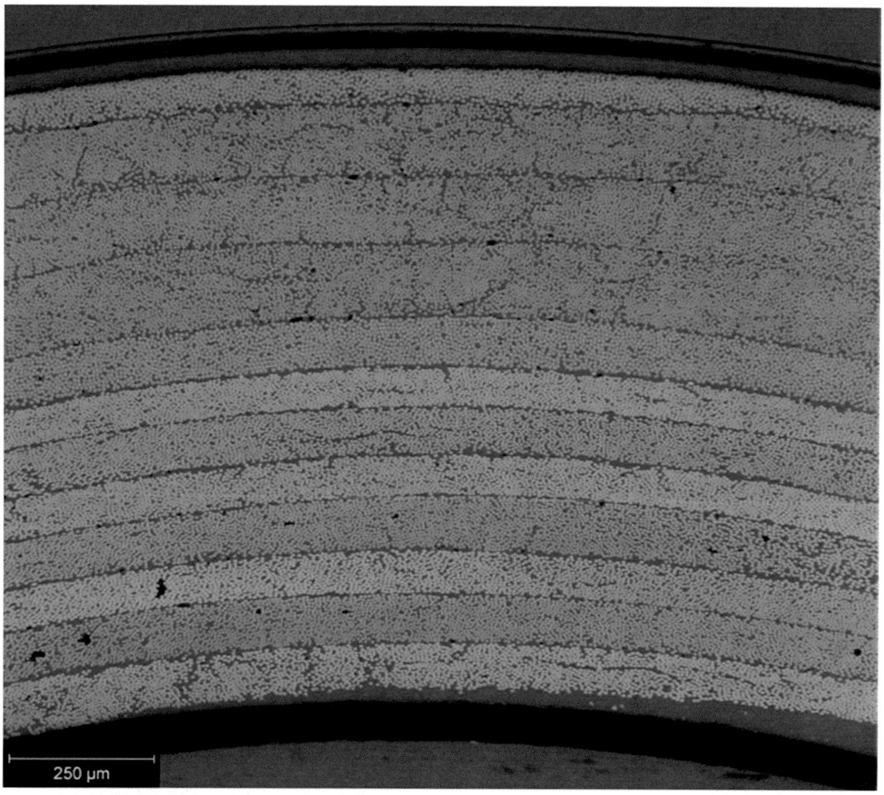

Fig. 7.5 Micrograph of a composite part made with two unidirectional carbon fiber prepreg materials having different thicknesses (areal weights) and fiber types. Slightly uncrossed polarized light, 10× objective

Fiber Orientation Verification

As shown in Chapter 5, "Viewing the Specimen Using Reflected-Light Microscopy," optical microscopy can be used to determine the fiber angles in composites made from unidirectional fiber prepreg materials. Also, it was shown that reflected polarized light can be used to further enhance fiber angle differences. Usually, a composite material can be cross sectioned and verification of the fiber angles can be determined quite easily if the orientation for the lay-up is known (Fig. 7.7). However, if the composite is cross sectioned in only one direction, it may be difficult to accurately determine the fiber orientation of the other plies if they are unknown. In this instance, the composite can be cross sectioned perpendicular to each assumed fiber (ply) angle by reproducing a pattern on the surface of the composite (Fig. 7.8). The fibers should be exactly the same aspect ratio in all plies of the same orientation in this section. This technique can work well with only a few fiber angles. Unfortunately, if the fiber angles are not known or there are many fiber angles, this technique may be too time-

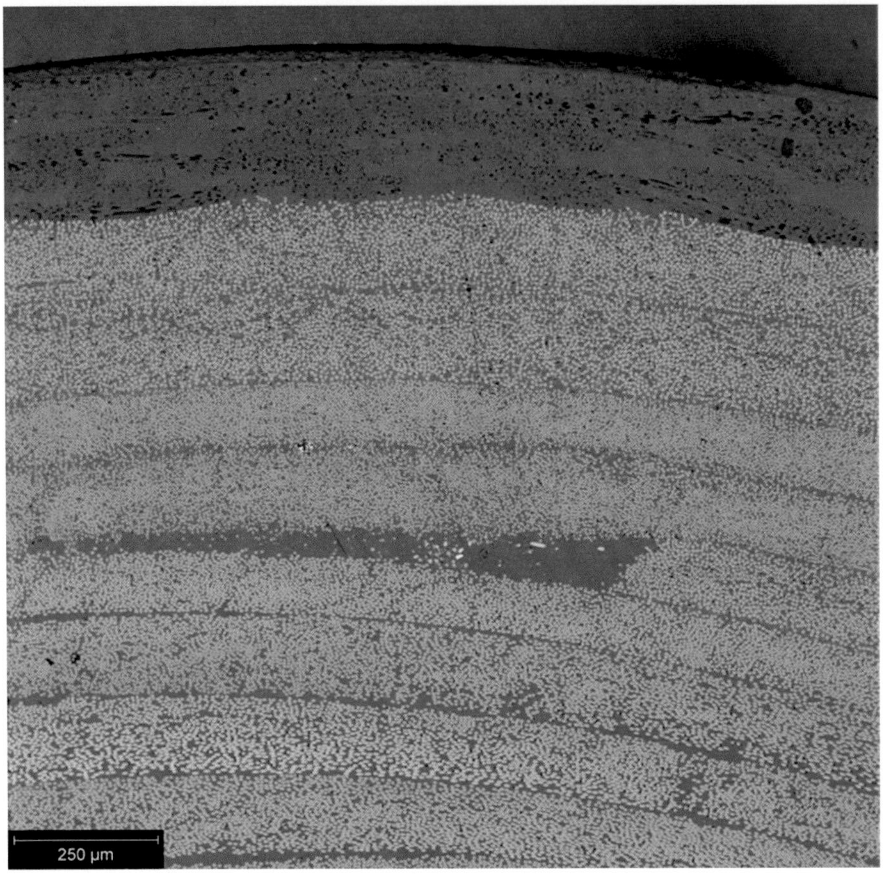

Fig. 7.6 Cross section of a composite part made with glass fabric prepreg and two unidirectional carbon fiber prepreg materials having different thicknesses (areal weights) and fiber types. Slightly uncrossed polarized light, 10× objective

consuming. A much better method can be used to identify fiber angles by sectioning and polishing the specimen at 45 degrees to the plane view (Fig. 7.9) (Ref 2). This creates elliptical images of the fibers in which the angle of the fibers can be determined based on the aspect ratio of the fibers and simple geometry. This is an easy method to determine ply angles using bright-field illumination, but the fiber diameter must be accurately known and, for high precision, preferably should be cylindrical instead of bean shaped. Fiber orientation analysis in tubular composites can be challenging if the part is only cross sectioned through the circumference and viewed normal to the radial direction. A sectional view normal to the axial direction using the same sample may provide better insight into the fiber orientation as well as the morphology of the part. For complete analysis, a tubular part may also need to be viewed normal to the circumference. Figure 7.10 shows the various sections that may be required for analyzing

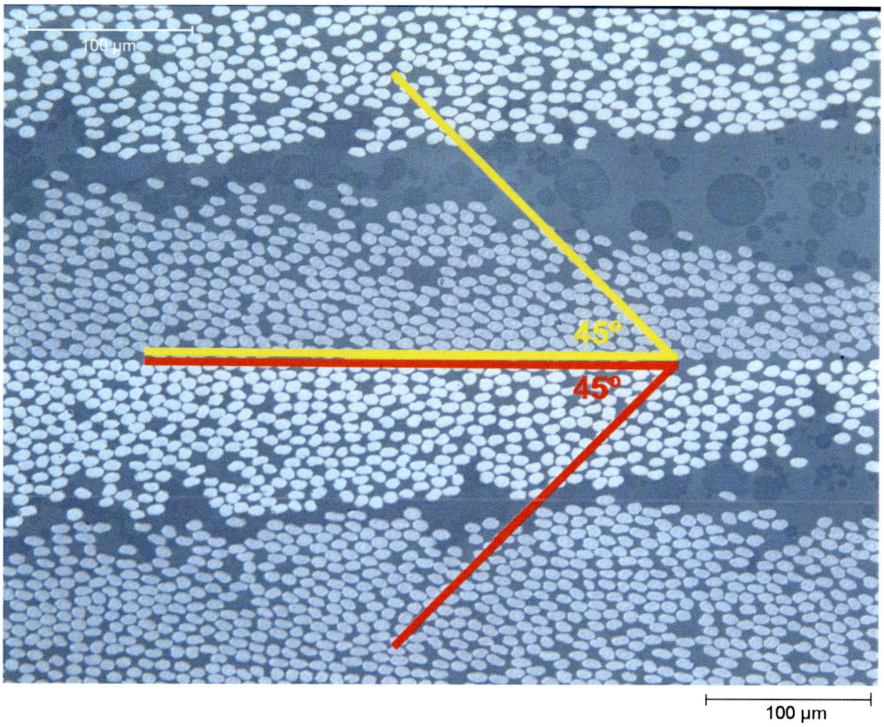

Fig. 7.7 Micrograph of an area in a composite part made from unidirectional carbon fiber prepreg having ±45° ply (fiber) angles. Slightly uncrossed polarized light, 25× objective

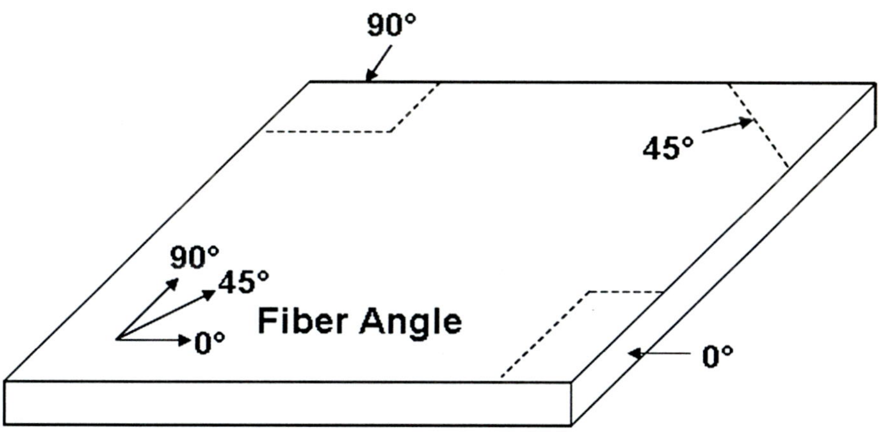

Fig. 7.8 Schematic showing the sections required for determining ply orientations of a quasi-isotropic laminate

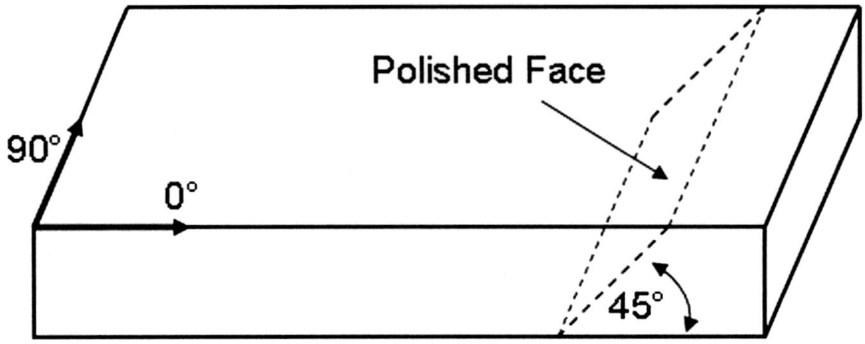

Fig. 7.9 Schematic showing the sectioning of a composite part on an angle through the thickness. This is done to elongate the fibers for easier determination of the ply angles.

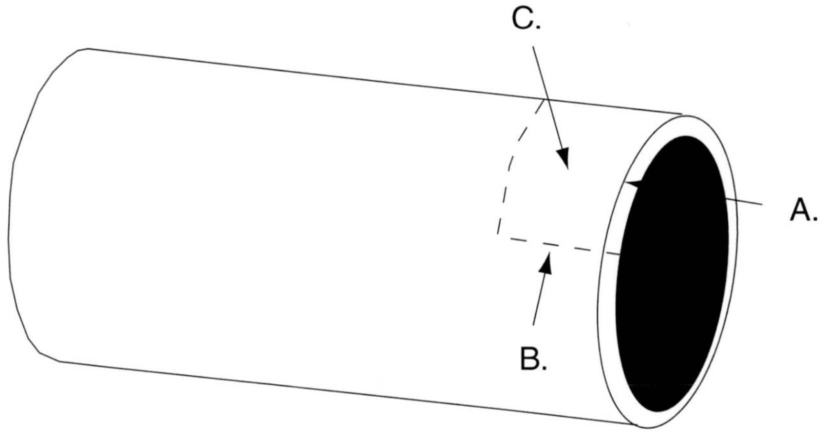

Fig. 7.10 Schematic showing the sections required for identification of the ply orientation and structure of a tubular composite part. A: view normal to the radial direction; B: view normal to the axial direction; and C: view normal to the circumference

tubular composite parts. Figure 7.11(a and b) show polished sections of a tubular composite that are viewed normal to the radial direction (A) and normal to the circumference (C), respectively. As can be seen in Fig. 7.11(b), polishing of a section of the tube surface provides an easy identification of the fiber angle. If the ply orientation changes throughout the thickness of a tube, the polished section can be ground down further and then polished to view the layers below. In Fig. 7.11(a), the fiber angle would be very difficult to determine from only a radial cross section, because the fiber angle is quite low. Errors in ply and fiber orientation analysis are usually found with fiber angles lower than 20 degrees.

Ply orientation is usually not as difficult to determine with woven fabric composites, as compared to composites made from unidirectional fiber

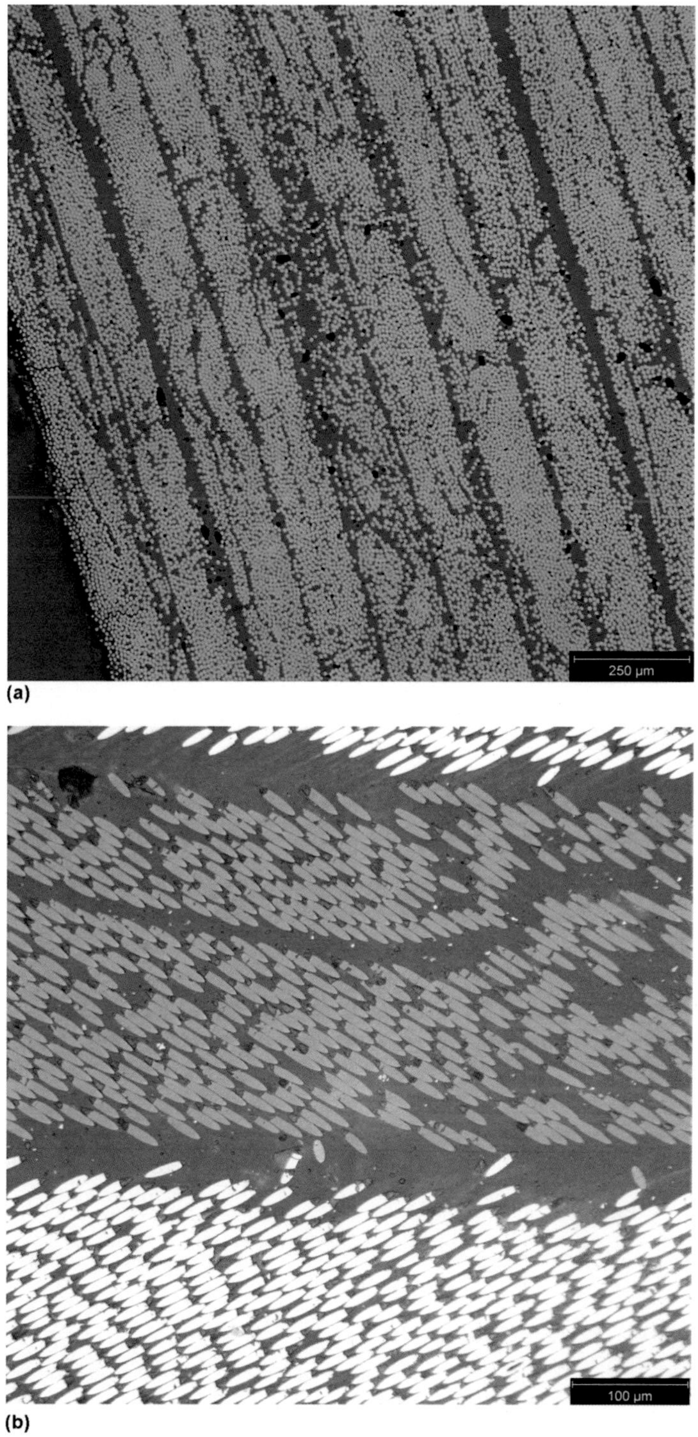

Fig. 7.11 Sections taken from a tubular composite showing the fiber orientation. (a) Viewed normal to the radial direction. Slightly uncrossed polarized light, 10× objective. (b) Viewed normal to the circumference. Slightly uncrossed polarized light, 25× objective

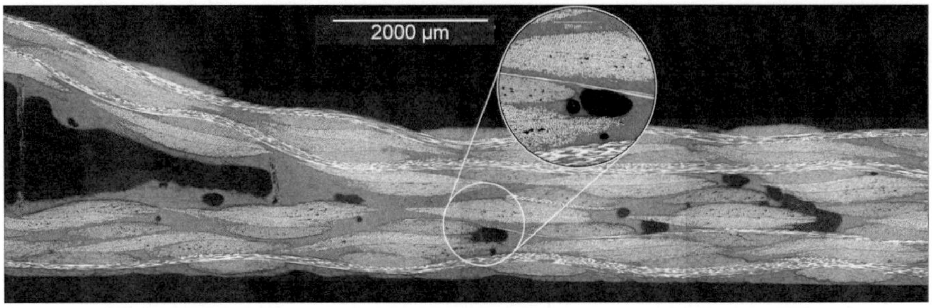

Fig. 7.12 Montage of a chamfer area in a honeycomb composite part made with carbon fabric prepreg skins. Bright-field illumination, 5× objective. A magnified view of an area having a void is also shown.

prepregs, because usually there are only one or two orientations. However, care must be taken in the analysis if the lay-up includes a design where the plies are alternately oriented in the warp and fill directions. Figure 7.12 shows a montage of a chamfer area of a honeycomb composite structure. In this figure, the 3k-70 plain weave carbon fabric plies are easily observed, as are voids in the structure.

REFERENCES

1. B.T. Astrom, Introduction to Manufacturing of Polymer-Matrix Composites, *Composites,* Vol 21, *ASM Handbook,* ASM International, 2001, p 419
2. J. Houser, Method for Determining Ply Count and Orientation of Laminates in Carbon Fiber Parts, *SAMPE J.,* March/April 1987, p 55

CHAPTER 8

Void Analysis of Composite Materials

ACHIEVING THE BEST-PERFORMING COMPOSITE PART requires that the processing method and cure cycle create high-quality, low-void-content structures. If voids are present, the performance of the composite will be significantly reduced. This is usually noticed by significant reductions in resin-dominated properties, such as compression and shear strength (Ref 1, 2). Voids in composite materials are areas that are absent of the composite components: matrix (resin) and fibers. The void areas in composite materials may be spherical or oblong, but other complex void morphologies may also be observed. It is known that the morphology of the void and the location in the composite significantly affect the mechanical properties, as does the size, number, and volume of the voids. There are multiple causes of voids in composite materials, and these have been generally categorized as voids that are due to volatiles (such as solvents, water) or voids that result from entrapped air (Ref 3). The void formation and void morphology in composite parts are a function of the materials and processing conditions.

Volatiles and Void Content

The curing of high-performance, fiber-reinforced, thermoset-matrix composites requires the application of pressure and temperature to cause resin to flow, providing compaction of the part (Ref 4). If water or solvent is contained in the matrix, and the vapor pressure of these materials exceeds the hydrostatic resin pressure before gelation, voids can form in the composite (Ref 5, 6). The hydrostatic resin pressure in a composite is a function of many variables, including the resin viscosity, gelation time, fiber bed permeability and architecture, lay-up, impregnation, ramp rate,

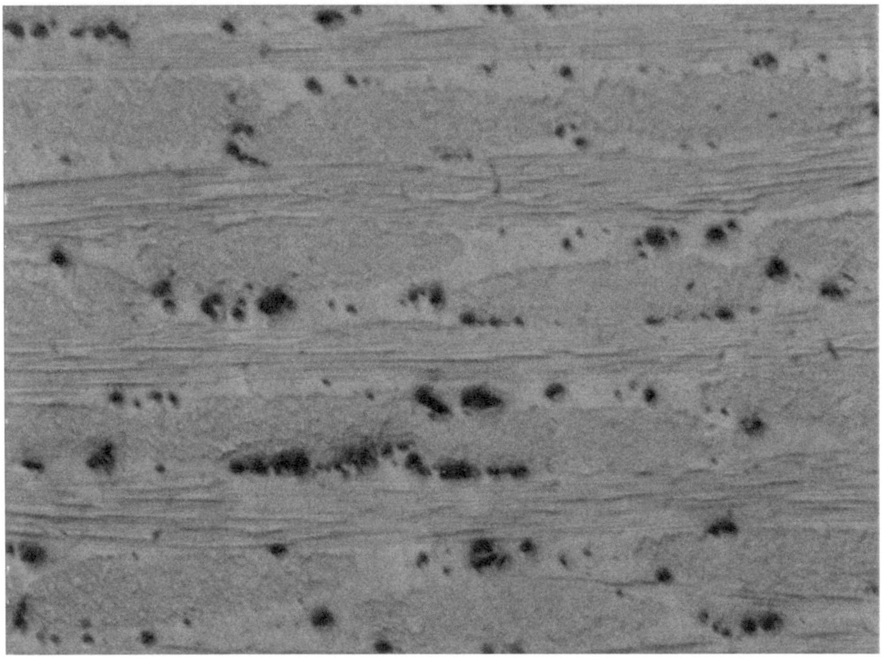

Fig. 8.1 Glass fabric composite that has high void content. Void areas are due to residual solvent from the prepregging process. Bright-field illumination, 10× objective

cure cycle, and applied pressure. Depending on these variables, the hydrostatic resin pressure may vary over the duration of the cure cycle and throughout the laminate thickness (Ref 7). It is commonly found that water in the formulation is the volatile compound that produces voids in the composite part. It has been shown that as-received epoxy resins contain between 0.3 and 0.7 wt% water (Ref 8). Furthermore, many prepreg-matrix formulations rapidly absorb moisture from ambient humidity during lay-up. Therefore, it is frequently observed that more voids are found in parts that require longer lay-up times. Many studies have found that absorbed water is the main factor in void production in specific prepreg systems (Ref 6, 9). In addition to water, solvents may be used in the formulation stage or during the processing of prepreg materials (Fig. 8.1) (Ref 10). These volatile materials are difficult to completely remove and, in many cases, cause voids in the cured parts.

Voids due to Entrapped Air

Voids are also commonly found in composite materials due to entrapped air from resin mixing (such as bubbles in the resin) that is not removed before curing. The entrapped air may be found in resin films or liquid resins, depending on the processing method. High-viscosity, controlled-flow resins and resins having high thixotropy can almost completely resist degassing. During the lay-up of a composite part using prepreg materials, air may be entrapped between the adjacent plies, resulting in voids in the cured structure (Ref 11, 12). The quantity and location of the voids depends on many factors, including the tack, prepreg impregnation, surface morphology, lay-up and nesting, thickness of lay-up, debulking stage, and cure parameters. The voids in composite materials may also have some water vapor content, which increases the size of the void (Ref 9, 13). The advancement of resin matrices can also cause increased void content in prepreg-based composite materials. An increase in the conversion of the resin system during storage or from long lay-up time may reduce the resin flow to the point that the part does not completely consolidate, and therefore, voided areas are formed. This is especially noticed at ply-drops. Additionally, inadequate compaction pressure, fiber bridging, and excessive resin bleed lead to higher void contents. Voids may be located throughout the composite or near the bottom of the part, due to lower pressure near the tool surface (Ref 6, 14, 15). Figure 8.2 shows an autoclave-cured composite part made with plies of woven carbon fabric prepreg and unidirectional carbon fiber prepreg. Voids were found mainly at the bottom of the lay-up and in the woven fabric prepreg, most likely due to lower resin pressure.

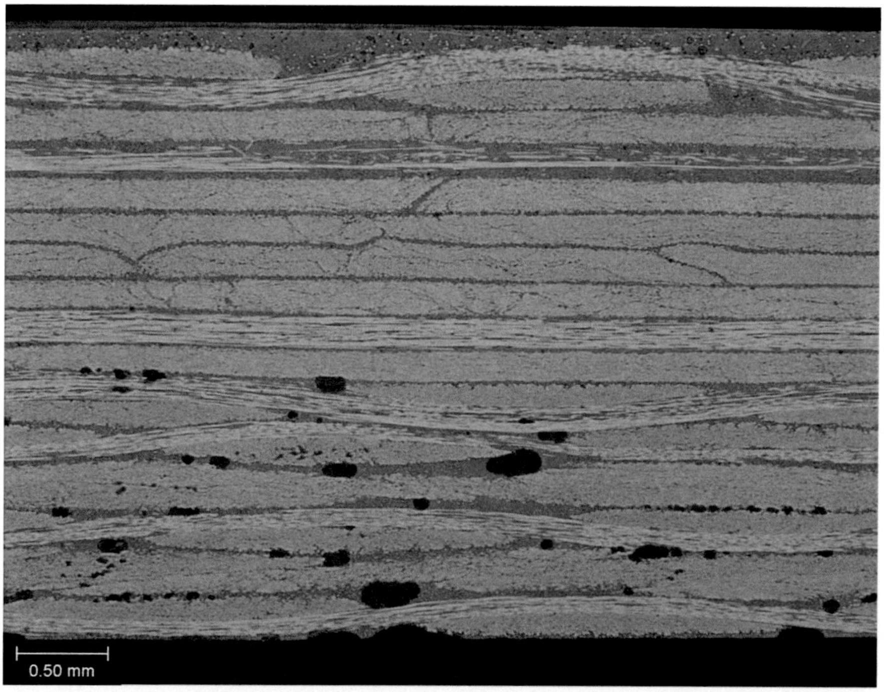

Fig. 8.2 Laminate made with unidirectional carbon fiber prepreg and woven carbon fabric prepreg plies. Voids are shown in the woven fabric area at the bottom of the composite part that was against the tool surface during cure. Bright-field illumination, 65 mm macrophotograph

Although most composite materials are cross sectioned through the thickness to determine the void content, more insight may be gained into the morphology by sectioning and polishing at a slight oblique angle to the composite surface. Figure 8.3 demonstrates the different morphology of voids, with a view of a 3k-70 plain weave carbon fiber composite sectioned parallel to the fabric plane. The voids are very irregular and located mainly in the interstitial regions in the woven fabric.

Some processing methods and lay-ups are more susceptible to air entrapment. Parts having tight radii and complex shapes are more apt to have voids located in the low-pressure areas. Likewise, tubular composite parts that have thick cross sections and high ply angles can cause large constraints during the compaction process that can resist even the highest applied pressures. Accordingly, these types of composites may contain more voids. This is often found in tubular composite parts that are wrapped with shrink tapes or are bladder molded. Figure 8.4 shows voids in a cross section taken from a tubular composite part. The majority of the voids are located in the interlayer region. Grinding and polishing the composite at a tangent to the tube surface provides another view of the voids and a more complete understanding of the morphology (Fig. 8.5a, b). The irregular-shaped voids were found to exist predominantly in the interlayer areas between the angled plies, but these voids also extend into the intraply area.

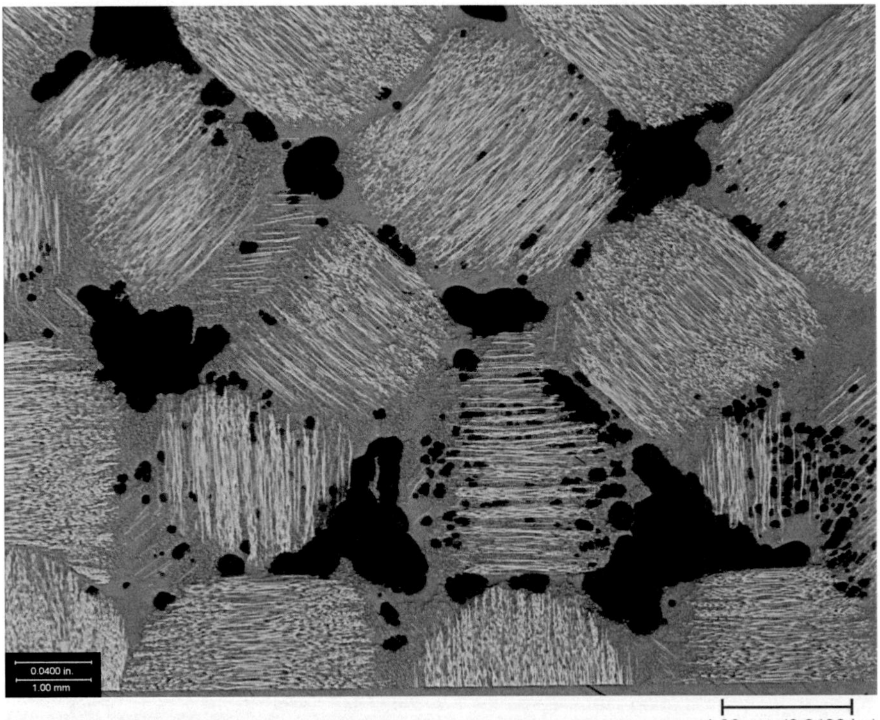

Fig. 8.3 Voids in the interstitial areas of a plain weave carbon fiber composite. Bright-field illumination, 65 mm macrophotograph

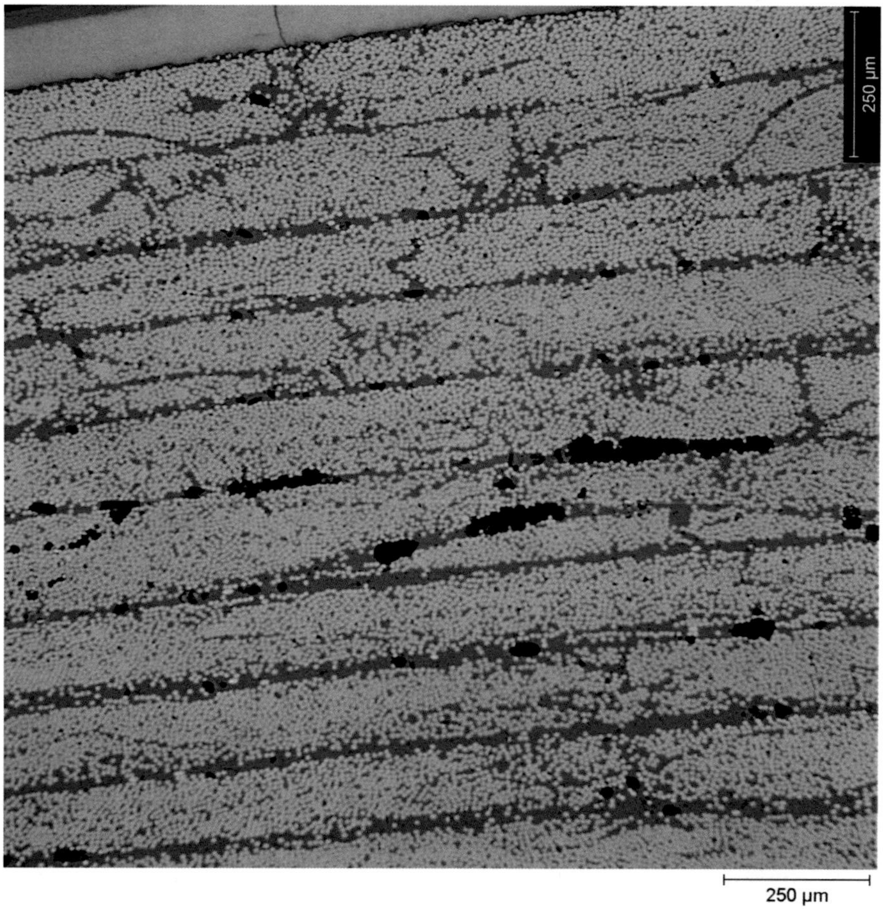

Fig. 8.4 Entrapped air voids in a tubular composite part made with unidirectional carbon fiber prepreg. Slightly uncrossed polarized light, 10× objective

Voids at Ply-Drops

The design and construction of a composite part dictates to a large extent if voids will be found in a composite and where the voids will most likely exist. It is common for composite parts to be designed with prepreg plies terminated at locations throughout the thickness of the lay-up. The junctures of these ply-drops contain areas where air may be trapped. Furthermore, at the edge of a ply-drop, a low-pressure area may be created due to bridging of the plies, above which can restrict compaction of the composite. Figure 8.6 shows voids in a shrink-wrapped, processed composite part made from unidirectional carbon fiber prepreg. The ply-drop near the surface of the part is shown to provide a good transition, while the ply-drop area at the bottom of the composite has a void at the end of the ply. This is due to the lower pressure near the tool surface, as are the small intraply and interlaminar voids in this area of the composite part. The pres-

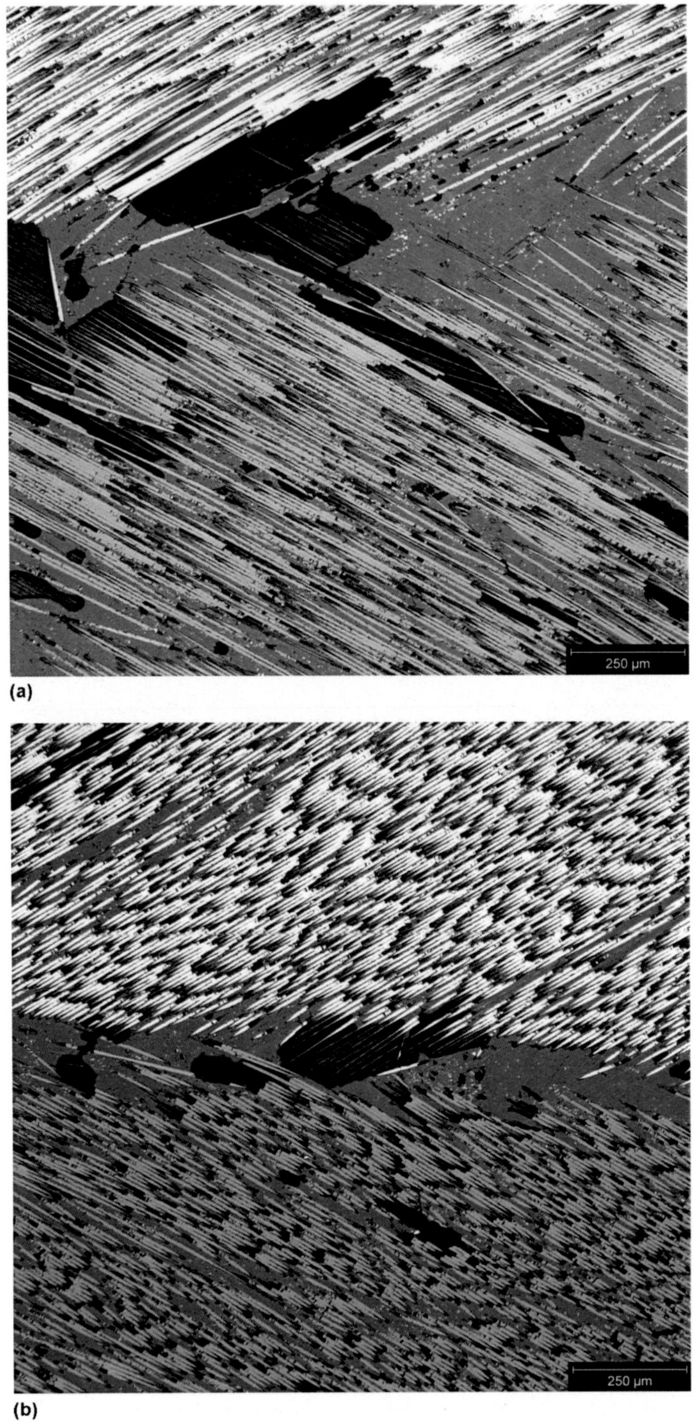

Fig. 8.5 Voids due to entrapped air shown in the interlayer region of a tubular composite part. (a) Section polished on a tangent to the outer radius. Slightly uncrossed polarized light, 10× objective. (b) Section polished on a tangent to the interior radius showing a similar fiber angle. Slightly uncrossed polarized light, 10× objective

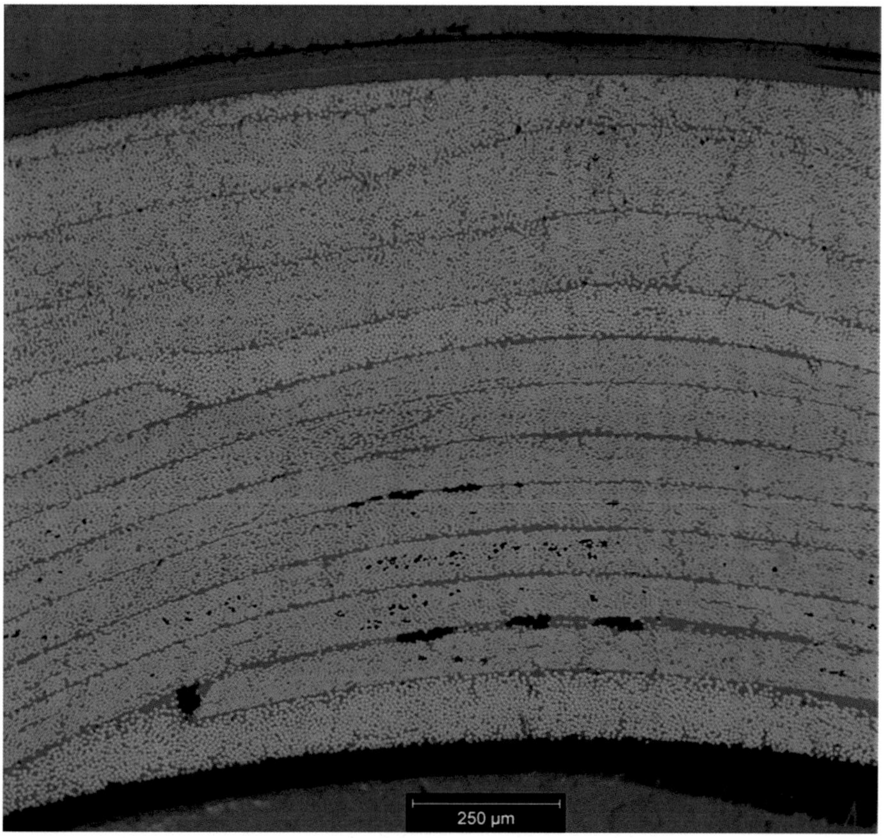

Fig. 8.6 Voids in the interlayer region and at the ply-drop in the interior of a tubular composite part. Bright-field illumination, 10× objective

ence of voids at ply-drops is even more of an issue when two plies are terminated together in a composite structure (Fig. 8.7).

Voids due to High Fiber Packing

The development of composite materials can be made with very high fiber packing and interfiber spacing, so that air can be entrapped and not removed during processing (Ref 3). Furthermore, the tight fiber packing can obstruct resin flow into unimpregnated areas during cure. This is sometimes found in pultruded composites but also in prepreg-based composites. Prepreg materials developed with high fiber tensions can reduce the fiber tow permeability. In this situation, the initial penetration of the resin is restricted by the fiber packing and is not relaxed after prepreg processing. Figure 8.8(a and b) shows voids in a unidirectional carbon fiber composite material having a high fiber volume and tight fiber packing. The two figures are taken from different axes of the composite to show the morphology of the voids. Frequently, it is assumed that the voids are

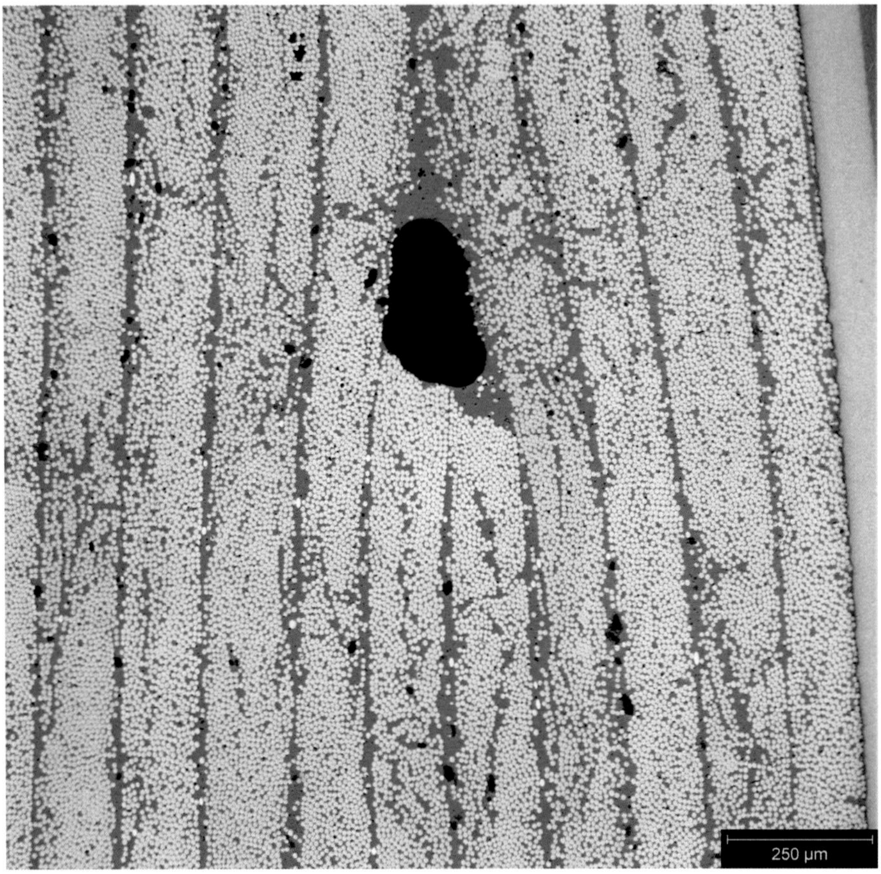

Fig. 8.7 Large void at the termination of two prepreg plies. Slightly uncrossed polarized light, 10× objective

spherical in nature due to only sectioning the composite perpendicular to the fiber orientation, as shown in Fig. 8.8(a). However, this is often not the case, as revealed in Fig. 8.8(b). This section was taken parallel to the fiber direction and shows very long voids throughout the composite. The spherical morphology is more common due to solvent-induced voids, whereas irregular void shapes are often found from entrapped air and areas that were initially not impregnated with resin.

Voids in Honeycomb Core Composites

The lay-up of honeycomb parts often results in voided areas in the composite structure as a result of low pressure in the facesheets and vacuum in the core during manufacturing. Voids are often found in the fillet regions because, in many cases, the core environment is under vacuum. Solvents can more easily vaporize and entrapped air can grow to a larger volume

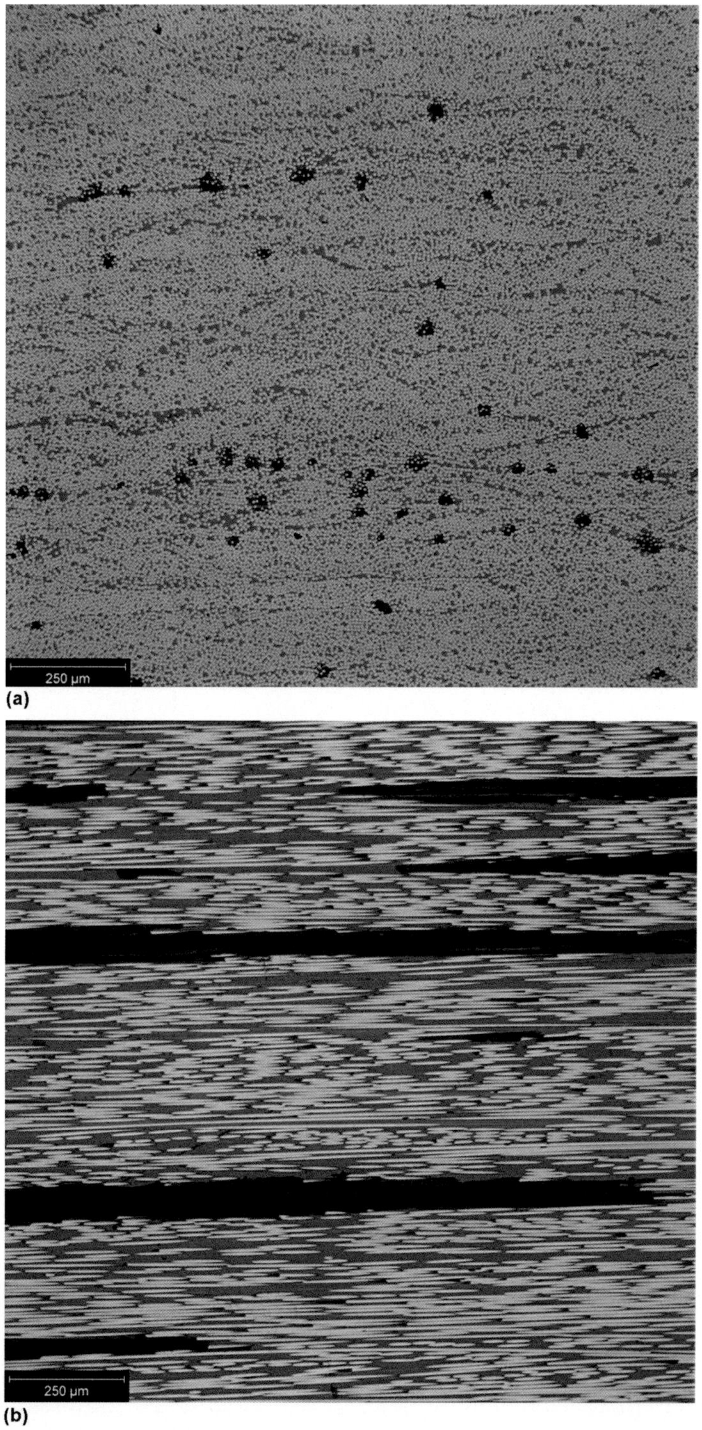

Fig. 8.8 Voids in a high-fiber-volume unidirectional carbon fiber composite part. (a) Sectioned and polished perpendicular to the fiber direction. Bright-field illumination, 10× objective. (b) Sectioned and polished parallel to the fiber direction. Bright-field illumination, 10× objective

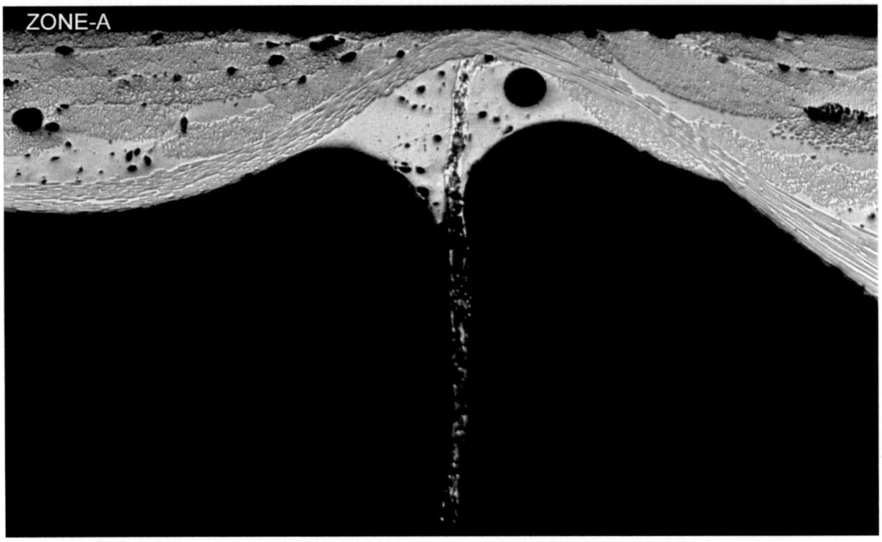

Fig. 8.9 Glass fabric prepreg honeycomb core composite with voids throughout the structure. Bright-field illumination, 5× objective

under reduced pressure and increased temperature during cure. Figure 8.9 shows voids throughout a glass fabric prepreg facesheet on a honeycomb core. Voids are shown in the fillet region as well.

Void Documentation

The void content in composite materials is commonly determined through nondestructive inspection (NDI) techniques or volumetrically by using density/specific gravity measurements of composite materials. The NDI techniques, including through-transmission ultrasound (C-scan) and x-ray radiography methods, are excellent methods for determining the void area in composite materials but are limited to the through-thickness direction and lack morphological detail of the voids throughout the thickness (Ref 16). Volumetric void analysis techniques, such as ASTM D 2734, work well for determination of the void volume if the density of the matrix resin and fibers are known as well as the exact ratio of these materials. With this technique, the morphology of the voids is also unknown. As a complementary technique, optical microscopy can be used to determine the morphology of the voids, which is, in many cases, as important as knowing the void volume. This is critical to understanding the relation of the voids to the mechanical property measurements. To completely understand the void morphology, the composite material should be viewed in two and, most likely, all three principal axes, because one or two planes usually do not provide the required detail. The number of planes that require analysis is highly dependent on the lay-up of the composite part.

After a face of a composite part is polished, it can be documented using image analysis software and the void area and number documented for a specific cross-sectioned area. This documentation, along with other NDI techniques and void volume information developed from density/specific gravity measurement methods, provides a complete analysis of the voids in a given composite structure.

REFERENCES

1. J.M. Tang, I.W. Lee, and G.S. Springer, Effects of Cure Pressure on Resin Flow, Voids, and Mechanical Properties, *J. Compos. Mater.,* May 21, 1987, p 421–440
2. P. Olivier, J.P. Cottu, and B. Ferret, Effects of Cure Cycle Pressure and Voids on Some Mechanical Properties of Carbon/Epoxy Laminates, *Composites,* Vol 26, 1995, p 509–515
3. D. Purslow and R. Childs, Autoclave Moulding of Carbon Fiber-Reinforced Epoxies, *Composites,* Vol 17, April 2, 1986, p 127–136
4. A.C. Loos and G.S. Springer, Curing of Epoxy Matrix Composites, *J. Compos. Mater.,* March 17, 1983, p 135–169
5. J.L. Kardos, M.P. Dudukovic, and R. Dave, Void Growth and Resin Transport during Processing of Thermosetting-Matrix Composites, *Epoxy Resins and Composites IV, Advances in Polymer Science,* Vol 80, Springer, Berlin/Heidelberg, 1980, p 102–122
6. R.A. Brand et al., "Processing Science of Epoxy Resin Composites," AFWAL-TR-83-4124, final report, General Dynamics Convair Division, AFWAL/MLBC, Wright-Patterson AFB, OH, Jan 1984, p 169–175
7. F.C. Campbell, A.R. Mallow, and C.E. Browning, Porosity in Carbon Fiber Composites—An Overview of Causes, *J. Adv. Mater.,* July 1995, p 18–33
8. S.R. White and Y.K. Kim, Staged Curing of Composite Materials, *Composites Part A,* Vol 27 (No. 3), 1996, p 219–227
9. M.A. Grayson, Chap. 32, Measurement of the Distribution of Water in a Graphite Epoxy by Precision Abrasion Mass Spectrometry, *Resins for Aerospace, ACS Symposium Series,* Vol 132, Aug 28, 1980, p 449–457
10. B.S. Hayes, J.C. Seferis, and R.R. Edwards, Self-Adhesive Honeycomb Prepreg Systems for Secondary Structural Applications, *Polym. Compos.,* Vol 19 (No. 1), Feb 1998, p 54–64
11. J.F. Harper, N.A. Miller, and S.C. Yap, The Influence of Temperature and Pressure during the Curing of Prepreg Carbon Fiber Epoxy Resin, *Polym.-Plast. Technol. Eng.,* Vol 32 (No. 4), 1993, p 269–275
12. W.-B. Young, Compacting Pressure and Cure Cycle for Processing of Thick Composite Laminates, *Compos. Sci. Technol.,* Vol 54, 1995, p 299–306

13. R. Dave, J.L. Kardos, S.J. Choi, and M.P. Dudukovic, Autoclave vs. Non-Autoclave Composite Processing, *32nd Int. SAMPE Symposium,* April 6–9, 1987, p 325
14. A.O. Kays, "Exploratory Development on Processing Science of Thick-Section Composites," Air Force Contract F33615-82-C-5059, final report, AFWAL-TR-85-4090, Sept 1982 to May 1985
15. B.S. Hayes, L.M. Gammon, and J.C. Seferis, The Influence of Prepreg Processing Conditions on Final Part Porosity, *Proceedings of the 30th Annual International Metallographic Society, Inc.,* July 13–16, 1997 (Seattle, WA)
16. Nondestructive Testing, *Composites,* Vol 21, *ASM Handbook,* ASM International, 2001, p 699–725

CHAPTER 9

Microcrack Analysis of Composite Materials

THE FORMATION OF MICROCRACKS in composite materials may arise from static-, dynamic-, impact-, or fatigue-loading situations and also by temperature changes or thermal cycles (Ref 1, 2). Impact loading is discussed in Chapter 11, "Impact Response of Composites," in this book. Other environmental factors, such as water, solvent, and chemical absorption, can also cause microcracks to form and expand in composite materials. The initiation of microcracking, and the propagation of the microcracks, usually occurs more rapidly in a composite material if a combination of these events occurs simultaneously. For example, the exposure to fatigue loading in addition to thermal cycling may initiate faster microcrack formation, greater propagation, and subsequent damage. This is due to enhanced multiaxial stresses in the composite and is further complicated by the increase in the modulus of the matrix and reduction in strain at lower temperatures. Unfortunately, most composite materials are not subjected to combined mechanical-thermal loading situations before being put into service, and therefore, premature degradation or failure can result.

After the formation of microcracks, the mechanical properties are usually degraded, and will continue to degrade, as the microcracks become larger and increase in number. Often, this degradation is first noticed in the matrix-dominated properties. In response to mechanical loading, microcracks are most often found in the plies off-axis to the loading direction (Ref 1). Small microcracks in the surface may not be initially detrimental to performance, but these can propagate to form larger and/or wider cracks that can also result in delamination. It is important to determine when the first microcrack forms in a composite material from thermal or mechanical loading of the composite material. Even small microcracks can propagate in tough materials in the right environment, creating paths for water, solvents, and cryogenic liquids, which can cause further damage.

The composite constituents (such as resin matrix and fibers) and the ratio of these materials play a significant role in the microcracking behavior of the cured composite, along with the formed interphase during cure (Ref 3). Composite matrices are often toughened, which can increase the stress level that is necessary to initiate microcracking. The location of the toughness modifiers in the composite significantly affects the microcracking, as does the concentration of the toughener (Ref 4). Along with the matrix constituents, the fiber properties and fiber-matrix adhesion levels also influence the microcracking susceptibility of the cured composite (Ref 3, 5). In addition to the composite constituents, the lay-up and composite construction influences the microcracking response to applied mechanical loads and thermal cycles. The interlayer thickness and modulus, ply orientations, thickness of the ply layers, and the grouping of the plies affect the stress/strain levels at which microcracking occurs in fiber-reinforced composites (Ref 6 to 9).

The residual stresses in a composite material act to preload the composite and can initiate microcracking at much lower stress levels and/or fewer thermal cycles (Ref 10). Consequently, the thermal stresses, which are further increased as the temperature falls below the stress-free temperature, also influence to a large extent the microcracking behavior in a composite material. The stress-free temperature is usually near the cure temperature, and therefore, as the temperature is reduced even further below ambient, a greater propensity for microcracking occurs (Ref 11). In some thermosetting matrices, just the reduction in temperature from the cure temperature to ambient has caused microcracks to form (Ref 12). Residual stresses are caused by the anisotropic expansion coefficients of the adjacent plies and material constituents (resin/fiber) as well as the cure shrinkage of the matrix. It is commonly observed that modifications to the cure cycle, by changing the ramp rate or adding dwell times, can alter the chemical cure shrinkage in a thermoset matrix (Ref 13). However, the ultimate cure temperature and time dictates to a large extent the residual stress in composite materials.

Bright-Field and Polarized-Light Analysis of Microcracked Composites

In the preparation of composite specimens for analysis of microcracking, it is usually best to not mount the sample. If only hand polishing is available, it is best to bond the samples together so as not to impregnate the microcracks. If mounting is necessary due to large-scale damage and fiber fracture, such as arising from impact loading, the composite may need to be mounted to preserve all of the damage features. In this case, a contrast dye should be added to the casting resin, as described in Chapter 2, "Sample Preparation and Mounting," in this book.

Most analysis of microcracks in composites can be performed using unmounted samples and reflected-light optical microscopy techniques (Chapter 5, "Viewing the Specimen Using Reflected-Light Microscopy"). Figure 9.1(a and b) shows areas of a composite material where microcracks formed after thermal cycling. In Fig. 9.1(a), microcracks are found in the resin-rich areas between the tows and also in the fiber tows. It is common for microcracks to initiate in the resin-rich regions because there is greater resin cure shrinkage due to less fiber constraint. However, it has also been shown that initiation of microcracking is increased when the individual fiber spacing is decreased (or clustered) and/or there is a higher modulus interphase (Ref 14). These factors, along with lower fiber-matrix adhesion, increase the microcracking in thermally cycled composite materials (Ref 5). The microcracks are commonly wider where the cracking initiated, and therefore relieved more stress, and narrower where the crack is propagating. It is often that the propagating microcrack is constrained by adjacent plies, perpendicular to the propagation. This can be seen at the bottom of Fig. 9.1(b). Also in this figure, the crack is found to propagate around a fiber tow and through other fiber tows. It is interesting that microcracks were not found in the resin-rich area containing the void but in the adjacent fiber tows. This may be attributable to the change in the stress field in this area due to the presence of the void. Figure 9.2 shows a micrograph of a composite cross section after many thermal cycles. In this figure, the microcracks are found to be the widest in the resin-rich areas, where the microcracking initiated. However, after additional thermal cycles, microcracks were also found in the fiber tows. As in Fig. 9.1, the off-axis plies are shown to constrain the microcracks. In Fig. 9.2, the microcrack propagation is also shown to be interrupted by the presence of the voids.

Microcracks may begin on the surface or in areas throughout the thickness of the composite. Where microcracks initiate is commonly a function of the lay-up, thickness of the ply groups, and how the load is applied to the composite (Ref 7, 9). It has been found that flaws in composites are a primary cause of a more rapid initial rate of microcrack development that is observed early on in thermal cycling (Ref 15). The rate of microcrack development then usually slows down and levels out when these stresses are relieved. In Fig. 9.3, a microcrack is shown to initiate on the surface of the composite and at the edge of a glass fabric tow. The glass fabric is very hard to see using slightly uncrossed polarized light, due to the low contrast. It can be seen, however, that the crack path was altered by the presence of a carbon fabric tow and propagated between the dissimilar fiber tows.

When a composite material is analyzed after thermal cycling, there may be areas that appear as voids in the tows that are in the polishing plane.

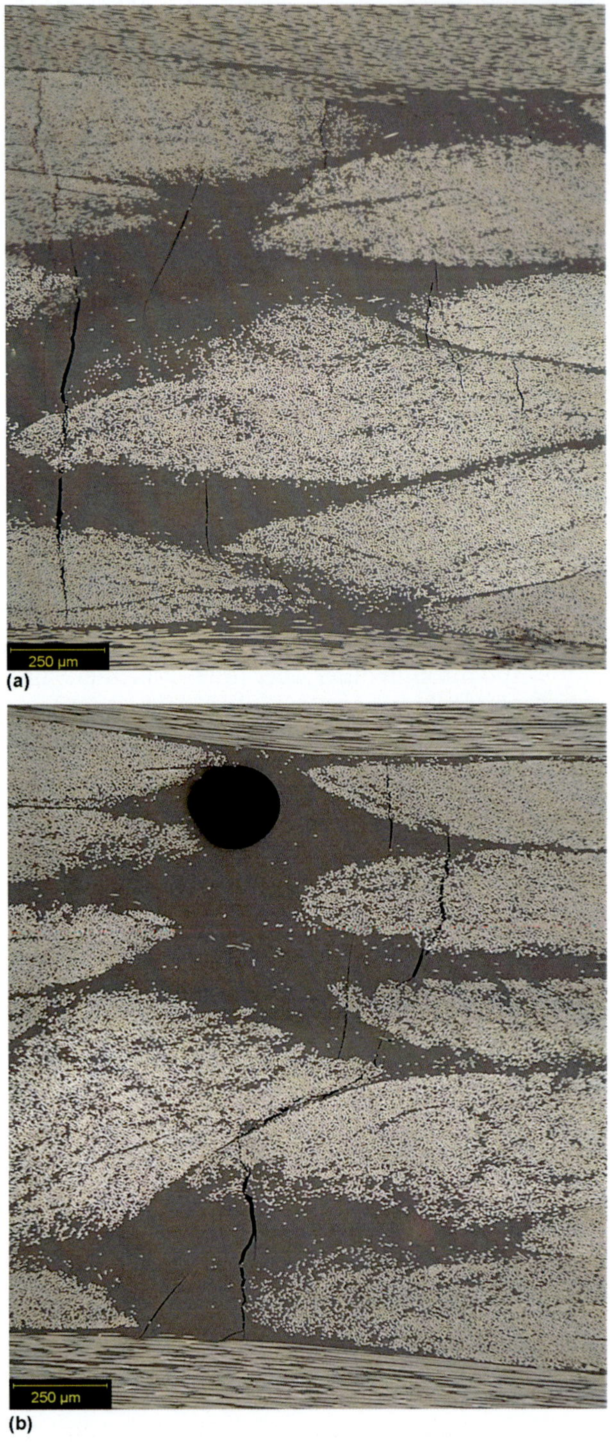

Fig. 9.1 Microcracks in a carbon fiber composite laminate due to thermal cycling. (a) Resin-rich region in the composite. Slightly uncrossed polarized light, 10× objective. (b) Resin-rich region containing a large void. Slightly uncrossed polarized light, 10× objective

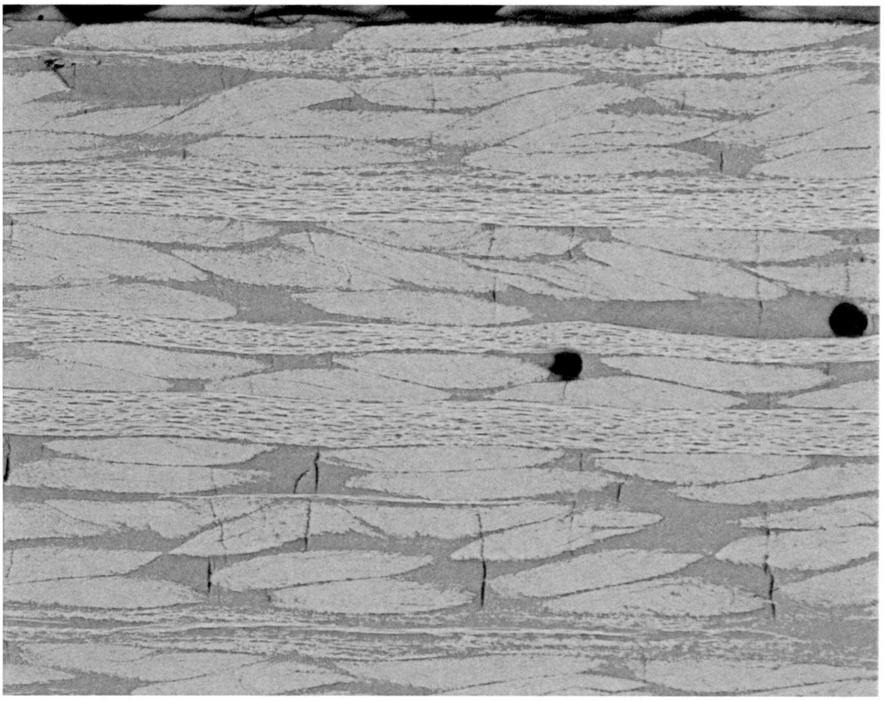

Fig. 9.2 Micrograph of a carbon fiber composite that microcracked during thermal cycling. Bright-field illumination, 65 mm macrophotograph

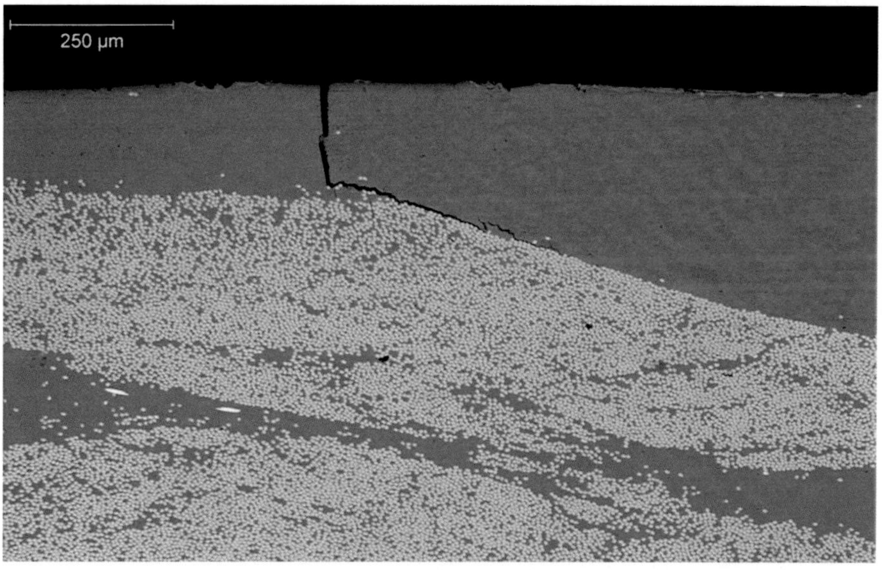

Fig. 9.3 Micrograph of a composite cross section showing a microcrack that initiated on the surface of the part. Slightly uncrossed polarized light, 10× objective

These voided areas are commonly microcracks that have formed in the tows (intraply area) from thermal cycling and only appear as long voids due to the polishing and viewing orientation. This can be observed in the tow at the bottom of Fig. 9.4, where the cracks appear as long, thin voids between the fibers that run parallel to the length of the tow. In contrast, the microcracks that appear at the top and center of this micrograph go through the cross section of the tows. If the viewing orientation was 90 degrees to the shown polishing plane, the same intratow features (such as long, thin voids) would be observed for the cracks at the top of this figure. Therefore, when counting microcracks in composites for analysis, the number should be determined separately in each axis and consist of those microcracks only going through the tow cross-sectional area. The microcracking may also be analyzed throughout the thickness of the composite, but this is usually limited to only one ply group, unless the section is cut and polished on an oblique angle from the surface.

In many cases, when the sample is polished, there may be microcracks just under the surface that can be observed using reflected-light illumination techniques. Figure 9.5 shows an example of a subsurface microcrack in a composite material where the reflective nature of the subsurface microcrack is easy observed. In many cases, the microcracks may be quite small or lack contrast and therefore will not be visible using bright-field

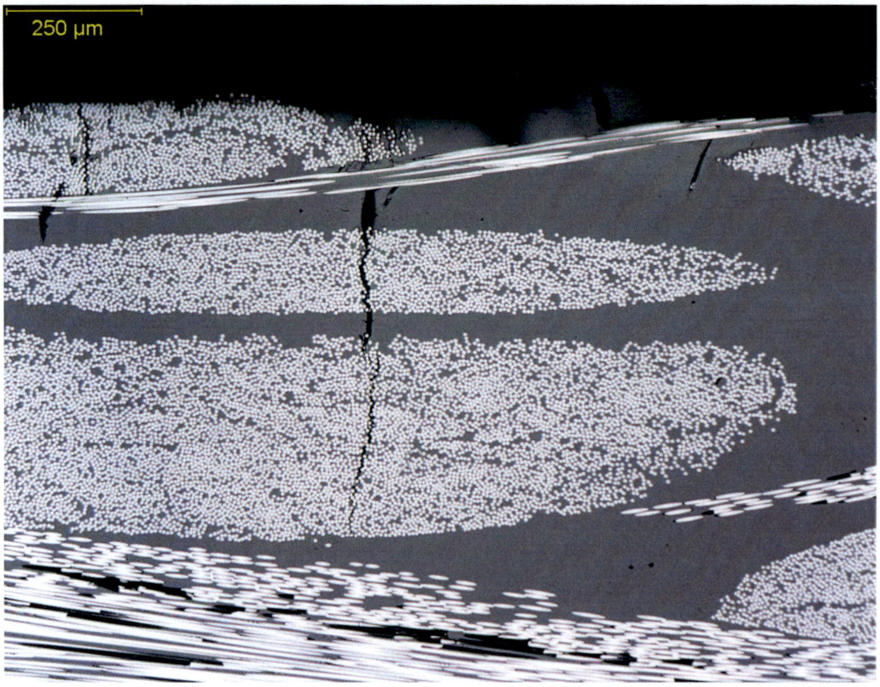

Fig. 9.4 Microcracked carbon fiber composite material illustrating the crack morphology in a fiber tow that is in the same plane as the polished surface. Bright-field illumination, 10× objective

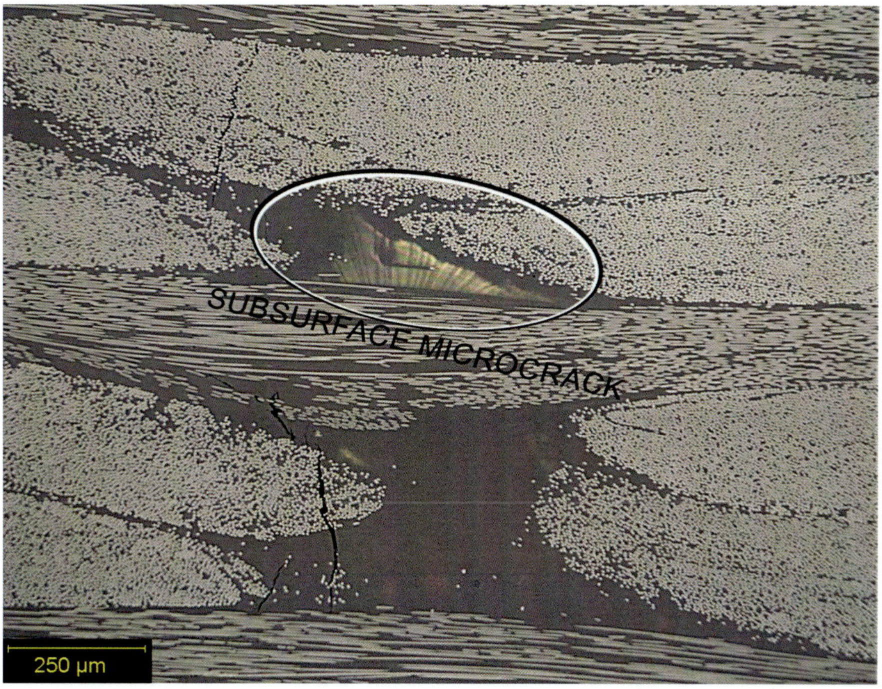

Fig. 9.5 Area of a carbon fiber composite that shows a subsurface microcrack. Slightly uncrossed polarized light, 10× objective

illumination or slightly uncrossed polarized light. In this case, a contrast dye and dark-field illumination or a laser dye and epi-fluorescence can be used to provide the necessary contrast to identify the microcracks.

Contrast Dyes and Dark-Field Analysis of Microcracked Composites

The application of dyes to the surface of polished composite cross sections is necessary to distinguish microcracks in composite materials that have low contrast. The most common types of composite materials where microcracks can be enhanced by the use of dyes are those composites having translucent or transparent fibers (such as glass, polyamide, polypropylene, and polyethylene). Figures 9.6 through 9.9 show a variety of microcracked composite cross sections with these types of fibers, where the application of dyes and epi-dark-field illumination was used to provide contrast. Before the red dye was applied to the polished composite surfaces, the dye was thinned with solvent (acetone) to lower the viscosity for better impregnation. After allowing time to infiltrate the microcracks, the excess dye was lightly wiped off the surface with a solvent-soaked cloth. These samples were then lightly back-polished before viewing. A composite made with plies of glass and thermoplastic fibers is shown in Fig. 9.6.

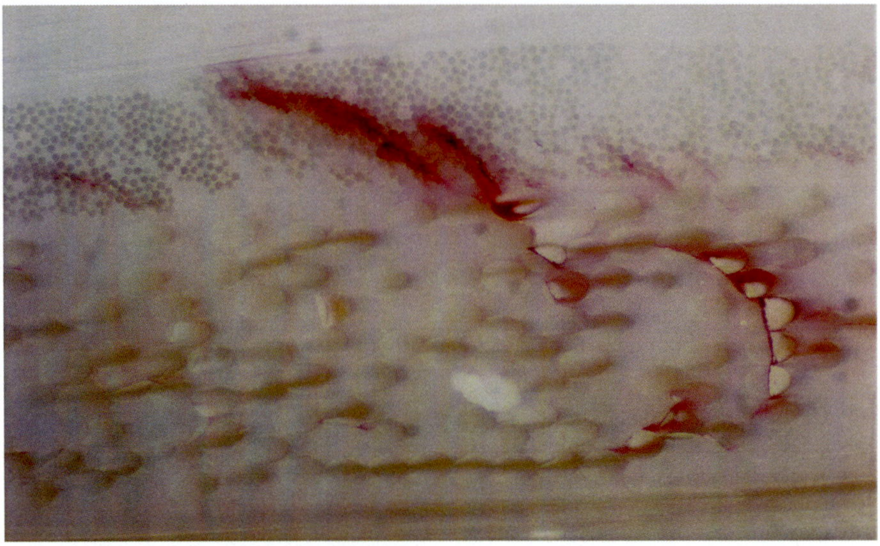

Fig. 9.6 Microcracks in a glass and thermoplastic fiber hybrid composite. Red penetration dye (Magnaflux Spotcheck SKL-H, Magnaflux Corp.), dark-field illumination, 25× objective

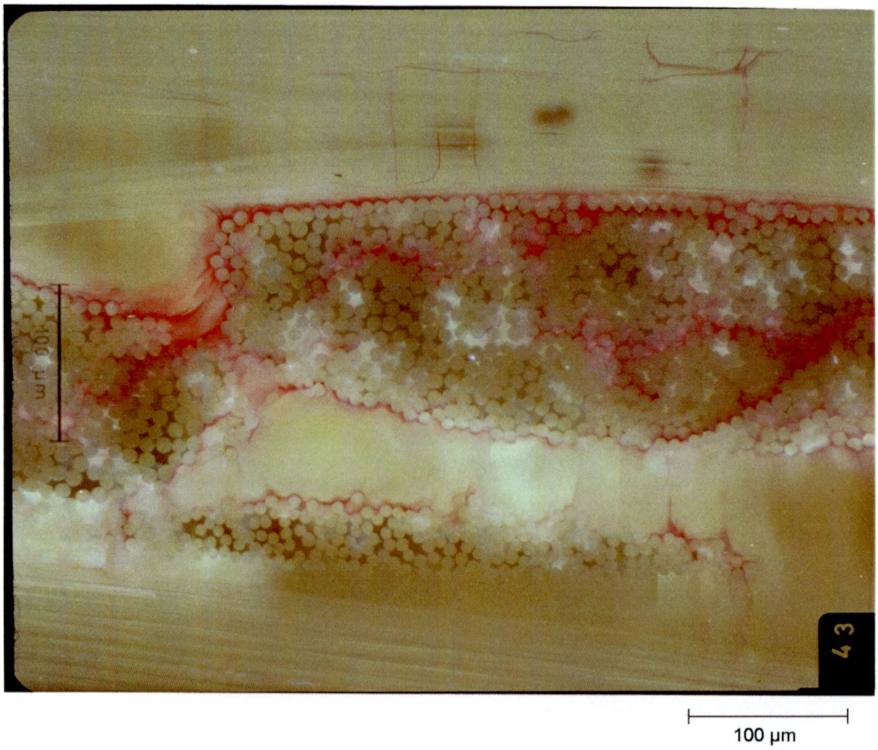

Fig. 9.7 Microcracks in a thermoplastic-matrix glass fiber composite. Red penetration dye (DYKEM Steel Red layout fluid, Illinois Tool Works, Inc.), dark-field illumination, 25× objective

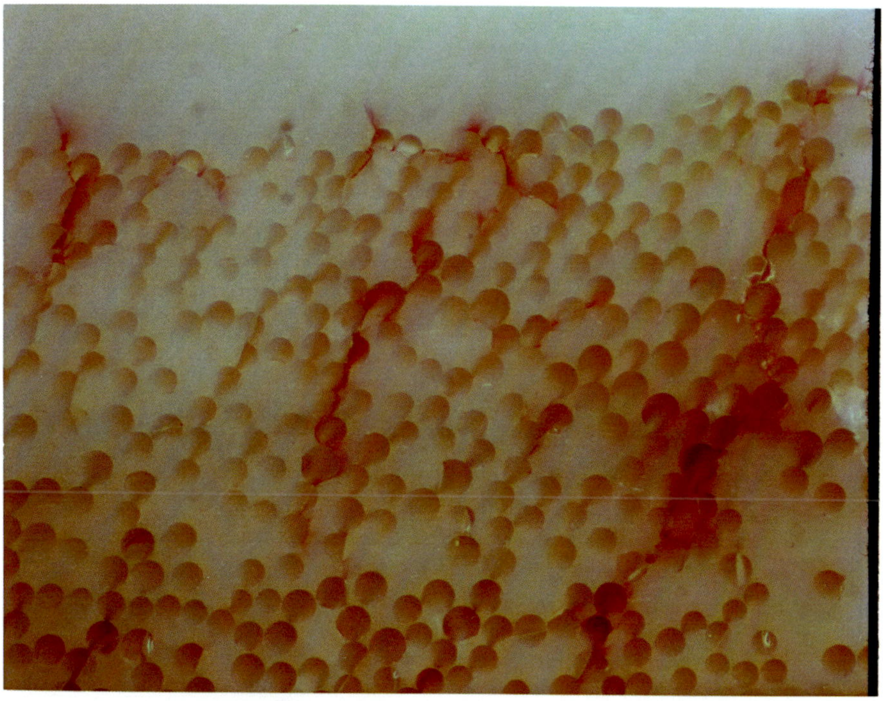

Fig. 9.8 Higher-magnification view of the microcracks in the intraply region of the thermoplastic-matrix glass fiber composite. Red penetration dye (DYKEM Steel Red layout fluid), dark-field illumination, 50× objective

The crack propagated between the smaller glass fibers but went through many of the thermoplastic fibers, indicating excellent fiber-matrix adhesion. In some composite materials, it is found that the microcracks form and propagate around the tows and at the fiber-matrix interface in the interlayer region. An example of this type of microcracking is shown in Fig. 9.7. The microcracks in this axis are constrained by the tows in the polishing plane. In another area of the composite, microcracks are shown in a glass fiber tow at a higher magnification (Fig. 9.8). The microcracks propagated at the interface and through the glass fibers. When dyes are used to enhance sample contrast, some details may be lost on the edges of the cracks due to the influence of the dye on the reflection of the absorbed incident light. Also, these dyes and their solvent carriers may spread on the surface or absorb into the substrate, causing a reduction in observable detail. These effects are shown in Fig. 9.9, where the microcracks are harder to distinguish due to dye absorption.

Dyes are also useful to identify microcracks in paint and primer layers on the surface of composite parts. Figure 9.10 shows a microcrack in a paint and primer layer on the surface of a composite material that was subjected to fatigue cycling.

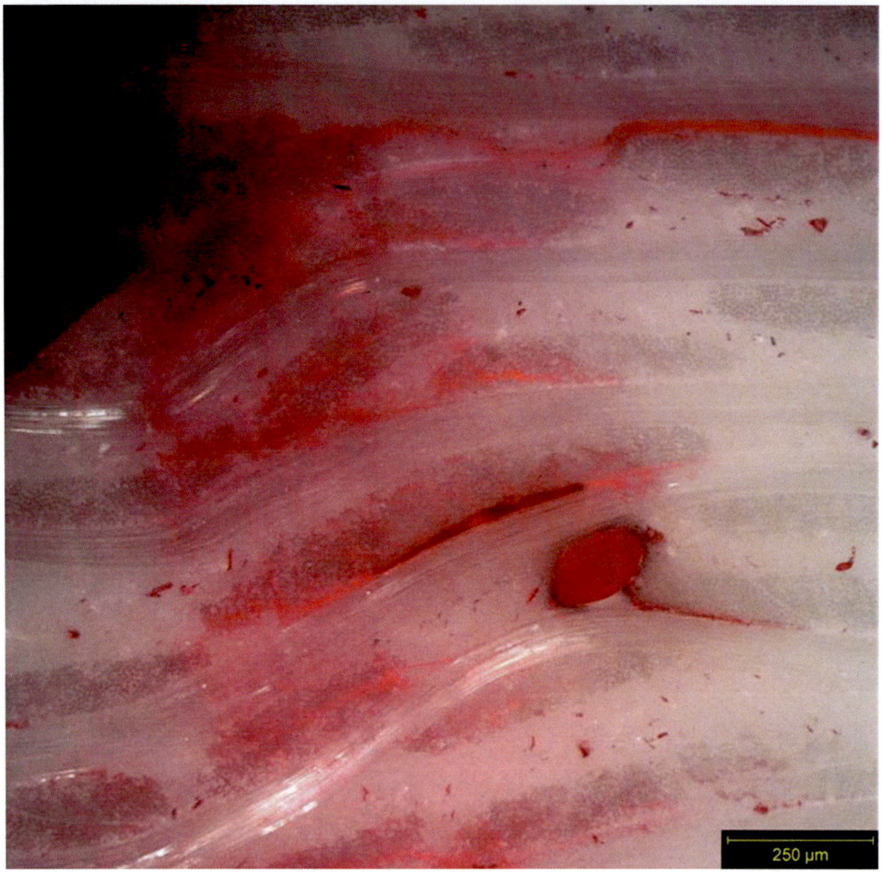

Fig. 9.9 Microcracked glass fiber composite showing a lack of detail due to absorbed solvent/dye by the matrix. Red penetration dye (DYKEM Steel Red layout fluid), dark-field illumination, 10× objective

Analysis of Microcracked Composites Using Epi-Fluorescence

The microcracks in composite materials made of carbon fibers and thermosetting matrices can most often be viewed using bright-field illumination, but in some cases, the microcracks are very small or lack adequate contrast. Accordingly, the use of epi-fluorescence may be necessary to determine if microcracks are present. Figure 9.11(a) shows that bright-field illumination of a thermally cycled carbon fiber composite did not indicate the presence of any microcracks. After application of a fluorescent dye to the polished surface, the composite specimen was viewed again using epi-fluorescence, which revealed a microcrack in an area that would not be possible to locate using bright-field illumination, due to the small size and location of the microcrack. Figure 9.11(b) shows a very small microcrack

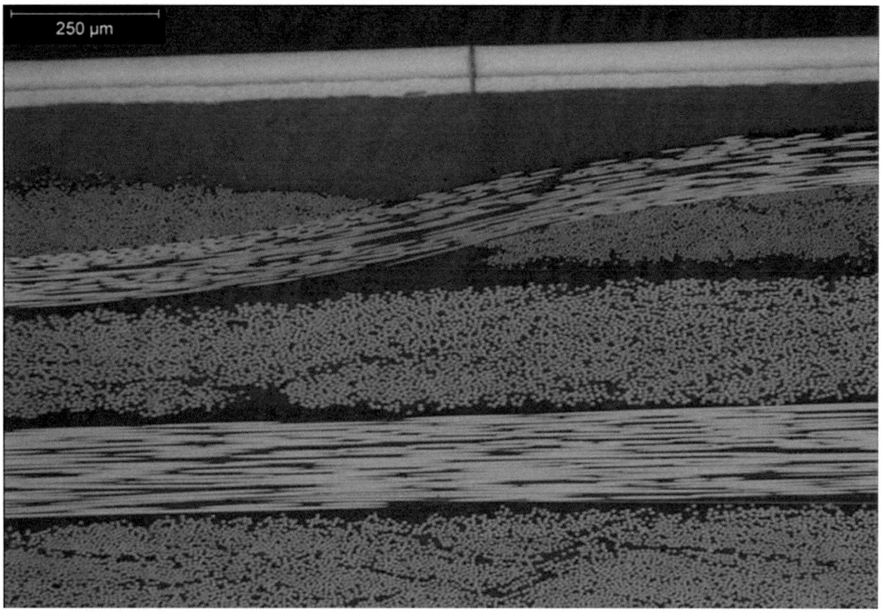

Fig. 9.10 Microcrack in the paint and primer layer on the surface of a composite as a result of fatigue cycling. Red penetration dye (DYKEM Steel Red layout fluid), slightly uncrossed polarized light, 10× objective

in this carbon fiber composite sample that is located and ends in a tow. Another area where microcracks were found in the tow region in this composite material is shown in Fig. 9.12. As with the previous figure, these microcracks would remain hidden without the use of epi-fluorescence. An area of large-scale microcracking of the same composite is shown in Fig. 9.13. The microcracking is found in the resin-rich areas and in the fiber tows. Microcracks can also be observed in the tow in the polishing plane. It was mentioned previously that these areas usually appear as voids with epi-bright-field illumination, but with the use of epi-fluorescence, the fine microcracks are more easily observed.

Sectioning and polishing a composite material on an oblique angle through the thickness provides another view of the microcracks in a composite material. Figure 9.14(a and b) compares the same area of a microcracked composite using bright-field illumination and epi-fluorescence. Again, without the epi-fluorescence, it would appear as though the composite did not have any microcracks.

The microcrack morphology can also be determined using epi-fluorescence. An example of this is shown in Fig. 9.15, where the crack bifurcated in the center of the carbon fiber intraply area. In analyzing the cracked area, it can be seen that many of the carbon fibers were separated from the matrix as a result of the crack propagation. In this material, the

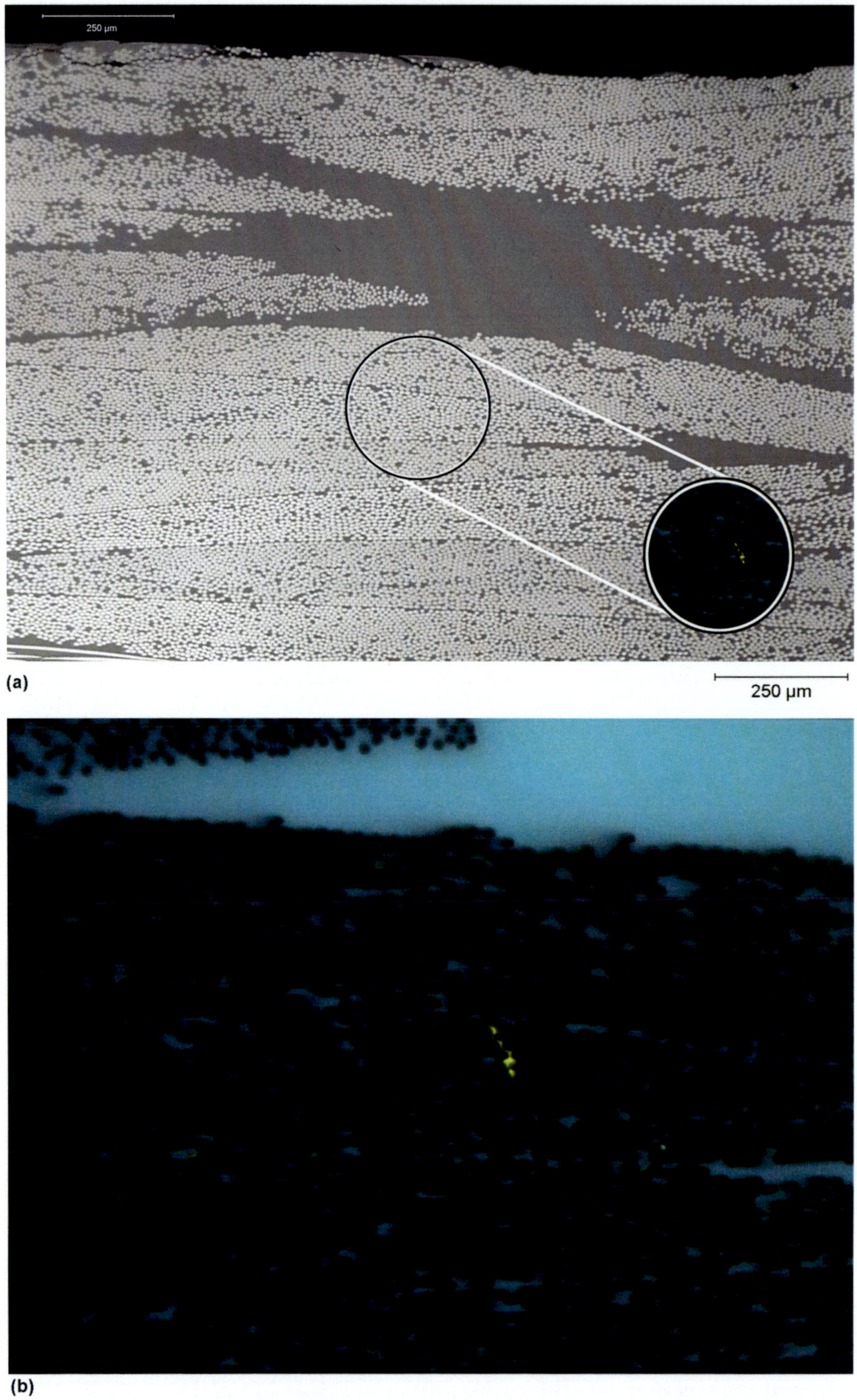

Fig. 9.11 Intraply microcrack in a carbon fiber composite that is difficult to observe using bright-field illumination but easily identified after the application of a fluorescing dye (Magnaflux Zylgo, Magnaflux Corp.) and epi-fluorescence. (a) Bright-field illumination, 10× objective, with an inset showing the crack using epi-fluorescence. (b) Epi-fluorescence, 390–440 nm excitation, 25× objective

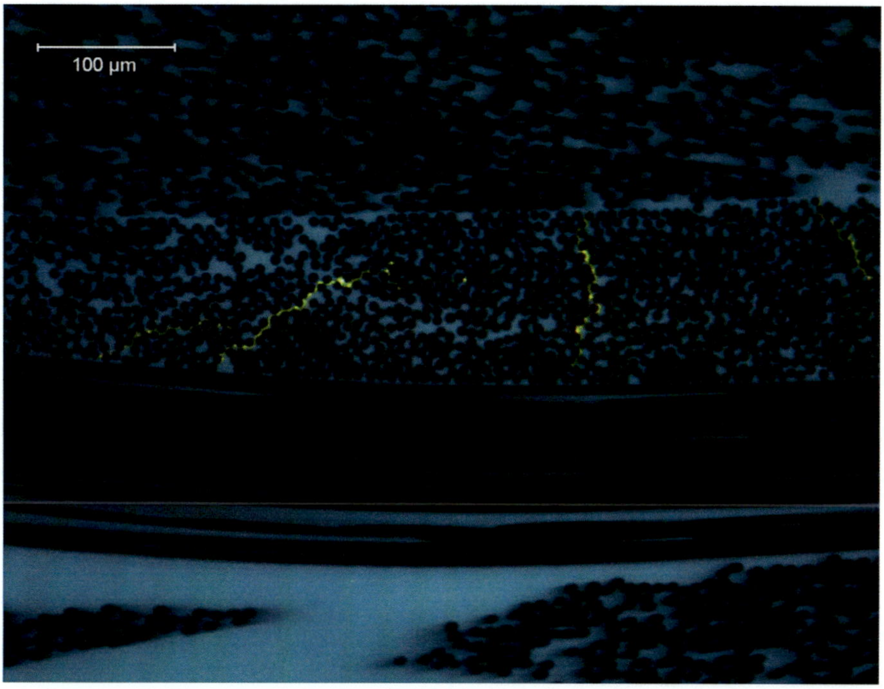

Fig. 9.12 Intraply microcracks in a carbon fiber composite material. Epi-fluorescence, 390–440 nm excitation, 25× objective

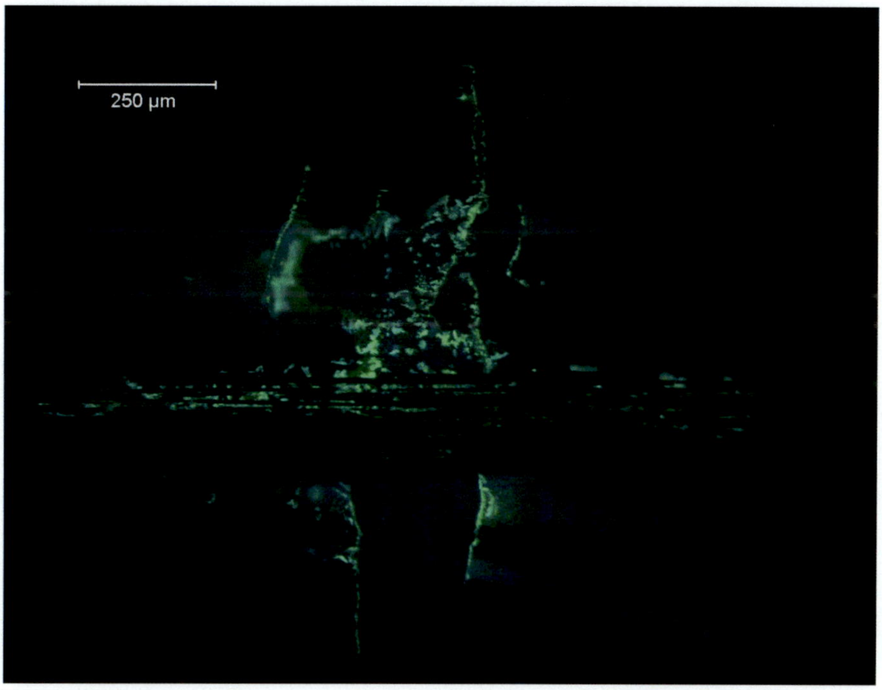

Fig. 9.13 Large-scale microcracking in a carbon fiber composite material. Epi-fluorescence, 390–440 nm excitation, 10× objective

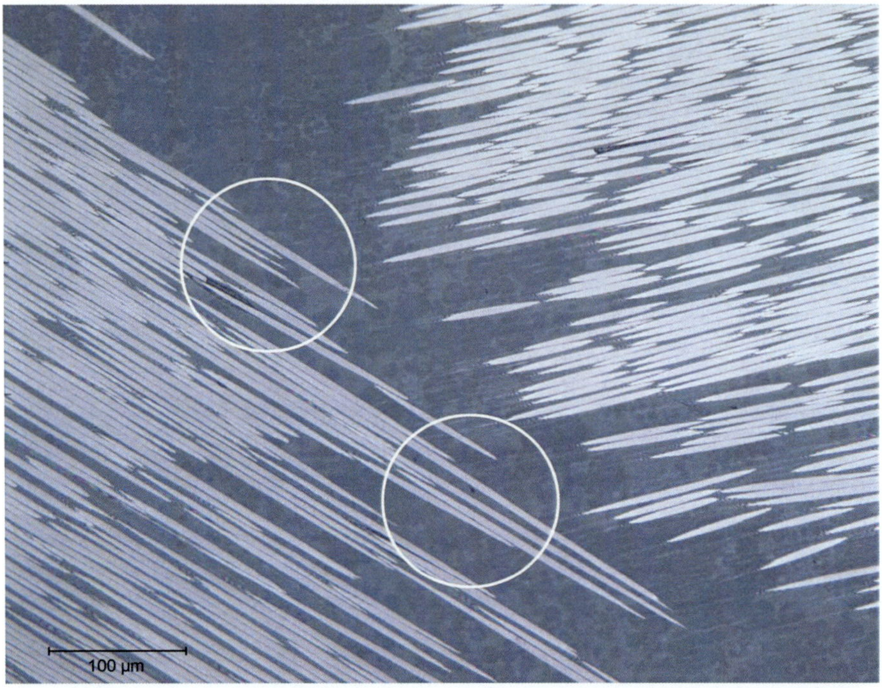

(a)

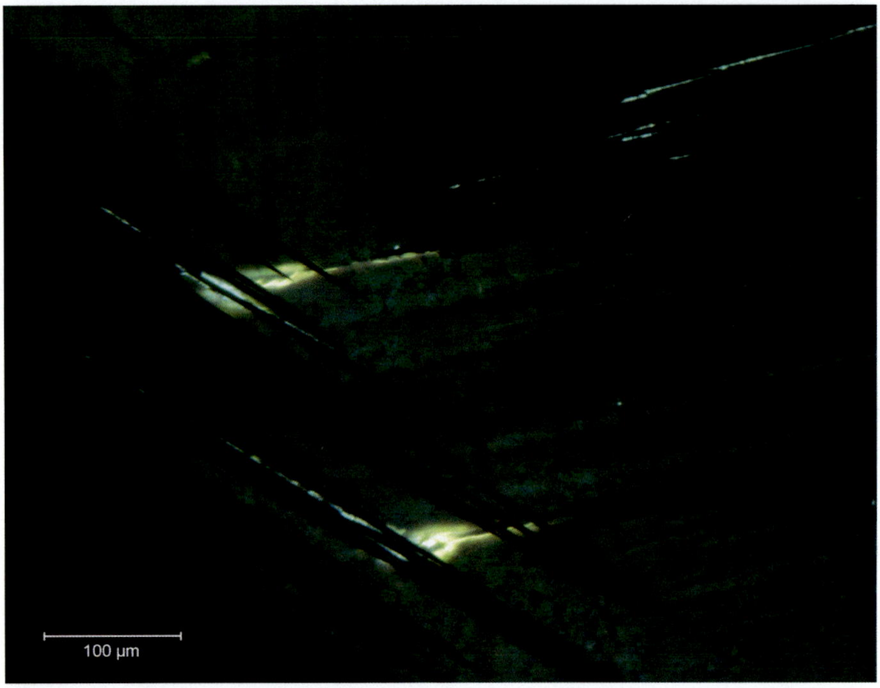

(b)

Fig. 9.14 Comparison of the same area in a carbon fiber composite using epi-bright-field illumination and epi-fluorescence. The microcracked area of the composite material was sectioned and polished at an oblique angle through the thickness to emphasize the interlayer region. (a) Bright-field illumination, 25× objective. (b) Epi-fluorescence, 390–440 nm excitation, 25× objective

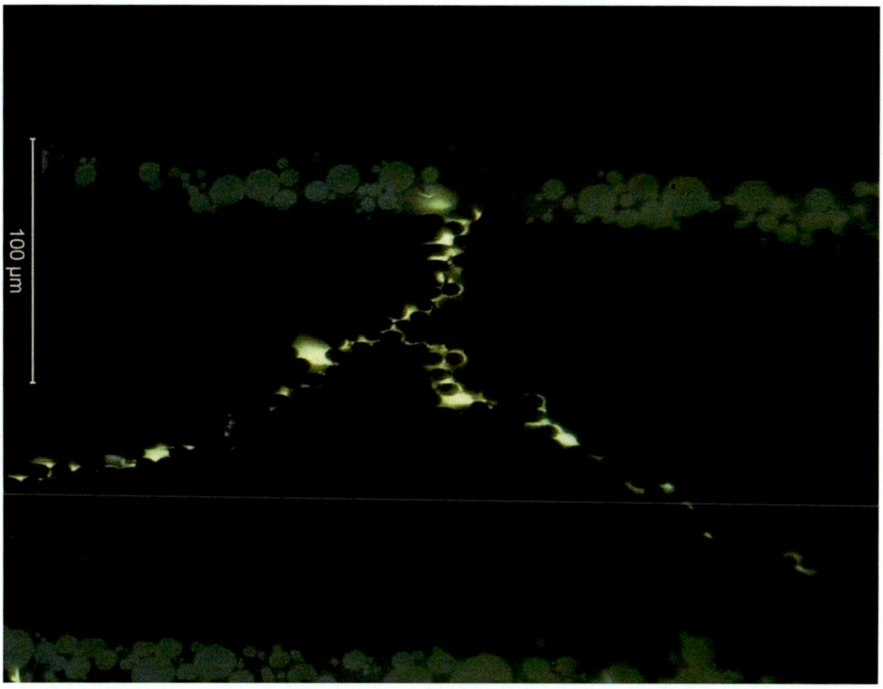

Fig. 9.15 Microcracks in the intraply region of an interlayer-toughened carbon fiber composite material that terminated at the interlayer region. Epi-fluorescence, 390–440 nm excitation, 25× objective

microcracks were found to not propagate into the particle-toughened interlayer region of the composite.

Determination and Recording of Microcracks in Composite Materials

The location of microcracks can be recorded and then the composite part subjected to further thermal or fatigue cycling to determine the propagation of the microcracks. It must be emphasized that sectioning of the composite and the polishing quality of the sample can affect microcrack formation and propagation. Therefore, it is best to subject, if possible, a larger- or full-scale composite part to the same thermal or fatigue cycling. Then, carefully section and polish a sample from different areas in the full-scale part to determine the extent of the microcracking. The sectioning and polishing of different areas in a complete part removes the effect of the free edge on the formation of microcracks and the possible stress relief by cutting or stress generation by poor sample preparation and polishing. The sectioning of the large part after exposure also has its drawback in the form of sectioning-induced microcracks and not being able to view the face before cycling. Both methods may be necessary to understand the effect of microcracking in a composite part.

REFERENCES

1. J.A. Nairn, Chap. 13, Matrix Microcracking in Composites, *Polymer Matrix Composites*, R. Talreja and J.A. Manson, Ed., *Comprehensive Composite Materials*, Vol 2, A. Kelly and C. Zweben, Ed., Elsevier Science, Amsterdam, 2000, p 403–432
2. N.L. Hancox, Thermal Effects on Polymer Matrix Composites: Part 1, Thermal Cycling, *Mater. Des.,* Vol 19, 1998, p 85–91
3. J.F. Timmerman, M.S. Tillman, B.S. Hayes, and J.C. Seferis, Matrix and Fiber Influences on the Cryogenic Microcracking of Carbon Fiber/Epoxy Composites, *Compos. Part A: Appl. Sci. Manuf.,* Vol 33, 2002, p 323–329
4. M. Nobelen, B.S. Hayes, and J.C. Seferis, Influence of Elastomer Distribution on the Cryogenic Microcracking of Carbon Fiber/Epoxy Composites, *J. Appl. Polym. Sci.,* Vol 90, 2003, p 2268–2275
5. J.F. Timmerman, B.S. Hayes, and J.C. Seferis, Cryogenic Microcracking of Carbon Fiber/Epoxy Composites: Influences of Fiber-Matrix Adhesion, *J. Compos. Mater.,* Vol 37 (No. 21), 2003, p 1939–1950
6. L.B. Ilcewicz et al., Matrix Cracking in Composite Laminates with Resin-Rich Interlaminar Layers, *Composite Materials: Fatigue and Fracture,* Vol 3, STP 1110, T.K. O'Brien, Ed., ASTM, Philadelphia, 1991, p 30
7. D.L. Flaggs and M.H. Kural, Experimental Determination of thc In-Situ Transverse Lamina Strength in Graphite/Epoxy Laminates, *J. Compos. Mater.,* Vol 16, 1982, p 103
8. D.S. Adams, D.E. Bowles, and C.T. Herakovich, Thermally Induced Transverse Cracking in Graphite-Epoxy Cross-Ply Laminates, *J. Reinf. Plast. Compos.,* Vol 5, 1986, p 152
9. T. Yokozeki, T. Aoki, T. Ogasawara, and T. Ishikawa, Effects of Layup Angle and Ply Thickness on Matrix Crack Interaction in Contiguous Plies of Composite Laminates, *Compos. Part A: Appl. Sci. Manuf.,* Vol 36, 2005, p 1229–1235
10. J.W. Lee and I.M. Daniel, Progressive Transverse Cracking of Cross-ply Composite Laminates, *J. Compos. Mater.,* Vol 24, 1990, p 1225–1243
11. K.S. Kim and H.T. Hahn, Residual Stress Development during Processing of Graphite/Epoxy Composites, *Compos. Sci. Technol.,* Vol 36, 1989, p 121–132
12. J.E. Lincoln, R.J. Morgan, and E.E. Shin, Fundamental Investigation of Cure-Induced Microcracking in Carbon Fiber/Bismaleimide Cross-Ply Laminates, *Polym. Compos.,* Vol 22 (No. 3), 2001, p 397–419

13. M.S. Madhukar, R.P. Kosuri, and K. Bowles, Reduction of Curing Induced Fiber Stresses by Cure Cycle Optimization in Polymer Matrix Composites, *Proceedings of the ICCM,* Aug 1995 (Whistler, BC), p III–157
14. D.L. Hiemstra and N.R. Sottos, Thermally Induced Interfacial Microcracking in Polymer Matrix Composites, *J. Compos. Mater.,* Vol 27 (No. 10), 1993, p 1030–1051
15. J.A. Nairn and S. Hu, Micromechanics of Damage: A Case Study of Matrix Microcracking, *Damage Mechanics of Composite Materials,* R. Talreja, Ed., Elsevier, Amsterdam, 1994, p 187–243

CHAPTER 10

Toughening Methods for Thermoset-Matrix Composites

THE PERFORMANCE of fiber-reinforced composite systems has evolved in response to new applications and designs. While there have been advances in fiber technology, the majority of new composite applications over the past few decades have been due to a redesigned balance of matrix properties with an emphasis on improved damage tolerance. First-generation advanced composites were not toughened in the sense that no additional modifiers were added to the base matrix composition to enhance the toughness. In impact situations, these systems suffered significant damage, reducing the postimpact strength of these materials to unacceptable levels for many applications. For example, as the desire grew to use composites in more damage-prone areas on aircraft, composites having greater damage tolerance were required. This led to the development of second-generation composite materials containing rubber or thermoplastic modifiers incorporated within the thermosetting matrix. These materials were added to increase the strain to failure of the primary phase and/or create a dispersed second phase, thereby enhancing the fracture toughness of the thermosetting matrix. These new matrices offered new design capabilities for composites in a variety of aircraft applications. To improve the damage tolerance of composite materials even further, an engineering approach to toughening was used to modify the highly stressed interlayer with either a tougher material (in this case, film or scrim) or through the use of preformed particles, leading to the third generation of composite materials. These interlayer-toughened composites provide the highest damage tolerance of current fiber-reinforced composite systems, with an excellent balance of mechanical properties.

Dispersed-Phase Toughening of Thermoset Matrices

The development of multiphase-structure thermosetting matrices using rubber and/or thermoplastic materials typically first involves the formation of an initially miscible solution of the modifier and uncured thermoset in which phase separation of the modifier occurs upon cure, forming distinct domains (Ref 1 to 5). One of the most widely used materials for dispersed-phase-toughening of epoxy matrices are reactive liquid polymers (RLP), such as carboxyl-terminated butadiene-acrylonitrile rubber (Ref 6 to 8). The average diameter of the dispersed phases in rubber-toughened thermoset-matrix composites commonly ranges from 0.5 to 10 μm. The size and degree of phase separation is a complex function of the formulation constituents, rubber or thermoplastic characteristics (such as solubility, molecular weight, and functionalization), and cure cycle. It is well known that phase separation is viscosity or mobility dependent and; therefore, affected by slight advancement or staging of the thermosetting resin. As a result, the phase separation of the rubber may be incomplete, which can lead to ambient- and elevated-temperature property degradation.

In the following figures, ultrathin sections were developed from the composite materials to use transmitted-light optical microscopy contrast techniques to determine the phases. Figure 10.1 shows a cross section of a

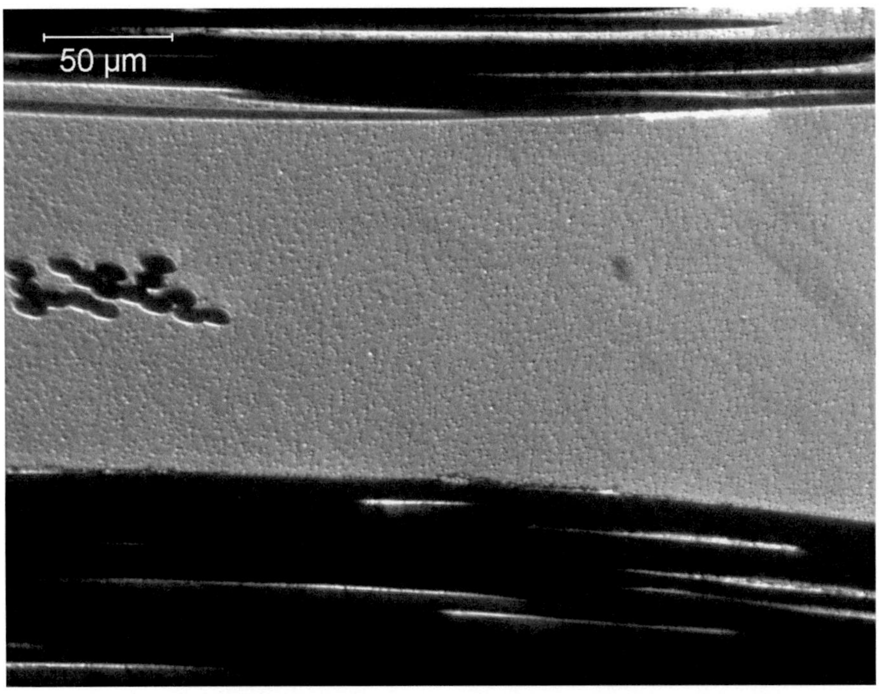

Fig. 10.1 Micrograph of a carbon fiber composite material that contains a very small dispersed-rubber phase in the matrix. Ultrathin section. Transmitted light, Hoffman modulation contrast, 40× objective

carbon-fiber-reinforced composite that has a very small dispersed-rubber phase in the matrix. The phase size in this composite was near the resolution of optical microscopy, approximately 0.5 μm. A cross section of a multiphase-matrix composite that was developed from two different rubber tougheners is shown in Fig. 10.2. These rubber tougheners are made from the same polymer backbone but differ in molecular weight. In the interply area, there is an irregular dispersed phase that was formed from the high-molecular-weight rubber. A smaller, spherical, dispersed-rubber phase is also seen in the figure and corresponds to the lower-molecular-weight rubber modifier. The size of this phase ranges from a little less than 1 to 5 μm in diameter. This phase was located more uniformly throughout the matrix and in the fiber tows. Larger phase sizes can be designed and developed so that they are forced to phase separate to a greater extent in the composite interlayer region due to the restriction of the fiber spacing in the intraply (Ref 9). However, in many cases, the dispersed phases are uniformly distributed throughout the matrix but are more easily identified in the interlayer region. The interlayer region can best be viewed by sectioning the composite on an angle through the thickness so the interlayer is expanded. In open-weave woven fabric composites, such as those made from 3k-70 carbon fiber fabrics, the best areas to view the morphology are usually in the interstitial regions. Figures 10.3 and 10.4 show cross sections of different multiphase-matrix composites that were sectioned, ground, and polished at between 10 and 20 degrees off parallel to the surface in order to view a larger area of the interlayer region. The multiphase morphology of these two systems is very complex. The dispersed phase is

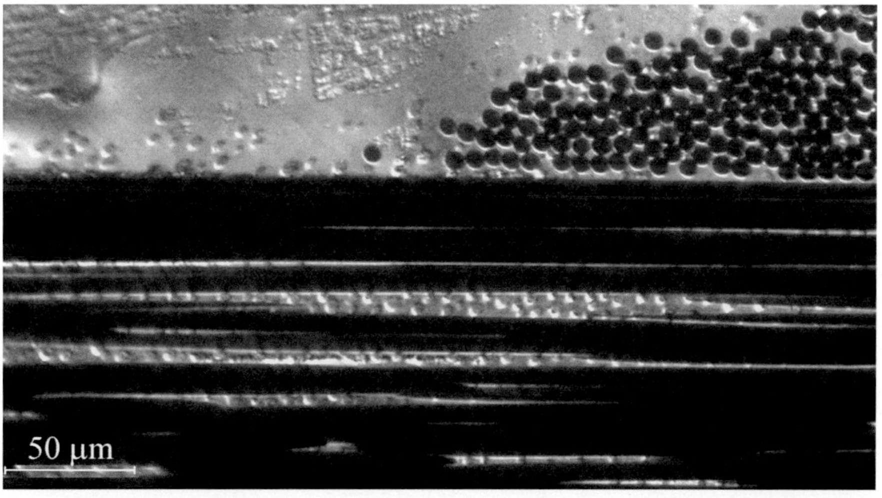

Fig. 10.2 Micrograph of a carbon fiber composite material that was toughened using two rubber materials of different molecular weight. Two different phase morphologies are observed, corresponding to the different tougheners. Ultrathin section. Transmitted light, Hoffman modulation contrast, 40× objective

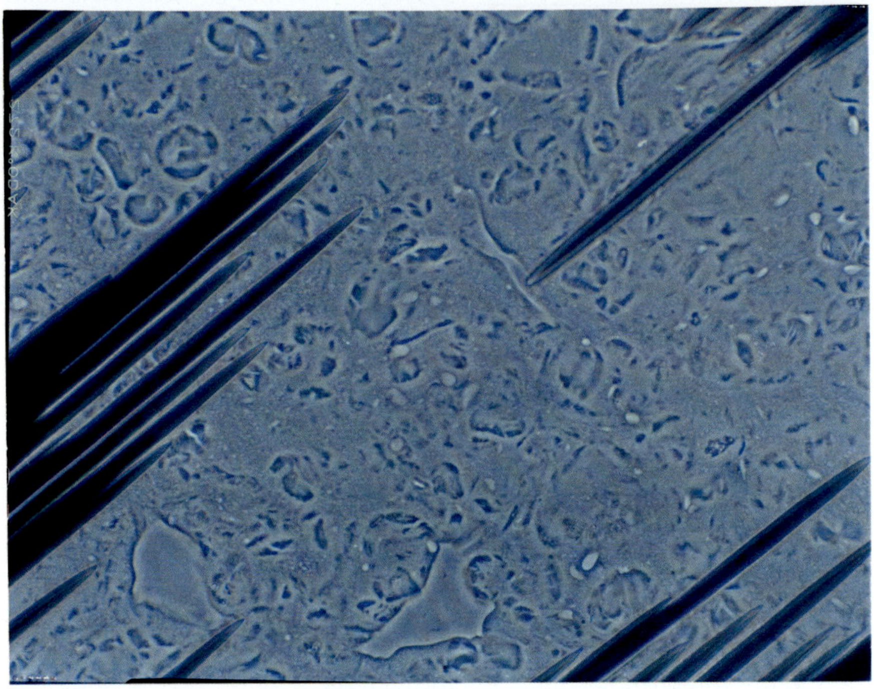

Fig. 10.3 Dispersed-phase-toughened carbon fiber composite material that was sectioned at an oblique angle to obtain a larger view of the interlayer region. Large, irregular phases, with some phases spherical and hollow, were found in the interlayer area and extended into the intraply area. Ultrathin section. Transmitted light, phase contrast, 40× objective

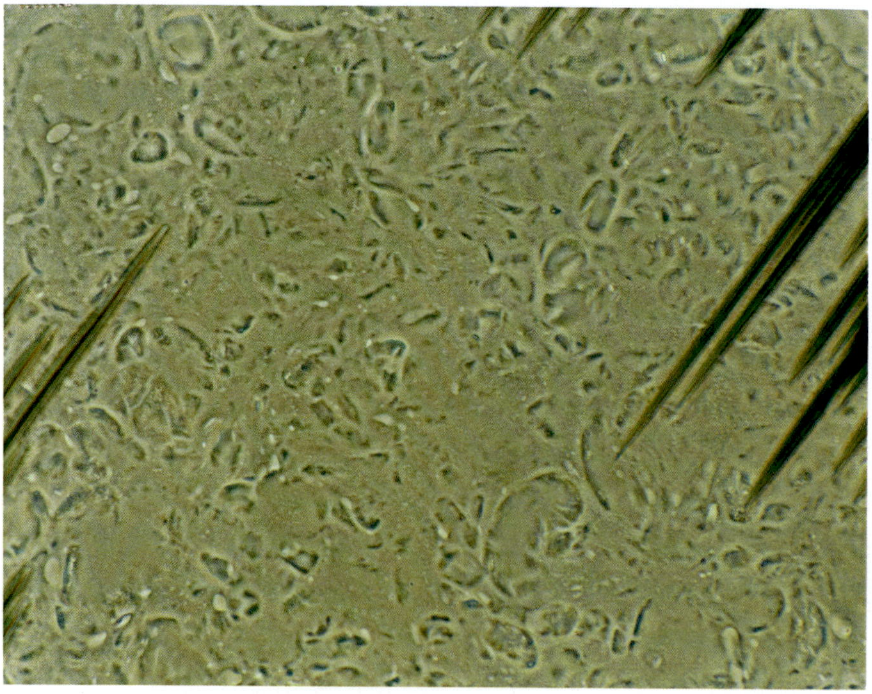

Fig. 10.4 Dispersed-phase-toughened carbon fiber composite material that was sectioned at an oblique angle to obtain a larger view of the interlayer region. A complex morphology was revealed, which was also present in the intraply area. Ultrathin section. Transmitted light, differential interference contrast, 40× objective

highly irregular, and the size varies from a few micrometers to as large as 15 μm.

Another technique that is used for developing multiphase thermosets involves the addition or formation of preformed particles in the uncured resin that are initially distinct particles and remain so after cure (Ref 10). As a result, the effect of the cure cycle, viscosity, and base matrix chemistry usually does not affect the final particle size. However, these particles may swell, or the outer shell may partially dissolve in or react with the thermosetting matrix. Materials that are commonly used to toughen fiber-reinforced composites are core-shell particles with a shell that is compatible and may also be reactive with the matrix, and a cross-linked rubber or thermoplastic core that does not dissolve. The size of the core-shell particles is usually smaller than the phases formed from RLP toughening. Core-shell particle sizes usually range from 50 to 500 nm but may be larger, depending on the product. Because of its resolution limits, optical microscopy may not be the best method for viewing these smaller particles. However, the effect on the fracture mechanism and the change in fracture morphology within a composite with core-shell-particle-type modification can be observed using optical microscopy. Another newer type of toughening material based on self-assembly of block copolymers has been developed for toughening thermosetting-matrix resins (Ref 11 to 13). These materials are capable of forming complex nanostructured morphologies from 10 to 500 nm within the thermosetting resins. As with core-shell particles in thermosetting matrices, the morphology of these materials is best viewed using other techniques, such as transmission electron microscopy.

Particle Interlayer Toughening of Composite Materials

A more recent method of toughening fiber-reinforced composites modifies only the highly stressed interlayer of the composite (Ref 10, 14 to 16). This engineering approach to toughening composites has been shown to reduce the propensity for delamination, which is a major cause of composite failure (Ref 17, 18). It is known that by increasing only the resin thickness in the interlayer region, the fracture toughness of composites can be significantly improved, providing a more unconstrained plastic zone during failure (Ref 19). Unfortunately, the improvement in damage tolerance by using this approach is limited by per-ply thickness constraints and complications in processing. These factors, along with handling issues (tack) and hot-wet property reductions, led to the development of preformed particle toughening of composite interlayers. The preformed particles used for interlayer modification are typically thermoplastic or rubber materials with average particle sizes between 20 and 50 μm (Ref 10). This toughening concept is most often applied to prepreg materials and, not until recently, resin transfer-molding applications through the use of particle-modified tackifiers or thermoplastic scrims in the interlayer (Ref 20).

In analyzing interlayer-modified composites, the sectioning plane and microscopy technique determine the level of detail that can be observed in the interlayer. Cross sections of various particle-toughened interlayer-modified composite materials are shown in Fig. 10.5. In Fig. 10.5(a), it is difficult but possible to observe the preformed particles in the interlayer region by cross sectioning the composite through the thickness and using slightly uncrossed polarized reflected light. Depending on the particle type and matrix, the phase may be more or less distinguishable using reflected-light techniques. The use of transmitted-light microscopy provides significantly greater detail of the morphology of polymer matrices containing different phases. In Fig. 10.5(b), it can be seen that there is a high particle concentration and high degree of packing of particles in the interlayer region of this interlayer-modified composite. Also, the particles in the interlayer region have been deformed by the restriction of the fiber bed. By grinding and polishing the cross section at a slightly oblique angle, the interlayer area can be expanded, as shown in Fig. 10.5(c). In this figure, the particles in the interlayer region are very spherical in nature and are shown to extend into the fiber tows, much like those observed in Fig. 10.5(a). An even larger view of a particle-modified interlayer region is shown in Fig. 10.5(d). This view was created from sectioning and polishing a quasi-isotropic carbon fiber composite at an even lower angle, so the interlayer region could be further expanded.

The two main methods for manufacturing prepregs that have preformed-particle-rich surfaces are single- and double-pass impregnation, although other methods have also been used (Ref 21). One method that has received much attention is double-pass impregnation, which involves first impregnating a homogeneous resin (no particles) into the inner fiber bed area, followed by selectively coating a particle-modified resin on the prepreg surface (Ref 22). Another technique for the development of prepreg having particle-modified surfaces is single-pass impregnation. This method involves filming a particle-modified resin and relying on the particles to filter out on the fiber tow surfaces or prepreg surface during the prepregging process (Ref 23). Figure 10.6 shows preformed particles in the interlayer region of a composite material that was developed using single-pass impregnation. Using epi-fluorescence, the preformed particles in the interlayer can be easily observed fluorescing green, as well as a smaller dispersed phase in the matrix that fluoresces yellow. It can be seen that most of the preformed particles are located in the interlayer region, but a few smaller preformed particles (hollow particles) are found in the intraply. If the large preformed particles are dispersed throughout the composite, there are typically disruptions in the fiber tows, and the interlayer-toughening effect will be diminished and mechanical properties reduced. It is known that as the particle size is decreased, especially when approaching the fiber

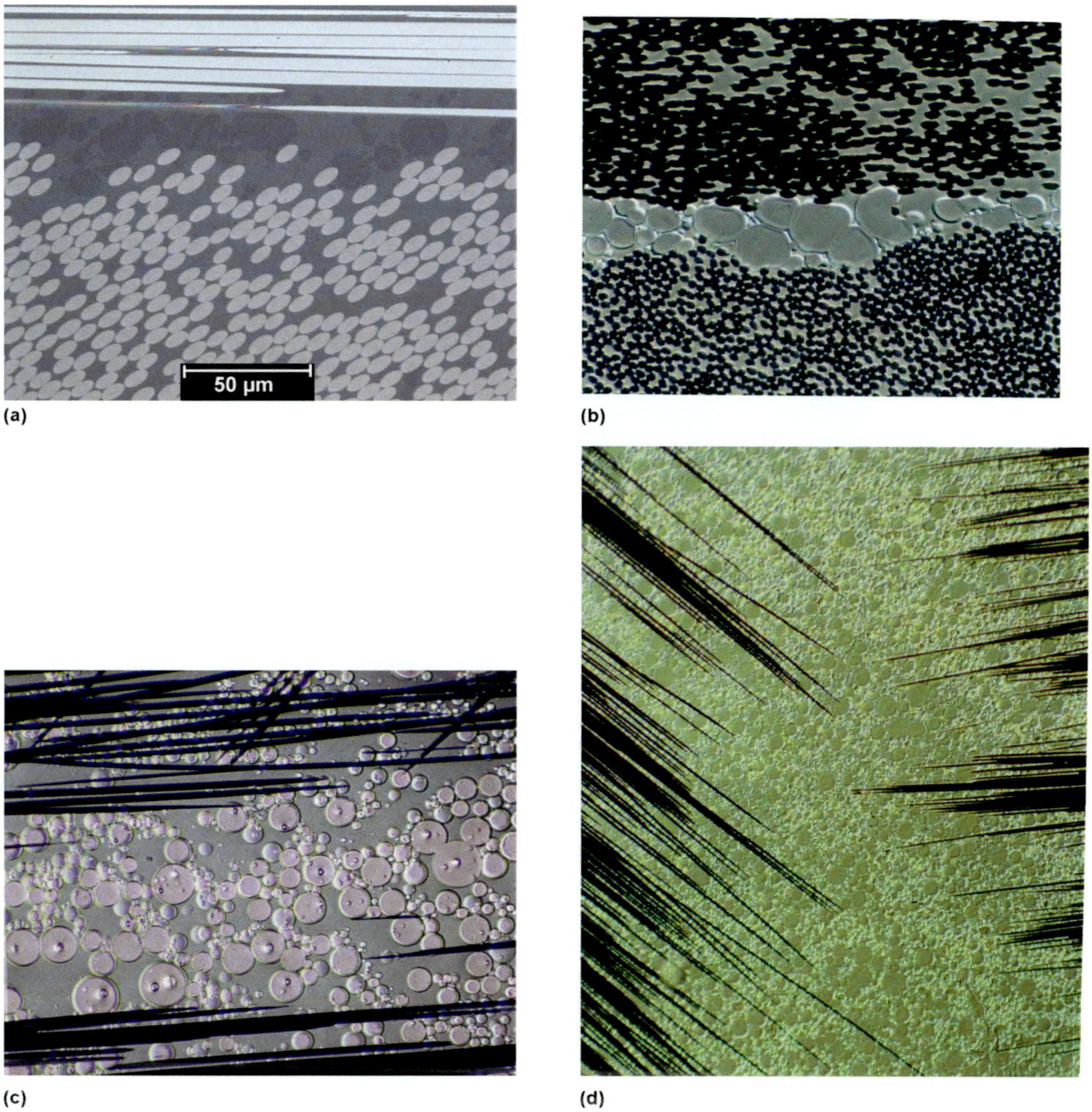

Fig. 10.5 Preformed-particle-modified interlayer regions of various carbon fiber composite materials showing differences in the optical analysis technique and the sample-preparation method. (a) Reflected-light optical analysis of an interlayer region showing particles residing in the interlayer region (i.e., light-gray circles). Slightly uncrossed polarized light, 50× objective. (b) Ultrathin section developed from a particle-modified toughened interlayer composite that was sectioned through the thickness. Transmitted light, Hoffman modulation contrast, 40× objective. (c) Ultrathin section of a particle-modified interlayer region that was developed from an oblique angle cross section of the composite to obtain a larger view of the interlayer. Transmitted light, differential interference contrast, 40× objective. (d) Larger field of view of a particle-modified interlayer region expanded by grinding and polishing at a low angle through the thickness. An ultrathin section was developed from the sample for transmitted-light analysis. Hoffman modulation contrast, 10× objective

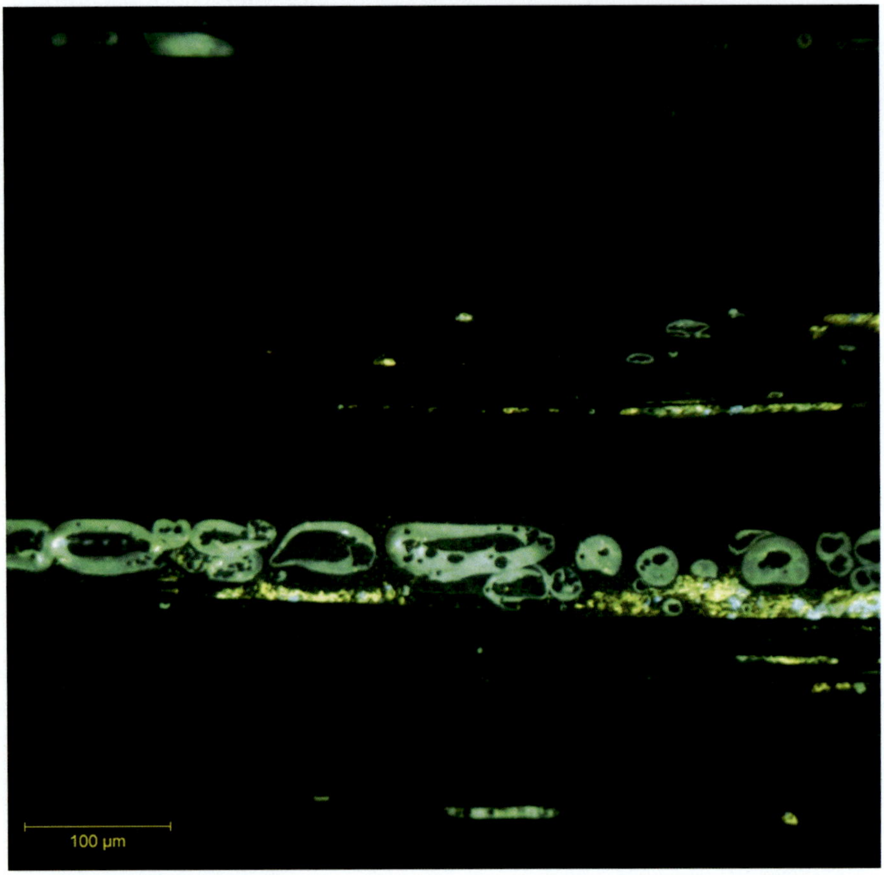

Fig. 10.6 Area of a preformed-particle-modified interlayer composite showing large, hollow particles fluorescing green in the interlayer region, with another smaller dispersed phase fluorescing yellow. To distinguish these particles, a fluorescing dye/solvent was wiped on the surface of the polished specimen, wiped off after 2 min, and cleaned with water. The solvent/dye preferentially absorbed into the particles and phases and not the thermoset matrix. Epi-fluorescence, 390–440 nm excitation, 25× objective

diameter, there is a greater probability that the particles will fail to filter out and instead will impregnate the fiber bed (Ref 24). This becomes even more of an issue for lower-fiber-areal-weight prepregs, because the tows may be spread farther apart before entering the impregnation zone, and greater impregnation pressure may need to be applied. Figure 10.7 shows a cross section of a composite material where the preformed particles did not filter completely in the interlayer region and were found extending into the fiber bed. The level of filtering of the preformed particles on the prepreg surface and therefore control of the cured composite interlayer dimensions is a function of the prepreg processing parameters and particle size distribution.

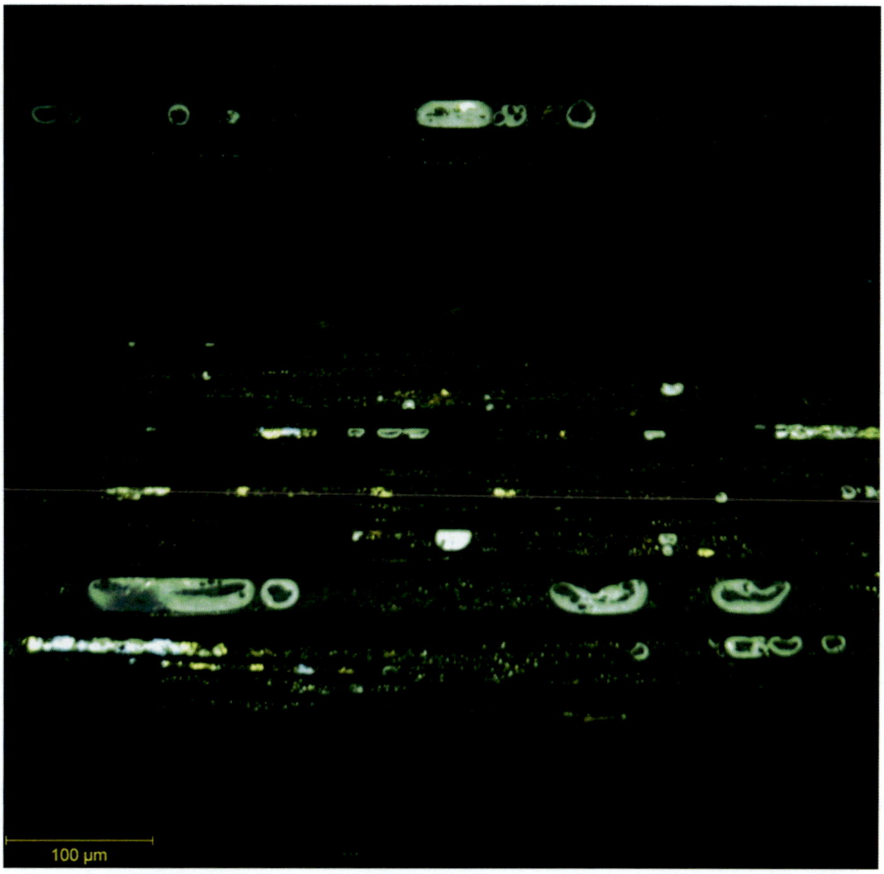

Fig. 10.7 Cross section of a preformed-particle-modified composite showing no distinct interlayer region. Due to the prepreg processing parameters, a particle-rich interlayer region was not developed. The hollow preformed particles fluoresce green and are shown compressed between fibers. Also present is a very small dispersed phase that fluoresces yellow. Epi-fluorescence, 390–440 nm excitation, 25× objective

It has been found that the size and distribution of individual resin phases and the morphology can directly affect the damage resistance of composites. In many cases, changes to the microstructure can be performed not only through the formulation but also in the manufacturing process. The capability to evaluate the morphology is critical in determining the material quality and in correlating the key microstructural features with material performance. Through three successive specimen-preparation techniques for the carbon-fiber-reinforced composite under analysis, the complex, multiphase resin system was completely evaluated using only reflected-light techniques. The reflected-light optical microscopy techniques that were used to enhance the contrast and show the morphology include the sample as polished, chemically etched, and using epi-fluorescence. All of these techniques were performed on the same cross section. In the as-polished condition using epi-bright-field illumination, it can be observed in Fig. 10.8 that there is a stratification of resin-rich areas between the fiber beds. Figure 10.8 also reveals a dispersed phase in between the fiber beds that is nearly circular (near spherical in three dimension) and varies in size from 3 μm to as large as 40 μm. With higher magnification, it can be seen that there are smaller circular phases that appear in the dispersed phases (Fig. 10.9a, b). In many previous microstructural investigations, this would have been the extent of the analysis, due to the limited techniques available to provide greater contrast.

Figure 10.10(a and b) is of the same composite, except that the polished specimen is etched for 30 seconds using a solution of CrO_3 (12.5 g), HNO_3 (50 mL), and H_2O (50 mL) (Table 5.3). Caution: This etch is hazardous. After using this etch, it can be observed that there is a subphase in addition to the dispersed phase in the matrix shown in Fig. 10.9(a and b). The use of this information allows for a visual characterization of the microstructure that can be correlated with material performance and formulation and processing changes. This also enables an understanding of how each phase contributes to damage resistance and fracture toughness.

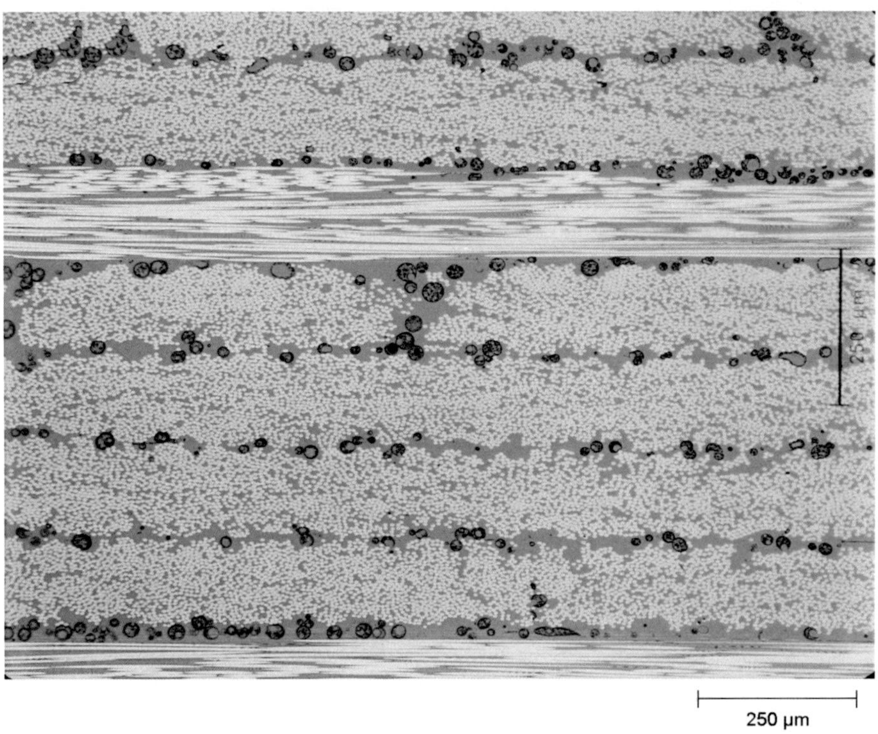

Fig. 10.8 Large view of a cross section of an interlayer-toughened composite showing multiple plies and interlayer regions. Bright-field illumination, 10× objective

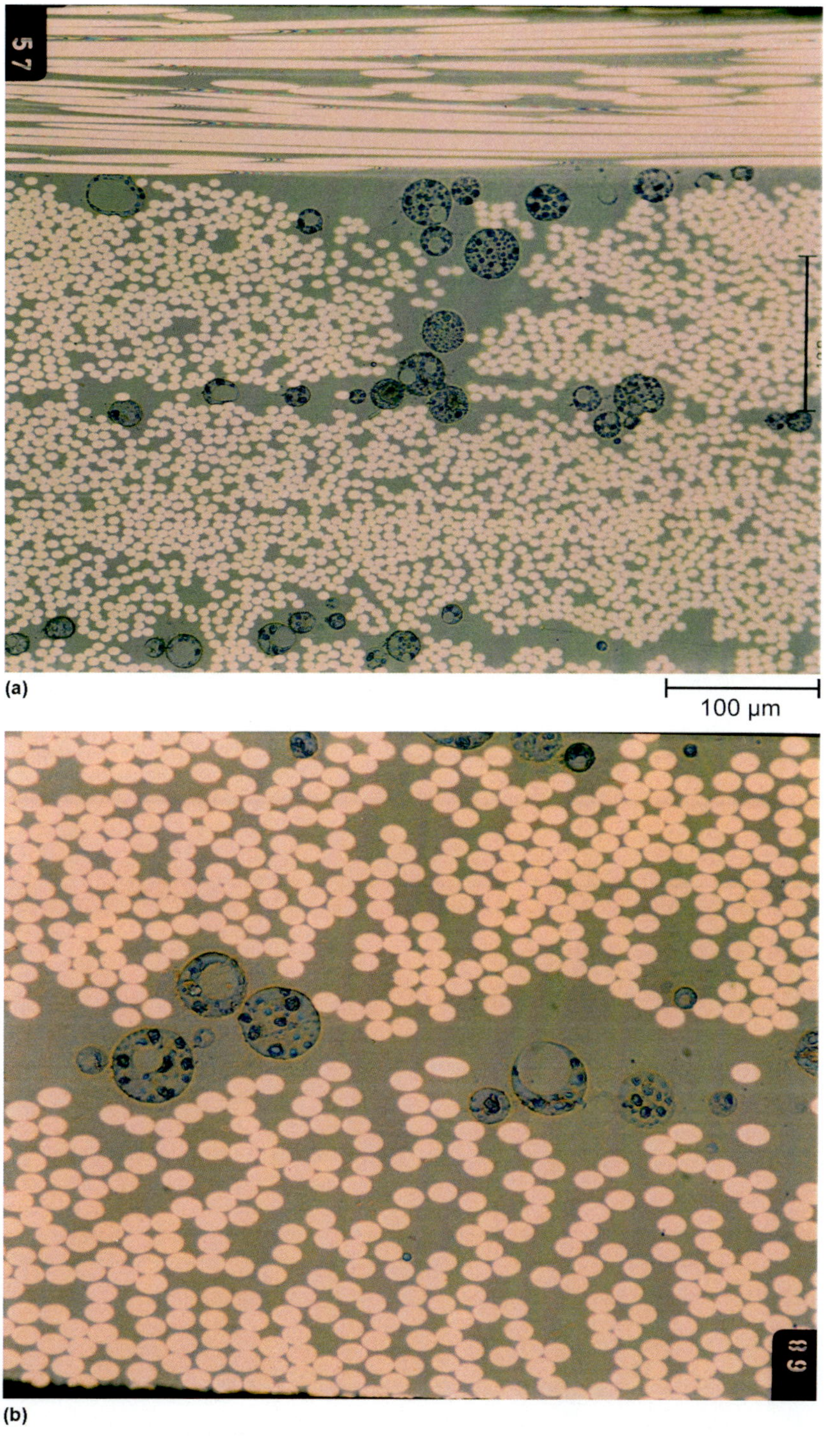

Fig. 10.9 Higher-magnification views of the composite cross section shown in Fig. 10.8. (a) Reflected light, phase contrast, 25× objective. (b) Bright-field illumination, 50× objective

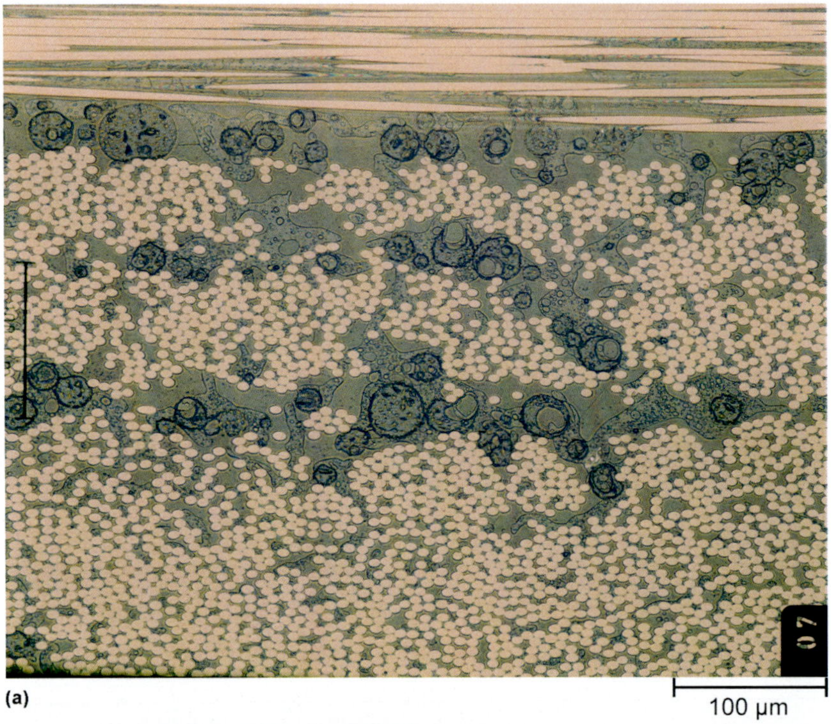

(a)

(b)

Fig. 10.10 After chemical etching, another phase can be observed in the matrix. (a) Reflected light, phase contrast, 25× objective. (b) Reflected light, phase contrast, 50× objective

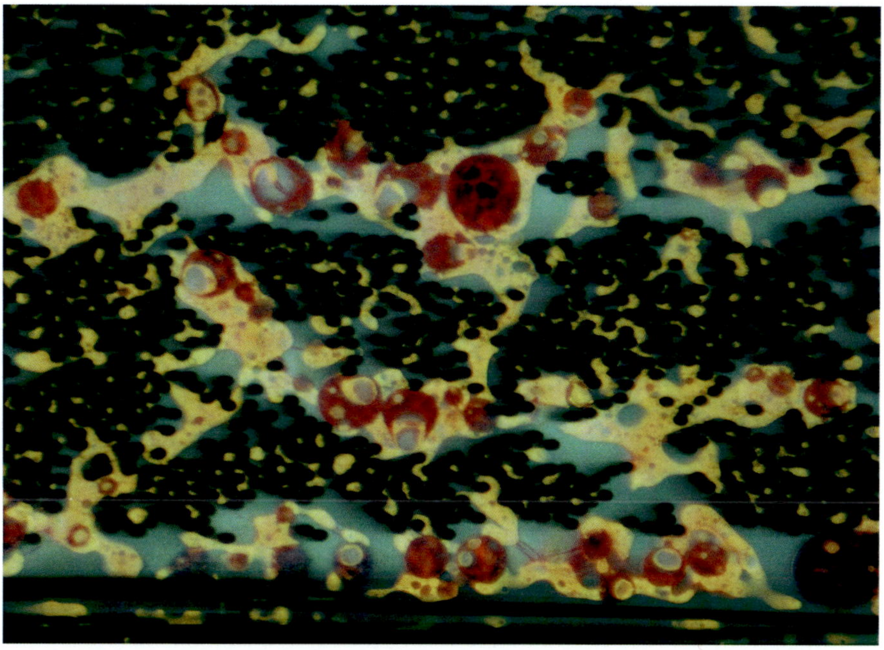

Fig. 10.11 Identification of the complex multiphase morphology of the matrix in the carbon fiber composite was further enhanced through the application of Rhodamine B dye/solvent to the polished cross section. Epi-fluorescence, 390–440 nm excitation, 25× objective

To determine if the different phases were formed from the same material, a dye was applied to the sample surface and viewed using epi-fluorescence. First, the sample was immersed in a solution of Rhodamine B dye (see Chapter 5, "Viewing the Specimen Using Reflected-Light Microscopy," for dyes). The sample was then ultrasonically cleaned in methylene chloride, followed by heating to 121 °C (250 °F) to remove additional solvent. Using epi-fluorescence microscopy with a 390 to 440 nm excitation range, each individual phase becomes distinctly colored, as shown in Fig. 10.11. This technique more accurately defines each phase and shows that each of the phases is a different chemistry. Because more contrast is possible, the morphology can be more quantitatively characterized using image analysis techniques.

REFERENCES

1. D. Verchere, H. Sautereau, J.-P. Pascault, C.C. Moschiar, S.M. Richardi, and R.J.J. Williams, Rubber-Modified Epoxies: Analysis of the Phase-Separation Process, *Toughened Plastics I,* C.K. Riew and A.J. Kinloch, Ed., American Chemical Society, Washington, D.C., 1993, p 335–363
2. X. Tang, L. Zhang, T. Wang, Y. Yu, W. Gan, and S. Li, Hydrodynamic Effect on Secondary Phase Separation in an Epoxy Resin Modified

with Polyethersulfone, *Macromol. Rapid Commun.,* Vol 25, 2004, p 1419–1424

3. Z. Zhang, J. Cui, S. Li, K. Sun, and W. Fan, Effect of Hydroxyl-Terminated Polyethersulfone on the Phase Separation of Polyetherimide-Modified Epoxy Resin, *Macromol. Chem. Phys.,* Vol 202, 2001, p 126–132
4. C.B. Bucknall and I.K. Partridge, Phase Separation in Crosslinked Resins Containing Polymeric Modifiers, *Polym. Eng. Sci.,* Vol 26 (No. 1), 1986, p 54–62
5. M. Akay and J.G. Cracknell, Epoxy Resin-Polyethersulphone Blends, *J. Appl. Polym. Sci.,* Vol 52, 1994, p 663–685
6. C.K. Riew, E.H. Rowe, and A.R. Siebert, in *Toughness and Brittleness of Plastics,* D.R. Deanin and A.M. Crugnola, Ed., American Chemical Society, Washington D.C., 1976, p 326
7. R.Y. Ting and R.J. Moulton, Fracture Properties of Elastomer Toughened Epoxies, *12th National SAMPE Technical Conference,* 1980, p 265
8. J.H. Klug and J.C. Seferis, Phase Separation Influence on the Performance of CTBN-Toughened Epoxy Adhesives, *Polym. Eng. Sci.,* Vol 39 (No. 10), 1999, p 1837
9. R.W. Hillermeier, B.S. Hayes, and J.C. Seferis, Processing of Highly Elastomeric Toughened Cyanate Esters through a Modified Resin Transfer Molding Technique, *Polym. Compos.,* Vol 20 (No. 1), 1999, p 155–165
10. B.S. Hayes and J.C. Seferis, Modification of Thermosetting Resins and Composites through Preformed Polymer Particles: A Review, *Polym. Compos.,* Vol 22 (No. 4), 2001, p 451–467
11. E. Girard-Reydet, J.-P. Pascault, A. Bonnet, F. Court, and L. Leibler, A New Class of Epoxy Thermosets, *Macromol. Symp.,* Vol 198, 2003, p 309–322
12. S. He, X. Wang, X. Guo, K. Shi, Z. Du, and Z.B. Zhang, Studies of the Properties of a Thermosetting Epoxy Modified with Block Copolymers, *Polym. Int.,* Vol 54, 2005, p 1543–1548
13. S. Ritzenthaler, F. Court, E. Girard-Reydet, L. Leibler, and J.P. Pascault, ABC Triblock Copolymers/Epoxy-Diamine Blends, Part 2: Parameters Controlling the Morphologies and Properties, *Macromolecules,* Vol 36 (No. 1), **2003**, p 118–126
14. N. Odagiri et al., T800H/3900-2 Toughened Epoxy Prepreg System: Toughening Concept and Mechanism, *Proceedings of the American Society of Composites, Sixth Technical Conference,* 1991, p 43
15. N. Sela and O. Ishai, Interlaminar Fracture Toughness and Toughening of Laminates Composite Materials: A Review, *Composites,* Vol 20 (No. 5), 1989, p 423–435
16. M.R. Groleau, Y.B. Shi, A.F. Yee, J.L. Bertram, H.J. Sue, and P.C. Yang, Mode II Fracture of Composites Interlayered with Nylon Particles, *Compos. Sci. Technol.,* Vol 56, 1996, p 1223–1240

17. K. Kageyama and I. Kimpara, Delamination Failures in Polymer Composites, *Mater. Sci. Eng. A,* Vol 143, 1991, p 167–174
18. F. Gao, G. Jiao, Z. Lu, and R. Ning, Mode II Delamination and Damage Resistance of Carbon/Epoxy Composite Laminates Interleaved with Thermoplastic Particles, *J. Compos. Mater.,* Vol 41 (No. 1), 2007, p 111–123
19. S. Singh and I.K. Partridge, Mixed-Mode Fracture in an Interleaved Carbon-Fibre/Epoxy Composite, *Compos. Sci. Technol.,* Vol 55, 1995, p 319
20. R.W. Hillermeier and J.C. Seferis, Interlayer Toughening of Resin Transfer Molding Composites, *Compos. Part A: Appl. Sci. Manuf.,* Vol 32, 2001, p 721–729
21. K. Jang, W.-J. Cho, and C.S. Ha, Influence of Processing Method on the Fracture Toughness of Thermoplastic-Modified, Carbon-Fiber-Reinforced Epoxy Composites, *Compos. Sci. Technol.,* Vol 59, 1999, p 995
22. M.A. Hoisington and J.C. Seferis, Process-Structures-Property Relationships for Layered Structured Composites, *Proceedings of the American Society of Composites, Sixth Technical Conference,* 1991, p 53–62
23. B.S. Hayes and J.C. Seferis, Toughened Carbon Fiber Prepregs Using Combined Liquid and Preformed Rubber Materials, *Polym. Eng. Sci.,* Vol 41 (No. 2), 1991, p 170–177
24. B.S. Hayes and J.C. Seferis, Influence of Particle Size Distribution of Preformed Rubber on the Structure and Properties of Composite Systems, *J. Compos. Mater.,* Vol 36 (No. 3), 2002, p 299–312

CHAPTER 11

Impact Response of Composites

AS FIBER-REINFORCED POLYMERIC COMPOSITES continue to be used in more damage-prone environments, it is necessary to understand the response of these materials when subjected to impact from foreign objects. One concern in the aircraft industry is low-velocity impacts, because they can cause barely visible impact damage, which can remain undetected (Ref 1 to 4). This type of damage may occur from such things as hail and runway debris or from maintenance operations. Higher-velocity impacts usually create a more catastrophic damage and can result in holes through the thickness of the composite, and therefore, visible damage. Accordingly, the mechanism of how composites fail is determined by the impact parameters, material characteristics, and composite design and manufacturing (Ref 5). The impact parameters that influence the damage mechanism are the area, velocity, and mass of the projectile that impinges on the composite part. Material characteristics that affect the damage tolerance and mechanism of failure are the type of fiber, fiber tow structure, fiber volume, weave and stitching, matrix properties and toughness, location and size of the toughening materials, interlayer thickness and properties, and the fiber-matrix interfacial properties. The design of the composite part, including the geometry, thickness, number of interfaces, and lay-up, affects the response of the composite to an impact event (Ref 6). Of the design considerations, the lay-up of the composite material adds an additional complexity due to the interfacial stresses that can develop during cure with more off-angled plies, essentially preloading the composite part. The manufacturing process can also affect the consolidation and void content in the cured part, altering the crack propagation and failure mechanism upon impact of the composite.

Analysis Methods for Impact-Damaged Composites

Upon impact, if the energy is not absorbed by the composite as elastic strain energy, then damage will occur in the composite (Ref 7). Common damage observed in fiber-reinforced composites involves crack formation in the matrix and intraply, delamination in the interlayer, fiber-matrix interface debonding, and fiber fracture. The damage that occurs in the composite will most likely be a combination of these damage events and will depend on the aforementioned factors. In composites made from unidirectional carbon fiber prepreg, it has generally been found that matrix splitting precedes delamination in the composite upon impact, and delamination follows the direction of the matrix splitting (Ref 8, 9).

There are a few nondestructive analytical methods used to examine composites after impact. Two methods used in the aircraft industry to analyze damage areas are ultrasound C-scan and x-radiography techniques. These analytical techniques can show the extent of damage in the composite but not the effect on the microstructure. Hence, the use of destructive analytical techniques can provide complementary information to nondestructive analytical methods. One technique that is usually overlooked is ply separation (Ref 10, 11). This timely technique involves the careful separation of the plies in the zone of the failure to determine the extent of failure within each interlayer. While these techniques are very useful for understanding the extent of damage in a composite, the details of the damage and the mechanism cannot be determined. As a complementary technique to the aforementioned analytical methods, optical microscopy is an invaluable tool. Additional information that can be gained from optical microscopy includes a complete view of the damage through the thickness and in the composite plane. Also, matrix strains and fracture morphology can be determined very effectively using optical microscopy. It is worth noting that a common error is for investigators to go directly to scanning electron microscopy, losing the detail that can be provided through the use of optical microscopy techniques.

Composite damage after an impact can be analyzed with optical microscopy to determine how the composite failed as well as the mechanism of failure. Epi-bright-field illumination can be effectively used to determine the origin and extent of the damage in the composite. To better identify the location of the damage in the intra- and interlayer areas, greater contrast may be required. In this case, epi-fluorescence can prove to be very valuable. Further information regarding the effect on the matrix microstructure, including matrix strains and fracture morphology, can be determined with the development of ultrathin sections and transmitted polarized light or one of many other contrast methods. Through the use of these techniques, a complete analysis of the damage response of fiber-reinforced composite materials to impact can be determined. This information can be used to develop composite materials and structural parts that are more resistant to damage from impact events as well as how to repair the damaged area to reduce the risk of further degradation in service.

The impact damage of a quasi-isotropic laminate made from unidirectional carbon fiber prepreg is shown in Fig. 11.1. This laminate was impacted at 3000 in.-lb/in. with a 0.625 in. spherical tup. Propagation of the cracks that were formed upon impact appears similar to a radial expanding staircase. The cracking and delamination can be easily observed using bright-field illumination. This sample was prepared with no mount, which allows for a quick assessment of the damage from impact. One issue associated with not mounting the sample is that the fragile fracture areas may be further damaged during sample preparation and may potentially cause artifacts. Details of the fracture may also be difficult to observe if the sample is not mounted. When mounting the sample using a casting resin, it is necessary to add a dye to distinguish it from the matrix resin. Figure 11.2 shows a cross section of an impacted composite laminate that was mounted with an epoxy casting resin containing Rhodamine B dye. Using only bright-field illumination, the details of the cracks are difficult to discern. This is where the use of epi-fluorescence is essential, so the details of the fracture can be observed.

Brittle-Matrix Composite Failure

The use of epi-fluorescence for analysis of the damage of a brittle-matrix composite after impact is shown in Fig. 11.3(a to e). A montage of the damage area after impact of the quasi-isotropic carbon fiber laminate is shown in Fig. 11.3(a). The montage allows a larger field of view to help capture the extent of damage but is time-consuming in development. It is easily observed by the red contrast where the Rhodamine-B-dyed casting resin impregnated the fracture areas during the mounting procedure. The matrix resin in the composite was found to naturally fluoresce light green when exposed to wavelengths in the range of 390 to 440 nm. The fracture pattern is typical of brittle failure of a composite material. From this large

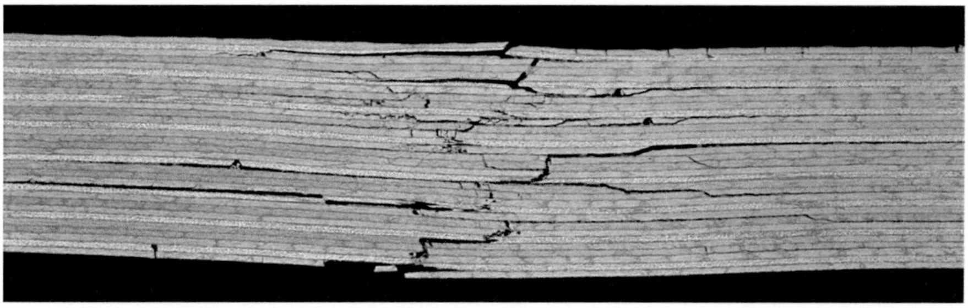

Fig. 11.1 Micrograph of impact damage of a 32-ply $(+45/0/-45/90)_{4S}$, high-temperature thermoplastic-matrix carbon fiber composite. Bright-field illumination, 65 mm macrophotograph

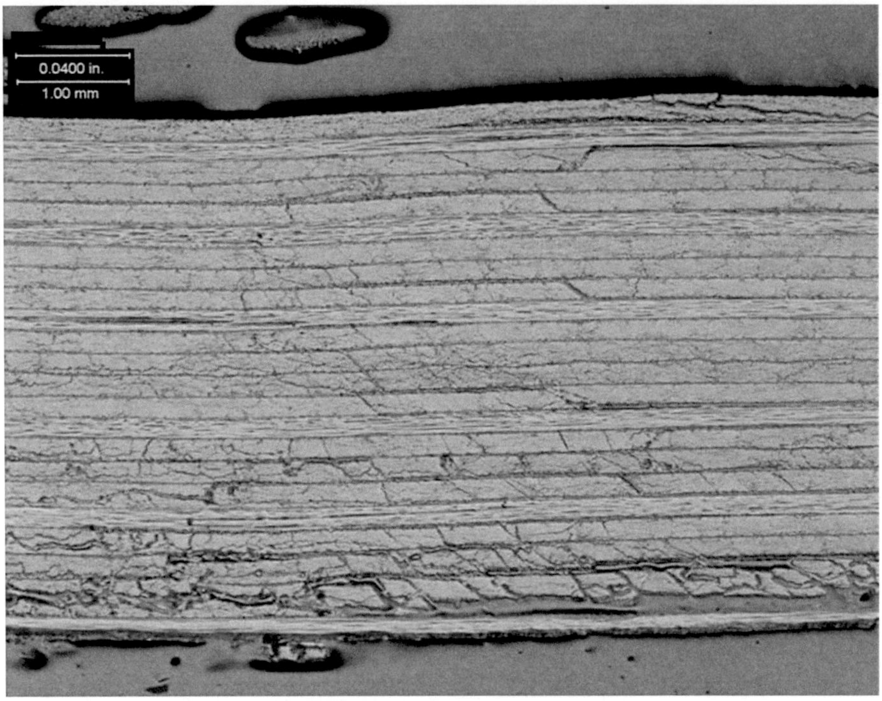

Fig. 11.2 Cross section of an impact-damaged carbon fiber composite that was mounted with Rhodamine-B-dyed epoxy casting resin. The use of epi-bright-field illumination does not allow the dye to fluoresce, and therefore, the cracks are hard to distinguish. Bright-field montage, 5× objective

sectional view, it can be seen that there is extensive interlayer delamination between almost every layer. Figure 11.3(b to e) is a series of micrographs taken with higher magnification from areas in Fig. 11.3(a). It can be seen that there is extensive fiber fracture (Fig. 11.3b) and mostly adhesive failure at the fiber-matrix interface, as shown in Fig. 11.3(c to e). In the resin-rich areas of the composite, shown in Fig. 11.3(d and e), cracks were found that propagated through the interlayer region between the plies. The brittle nature of this matrix, combined with the weak fiber-matrix interface, allowed the translation of the impact energy to destroy the composite laminate throughout the thickness.

Tough-Matrix Composite Failure

The impact damage of two toughened-matrix composites is shown in Fig. 11.4 and 11.5. In contrast to the composite shown in Fig. 11.3, significantly less damage is found in these laminates. Most of the damage in these composites was focused in the interlayer areas, with large ply groups still intact. While the damage progression in the composites shown in Fig. 11.4(a) and 11.5(a) is indicative of tough composite failure, there were less

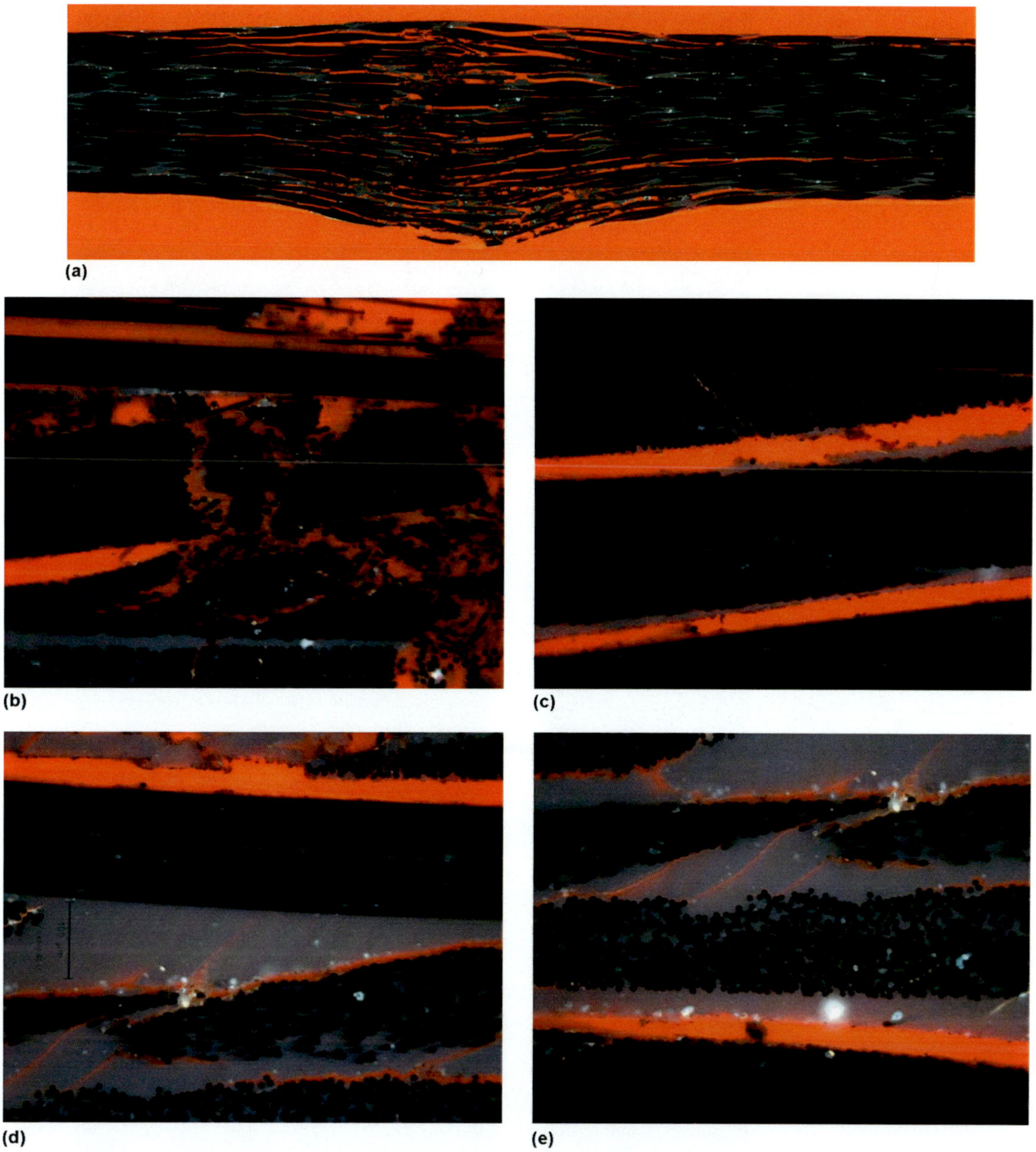

Fig. 11.3 Impact damage of a carbon fiber composite material that has a brittle matrix. (a) Montage of the impact area. Epi-fluorescence, 390–440 nm excitation, 5× objective. (b) Fiber fracture area in the composite. Epi-fluorescence, 390–440 nm excitation, 25× objective. (c) Fracture shown to occur extensively at the fiber (tow)-matrix interface and is mostly adhesive failure. Epi-fluorescence, 390–440 nm excitation, 25× objective. (d) Cracks shown at the fiber (tow)-matrix interface. Epi-fluorescence, 390–440 nm excitation, 25× objective. (e) Cracks spanning the resin-rich areas. Epi-fluorescence, 390–440 nm excitation, 25× objective

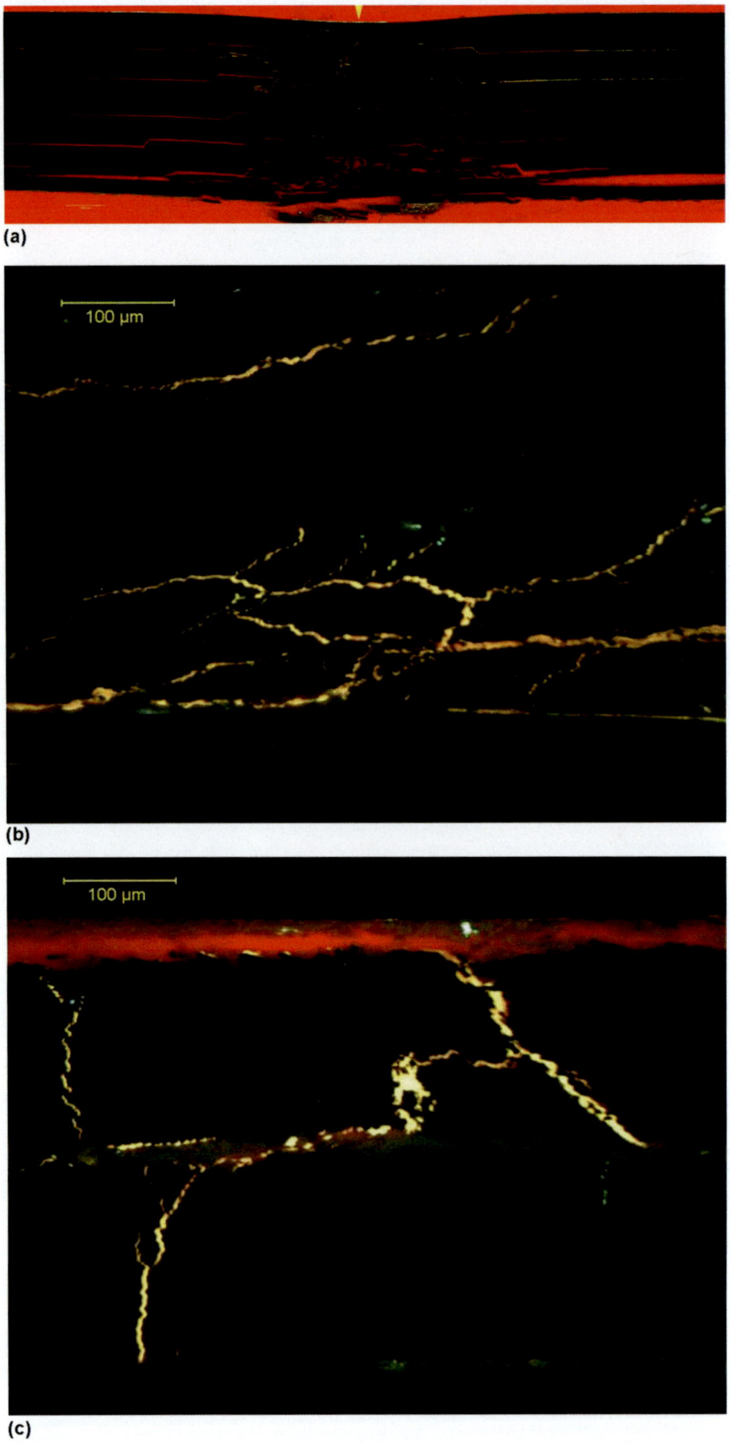

Fig. 11.4 Impact damage of a carbon fiber composite material that has a toughened matrix. (a) Montage of the impact area. Epi-fluorescence, 390–440 nm excitation, 5× objective. (b) Multiple intraply fractures. Epi-fluorescence, 390–440 nm excitation, 25× objective. (c) Adhesive failure at the fiber (tow)-matrix interface. Epi-fluorescence, 390–440 nm excitation, 25× objective

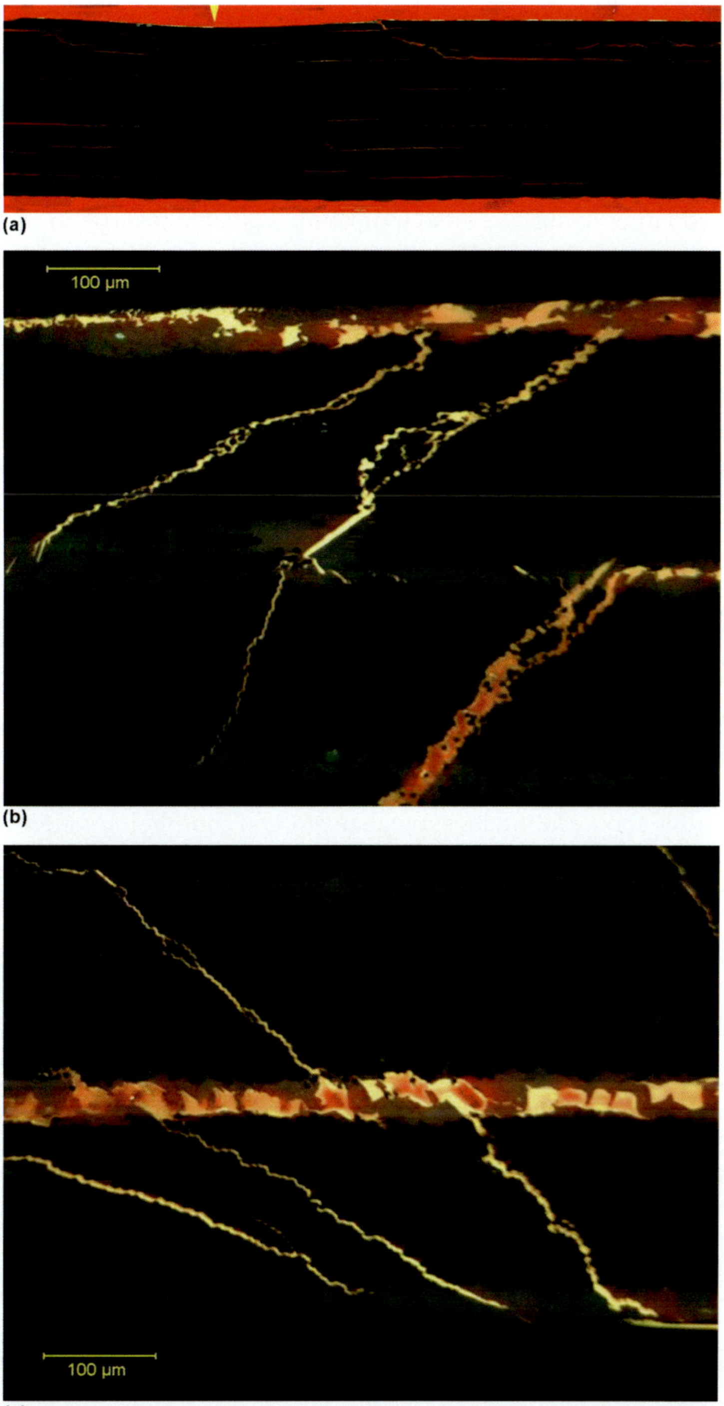

Fig. 11.5 Impact damage of a carbon fiber composite material that has a toughened matrix. (a) Montage of the impact area. Epi-fluorescence, 390–440 nm excitation, 5× objective. (b) Wide crack formation in the 45° plies and mainly cohesive failure in the interlayer. Epi-fluorescence, 390–440 nm excitation, 25× objective. (c) Failure in the interlayer showing the crack alternating between adjacent ply surfaces. Epi-fluorescence, 390–440 nm excitation, 25× objective

fiber breakage and intraply cracking in the composite shown in Fig. 11.5. This difference is especially noticeable directly under the impact zone. With higher magnification of these two composite materials, more quantitative differences in the fracture morphology were found. Figure 11.4(b) shows more extensive crack formation in the intraply region as compared to those in Fig. 11.5(b and c). The cracks in Fig. 11.5(b and c) are also wider and larger in surface area, indicating a greater absorption of energy. The micrograph in Fig. 11.4(c) shows an interlayer area with mainly adhesive fracture at the fiber-matrix interface. Without the red dye in the casting resin, it would be difficult to see if there is adhesive- or cohesive-type failure. The composite described in Fig. 11.5 had a significantly different failure mode in the interlayer area. Greater than 70 percent cohesive failure was found at the fiber-matrix interface, shown in Fig. 11.5(b and c). A comparison of the two different types of failures can help in explaining the differences in the residual strength after impact of these two composite materials.

Thermoplastic-Matrix Composite Failure Mechanisms

A macroview of the damage of a composite material after impact was shown in Fig. 11.1. This material was made from a tough, high-temperature thermoplastic resin and standard-modulus carbon fibers. To gain more information about the fracture mechanism and morphology after impact, an ultrathin section (less than 1 μm and up to 5 μm) was developed from a section in the montage. Various areas of the impacted specimen are shown in Fig. 11.6(a to c). Upon impact loading, shear deformation across the interply caused polymer alignment, resulting in strain birefringence (Fig. 11.6a). From analysis of the ultrathin section using transmitted cross-polarized light and a 530 nm interference filter, the birefringent-strained polymer regions appear white to yellow in color. A quantitative measurement of the strain can be made with a Berek-type compensator (Ref 12). Shear deformation along the interplies was the main energy-absorption mechanism. In contrast, the intraplies absorbed less energy due to fiber-spacing constraints. However, as seen in Fig. 11.6(a), spreading of the damage zone in the intraply region inhibited delamination.

As shear strains in the interply reached a critical value, the resolved tensile strains initiated cracks 45 degrees to the shear plane, becoming hackles as the crack tips linked together (Fig. 11.6b). The hackle spacing provides qualitative information about resin toughness and strength. A large interply deformation zone is found parallel to the intraply fracture. Fibers altered the local strain field by restricting polymer alignment. With the restriction of deformation and therefore reduced energy absorption, fracture occurred in the interlayer area (Fig. 11.6c).

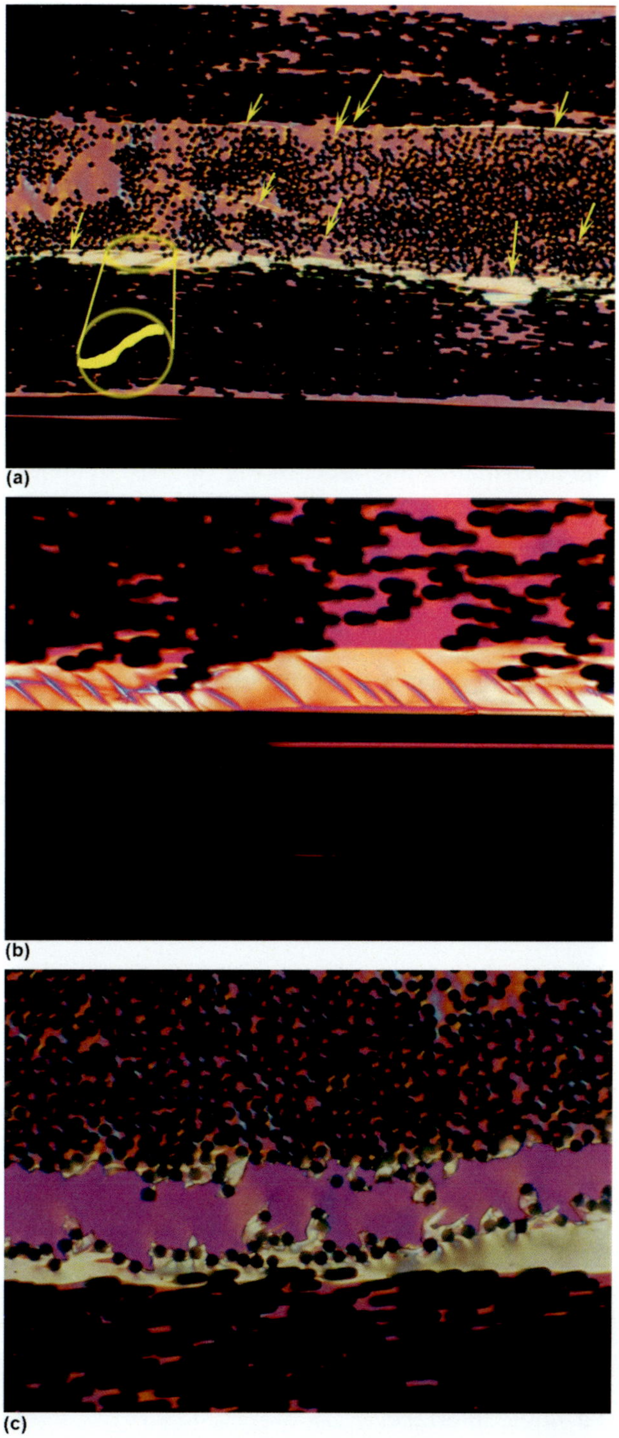

Fig. 11.6 Micrographs taken from ultrathin sections developed from areas of the high-temperature thermoplastic-matrix composite shown in Fig. 11.1. (a) Micrograph showing the matrix strains in the composite after impact. The arrows indicate areas of strain birefringence in the matrix. Transmitted polarized light, full wave plate, 20× objective. (b) Hackle formation in the interlayer region. Transmitted polarized light, full wave plate, 40× objective. (c) Fracture in the interlayer area. Transmitted polarized light, full wave plate, 40× objective

Untoughened Thermoset-Matrix Composite Failure Mechanisms

First-generation thermoset-matrix composites were untoughened in the sense that no additional materials, such as rubber or thermoplastics, were added to the base matrix composition to enhance the fracture toughness. This resulted in significant damage after impact, reducing the postimpact strength of these materials to unacceptable levels. In Fig. 11.7(a and b), areas taken from an untoughened composite after impact loading show the fracture morphology and the area ahead of the crack tip, respectively. The fracture morphology showed almost nonexistent hackle formation and very little strain birefringence. An adhesive-type failure was found in the interlayer region, where the crack was found to propagate around the fibers and at the fiber-matrix interface. In Fig. 11.7(b), it appears as though there was some strain birefringence ahead of the crack tip. Small areas of strain birefringence are seen in the intraply region as well, but this may be due as much to the effect of residual stress as it is to impact loading. As the desire increased to use composites in more damage-prone areas on aircraft, composites having greater damage tolerance were required. This led to the development of second-generation composite materials that contained rubber or thermoplastic modifiers incorporated within the thermosetting matrix.

Toughened Thermoset-Matrix Composite Failure Mechanisms

The damage resistance of composite materials is affected by the ability of the matrix to absorb impact energy and resist crack propagation. A cross section of a toughened, high-temperature thermosetting-matrix composite after impact is shown in Fig. 11.8. The formation of hackles in the fracture area is indicative of energy absorption by the matrix. However, the small size of the hackles and the more jagged appearance, compared to that shown in Fig. 11.6, imply less energy absorption. These features, and the observation that the deformation region did not span the interply, correspond to a more brittle matrix than that of the high-temperature thermoplastic matrix. As a comparison, Fig. 11.9 shows a cross section of an area of a toughened thermosetting-matrix composite that was subjected to impact testing, and the deformation region spanned the entire interlayer area. The fracture path was found to alternate between the interfaces of adjacent plies, confirming adequate fiber-matrix adhesion because the crack did not propagate at a single interface.

An optical analysis of the fracture mechanism of another toughened thermosetting-matrix composite after impact is shown in Fig. 11.10(a to c). Figure 11.10(a) shows an area in the composite where there was strain

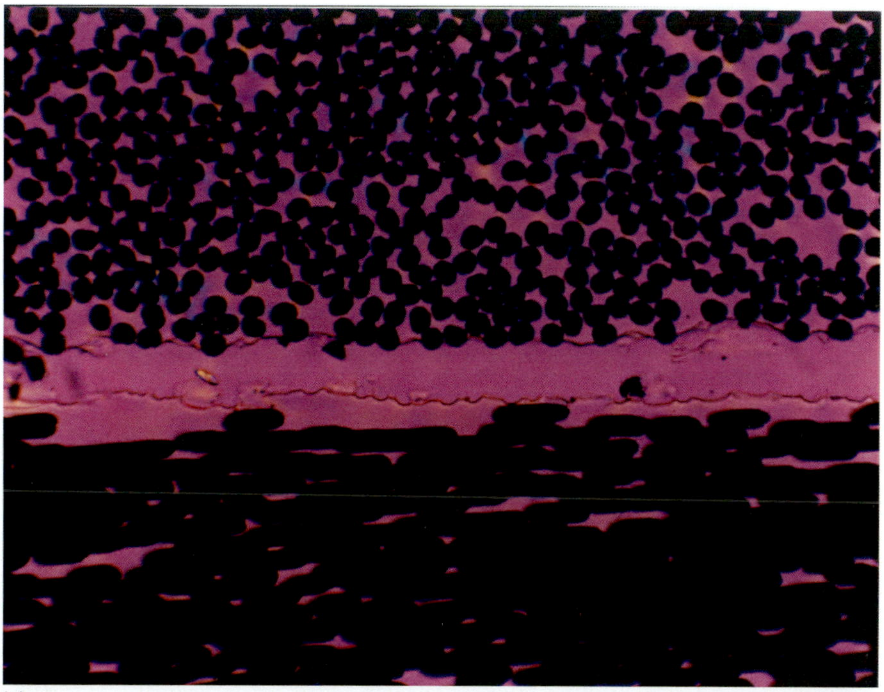

(a)

(b)

Fig. 11.7 Micrographs of an untoughened-matrix carbon fiber composite material after impact damage. (a) Fracture morphology showing no signs of hackle formation. Transmitted polarized light, full wave plate, 40× objective. (b) Area ahead of the crack tip. Transmitted polarized light, full wave plate, 40× objective

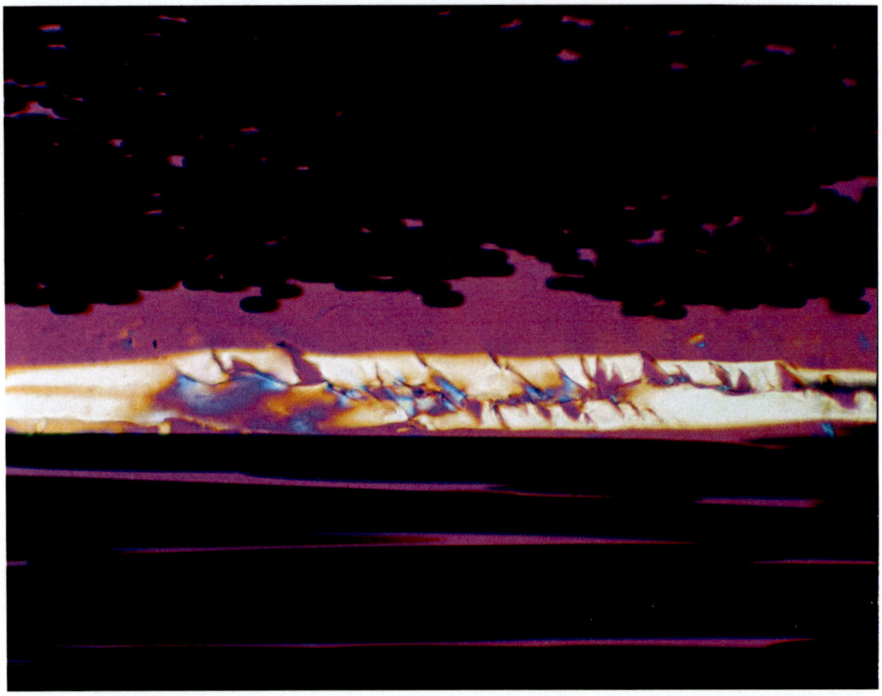

Fig. 11.8 Hackle formation in a high-temperature thermosetting-matrix composite after impact. Transmitted polarized light, full wave plate, 40× objective

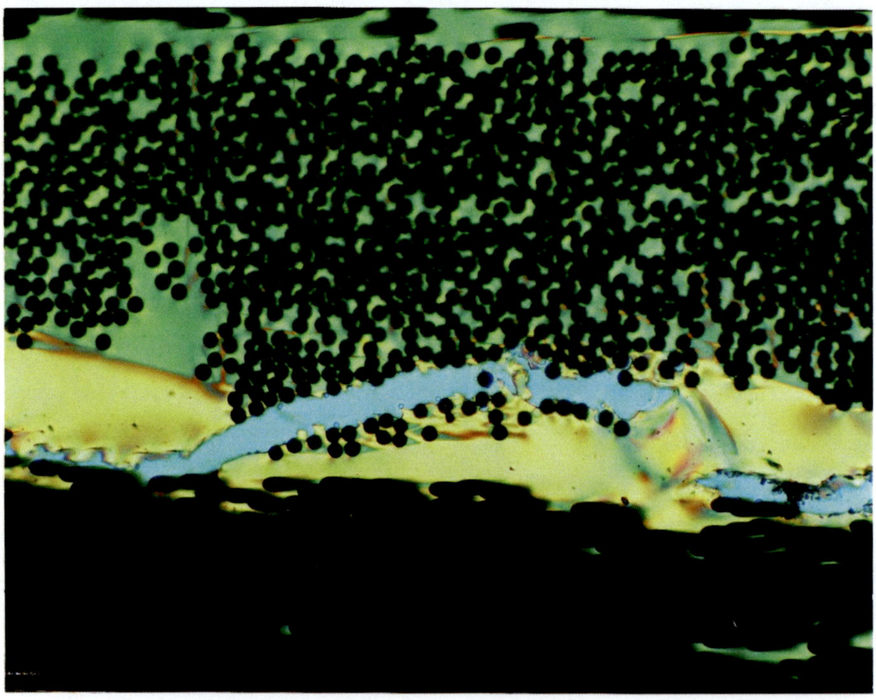

Fig. 11.9 Interlayer strain and fracture after impact of a toughened thermosetting-matrix composite. Transmitted polarized light, full wave plate, 40× objective

(a)

(b)

(c)

Fig. 11.10 Fracture morphology of a primary-phase-toughened matrix composite after impact. (a) Onset of hackle formation and strain in front of the crack tip. Transmitted polarized light, full wave plate, 40× objective. (b) Hackles in the interlayer region of the composite. Transmitted polarized light, full wave plate, 40× objective. (c) Fracture in the interlayer region. Transmitted polarized light, full wave plate, 40× objective

birefringence ahead of the crack tip, extending almost 100 μm. Also in this figure, the initiation of hackles is shown, but the hackle tips are not linked together. Moving further toward the area under the impact, the hackles are connected. These hackles are small and do not extend across the large interlayer. All of these images show little strain birefringence and therefore strain-induced variations within the matrix structure, as compared to Fig. 11.6. After complete fracture (Fig. 11.10c), cracks were found to have propagated from the interlayer region to the ply interfaces. These cracks were constrained by the fibers in the adjacent plies. The fracture path propagated intermittently from one interface to the next in the interply region.

Dispersed-Phase, Rubber-Toughened Thermoset-Matrix Composite Failure Mechanisms

The addition of a reactive rubber that phase separates uniformly throughout a thermosetting matrix upon cure is one of the oldest and most common toughening methods used for fiber-reinforced composite materials. Because the rubber has a higher coefficient of thermal expansion (CTE), it limits the shrinkage of the matrix during gellation and cure. This can result in lower cure shrinkage and lower residual stress in fiber-reinforced composites, along with increased fracture toughness. In dispersed-phase, rubber-toughened thermosets, the rubber particles are under triaxial stress after cure due to different CTEs for the rubber and the matrix. Upon impact, these rubber particles increase the toughness of the matrix and therefore the composite by one or more of the following major mechanisms:

- Cavitation of the particle phase and subsequent plastic void growth
- Particle bridging and tearing
- Creation of shear yielding between rubber particles (Ref 13 to16)

Figure 11.11 shows a crack in the interlayer region of a dispersed-phase, rubber-toughened matrix composite. After this composite was subjected to impact loading, the cross sections revealed that many of the particles had cavitated. Also, particles were found to be torn from the crack propagation. Ahead of the crack tip, the rubber particles were found to align and elongate in the crack propagation direction. These energy-absorbing mechanisms acted together to increase the damage tolerance of the composite material.

Particle Interlayer-Toughened Composite Failure Mechanisms

The toughening of composite materials through preformed thermoplastic or rubber particle modification of the interlayer is the most effective

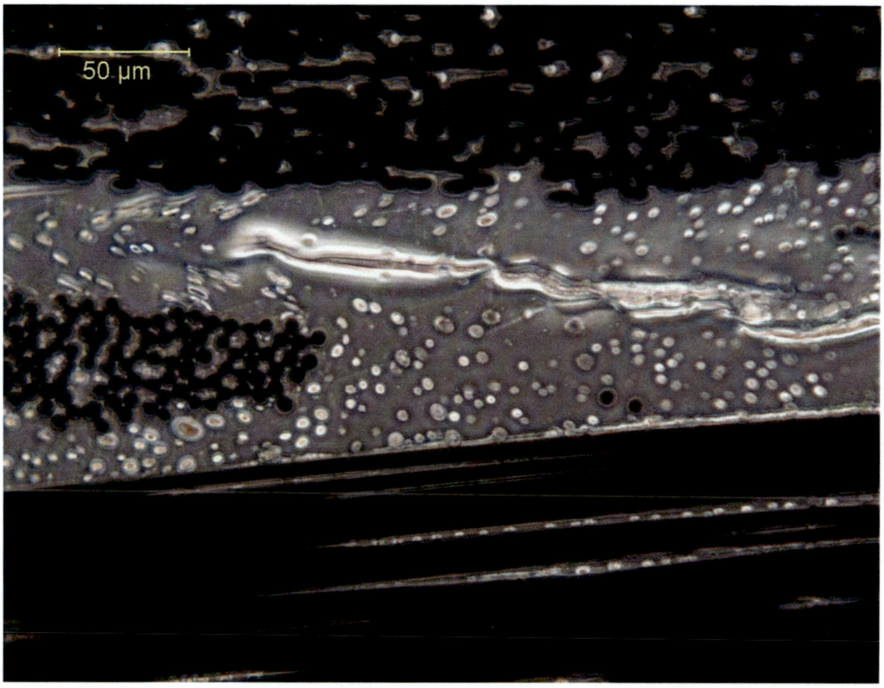

Fig. 11.11 Micrograph of crack propagation through a dispersed-phase, rubber-toughened thermoset-matrix composite after impact. Transmitted-light phase contrast, 40× objective

method for improving damage tolerance while providing an overall excellent balance of mechanical properties. This engineering approach to toughening composites focuses the particles in the highly stressed interlayer region. The impact damage of a particle-modified, interlayer-toughened composite material is shown in Fig. 11.12(a and b). In this composite system, the interlayer particles initiated cracking and plastic deformation of the matrix in the interlayer region at 45 degrees to the shear plane. Some of the cracks were found to propagate into the intraply region of the composite. Particles were also found deformed and torn where the crack propagated throughout the interlayer region (Ref 17). In localized areas, the particles were also found to elongate in the 45-degree direction of the cracking and to bridge the cracks. High strain birefringence was observed in some areas of the interlayer and was found more in the particles than in the matrix. There was no fiber-matrix interfacial failure found in any area of this composite.

The impact damage of another interlayer-modified composite is shown in Fig. 11.13. Different from the previous composite, the particles (phases) in the interlayer region are much larger and more irregular in shape. In contrast to the previous figure, the addition of the particles to the interlayer did not initiate microcracking, and most of the particle appeared unaffected by the fracture. These particles initiated hackle formation at their

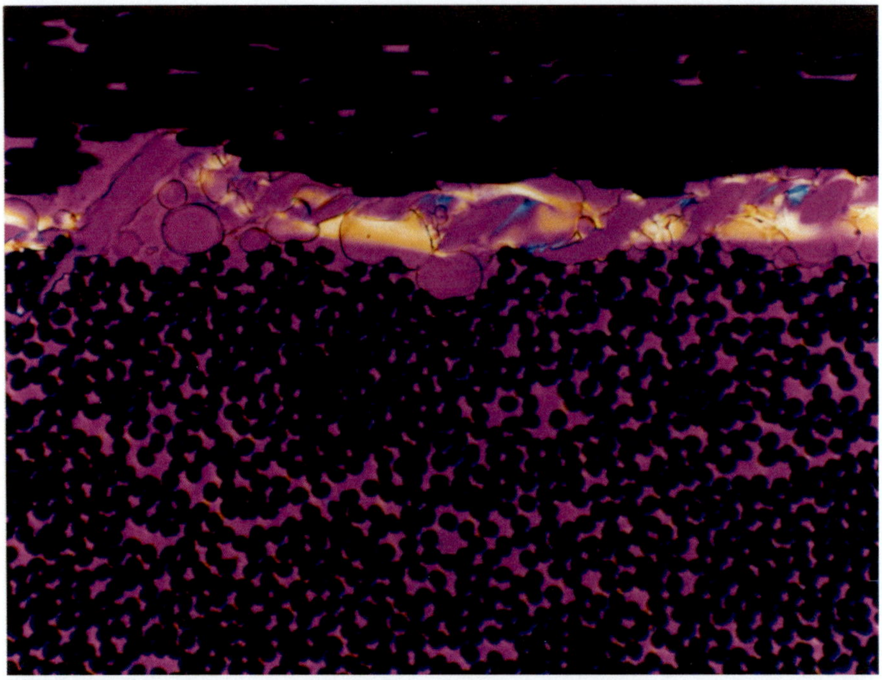

(a)

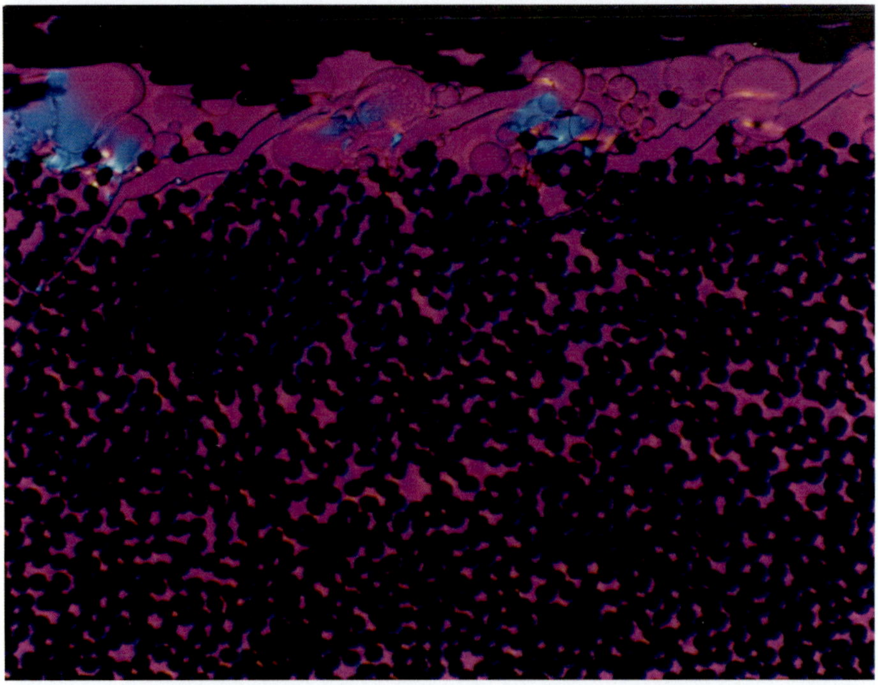

(b)

Fig. 11.12 Fracture morphology in a particle interlayer-toughened thermoset-matrix composite. (a) Strain birefringence in the interlayer particles. Transmitted polarized light, 20× objective. (b) Some of the particles are found to bridge the formed cracks, and some particles are torn. Transmitted polarized light, full wave plate, 20× objective

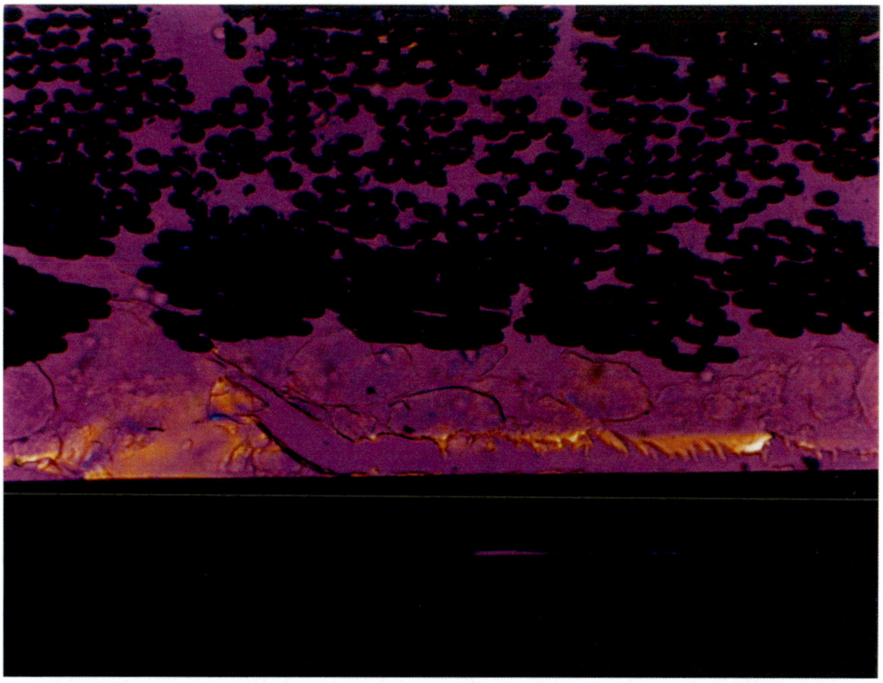

Fig. 11.13 Fiber-matrix interfacial failure in an interlayer-toughened thermoset-matrix composite. Transmitted polarized light, full wave plate, 20× objective

interface with the matrix resin. In some areas, the crack was found to propagate through the particles. Overall, the failure of the composite occurred at the fiber-matrix interface.

REFERENCES

1. P.O. Sjoblom, J.T. Hartness, and T.M. Cordell, On Low-Velocity Impact Testing of Composite Materials, *J. Compos. Mater.,* Vol 22, 1988, p 30–52
2. I.-I.T. Wu and G.S. Springer, Measurements of Matrix Cracking and Delamination Caused by Impact on Composite Plates, *J. Compos. Mater.,* Vol 22, 1988, p 518–532
3. D. Delfone, A. Poursartip, B.R. Coxon, and E.F. Dost, Non-Penetrating Impact Behavior of CFRP at Low and Intermediate Velocities, *Composites Materials: Fatigue and Fracture,* Vol 5, R.H. Martin, Ed., STP 1230, American Society for Testing and Materials, Philadelphia, PA, 1995, p 333–350
4. R.C. Madan, Influence of Low-Velocity Impact on Composite Structures, *Composites Materials: Fatigue and Fracture,* Vol 3, T.K. O'Brien, Ed., STP 1110, American Society for Testing and Materials, Philadelphia, PA, 1991, p 457–475

5. G. Zhou, The Use of Experimentally-Determined Impact Force as a Damage Measure in Impact Damage Resistance and Tolerance of Composite Structures, *Compos. Struct.,* Vol 42, 1998, p 375–382
6. L. Reis and M. de Freitas, Damage Growth Analysis of Low Velocity Impacted Composite Panels, *Compos. Struct.,* Vol 38 (No. 1–4), 1997, p 509–515
7. M. de Freitas and L. Reis, Failure Mechanisms on Composite Specimens Subjected to Compression after Impact, *Compos. Struct.,* Vol 42, 1998, p 365–373
8. C. Soutis and P.T. Curtis, Prediction of the Post-Impact Compressive Strength of CFRP Laminated Composites, *Compos. Sci. Technol.,* Vol 56, 1996, p 677–684
9. A. Hitchen and R.M.J. Kemp, The Effect of Stacking Sequence on Impact Damage in a Carbon Fibre/Epoxy Composite, *Composites,* Vol 26, 1995, p 207
10. E.G. Guynn and T.K. O'Brien, The Influence of Lay-Up and Thickness on Composite Impact Damage and Compression Strength, *Proc. 26th Structures, Structural Dynamics, Materials Conf.,* April 1985 (Orlando, FL), p 187–196
11. A. Sjogren, A. Krasnikovs, and J. Varna, Experimental Determination of Elastic Properties of Impact Damage in Carbon Fibre/Epoxy Laminates, *Compos. Part A: Appl. Sci. Manuf.,* Vol 32, 2001, p 1237–1242
12. M.W. Davidson, "The Berek Compensator," Polarized Light Microscopy, Olympus Microscopy Resource Center, http://www.olympusmicro.com/primer/techniques/polarized/berekcompensator.html
13. A.G. Evans et al., Mechanisms of Toughening in Rubber Toughened Polymers, *Acta Metall.,* Vol 34 (No. 1), 1986, p 79–87
14. Y. Huang and A.J. Kinloch, Modelling of the Toughening Mechanisms in Rubber-Modified Epoxy Polymers, Part II: A Quantitative Description of the Microstructure-Fracture Property Relationships, *J. Mater. Sci.,* Vol 27, 1992, p 2763–2769
15. A.J. Kinloch et al., Deformation and Fracture Behavior of a Rubber-Toughened Epoxy, Part 1: Microstructure and Fracture Studies, *Polymer,* Vol 24, 1983, p 1341–1354
16. S. Kunz-Douglass, P.W.R. Beaumont, and M.F. Ashby, A Model for the Toughness of Epoxy-Rubber Particulate Composites, *J. Mater. Sci.,* Vol 15, 1980, p 1109–1123
17. M.R. Groleau, Y.-B. Shi, A.F. Yee, J.L. Bertram, H.J. Sue, and P.C. Yang, Mode II Fracture of Composites Interlayered with Nylon Particles, *Compos. Sci. Technol.,* Vol 56, 1996, p 1223–1240

CHAPTER 12

Matrix Microstructural Analysis

MICROSTRUCTURAL ANALYSIS of the composite matrix is necessary to fully understand the performance of the part and its long-term durability. The microstructure of the matrix is dictated by the matrix component(s) and, in many cases, is influenced by the cure cycle. It is commonly thought that the microstructure of thermoplastic matrices is affected more by the cure cycle than the microstructure of thermosetting matrices. This is due to the semicrystalline nature of many engineering thermoplastics used as matrices in high-performance composites. However, the microstructure of thermosetting matrices can also be influenced by the cure cycle, where individual components may or may not phase separate in the matrix upon cure. As discussed in Chapter 10, "Toughening Methods for Thermoset-Matrix Composites," the use of various toughening materials in thermosetting matrices may alter the matrix microstructure and can be influenced by the cure cycle. Another factor that is sometimes disregarded is the fiber influence on the matrix microstructure in composite materials. In the following sections, the microstructural analysis of some engineering thermoplastic-matrix composites is discussed, followed by the analysis of a new bio-based thermosetting-matrix natural fiber composite system.

Crystalline Microstructures of Thermoplastic-Matrix Composites

It is well known that the crystallinity of thermoplastic-matrix fiber-reinforced composites is affected by the rate of cooling after being consolidated at high temperature and pressure. In this regard, thermoplastic matrices have characteristics similar to metals. For thermoplastic-matrix composites to have repeatable mechanical properties, the degree of crystallinity must be consistent throughout and for each part. This becomes an even greater challenge when trying to manufacture thicker parts in which skin-core effects can be present due to differences in heat transfer. The crystallinity that develops in thermoplastics depends on many factors, including the polymer backbone, structure, molecular weight, polydispersity, and secondary bonding (hydrogen, van der Waals) (Ref 1). This is further complicated by the addition of a reinforcing phase (Ref 2). The addition of discontinuous or continuous fibers and the volume fraction of the fibers affect the crystallinity in the matrix. In addition, differences in fiber surface roughness, size, surface activation, and sizing affect the nucleation and growth of crystals in many thermoplastic-matrix composites. While the degree of crystallinity can be determined from differential scanning calorimetry, x-ray analysis, or density measurements, these analytical methods provide little, if any, information on the origin and microstructure of the crystallinity in the composite. To determine the morphology of thermoplastic-matrix composites, transmitted polarized-light microscopy can be used very effectively.

The preparation of thermoplastic-matrix composites for optical microscopic analysis requires the creation of ultrathin sections so that transmitted polarized light can be used. Throughout the literature, this has often been performed using a microtome or similar instrument. However, this method often leads to destruction of the morphology and artifacts in the sample. In contrast, proper thin-section development through grinding and polishing alleviates these phenomena when performed correctly. The details of this preparation technique were described in Chapter 6, "Thin-Section Preparation and Transmitted-Light Microscopy," in this book. The thin sections developed in these studies have been feathered from as thick as 5 μm to less than 1 μm (1 μm sample thickness is optimal). Controlling the specimen thickness is essential to resolve spherulites, which can be stacked behind each other in thick sections, and also to avoid "ghostlike" images due to lack of refractive material in the thin sections. The thickness of the thin section may be determined geometrically, as indicated in Fig. 12.1, if problems develop with the analysis.

Cooling-Rate Effects on Thermoplastic-Matrix Crystallinity

The effects of cooling rate on the development of spherulites in a high-temperature thermoplastic-matrix carbon-fiber-reinforced composite material are shown in Fig. 12.2(a to c). In Fig. 12.2(a), it can be observed that the spherulites are significantly larger than those in Fig. 12.2(b). The reduction in size of the spherulites in Fig. 12.2(b) is a result of a faster cooling rate after consolidation of the thermoplastic composite, along with subsequent annealing. Although the spherulites are smaller in Fig. 12.2(b), there was no difference in the degree of crystallinity of the two materials as a result of the different cooling rates. When the sample was quenched

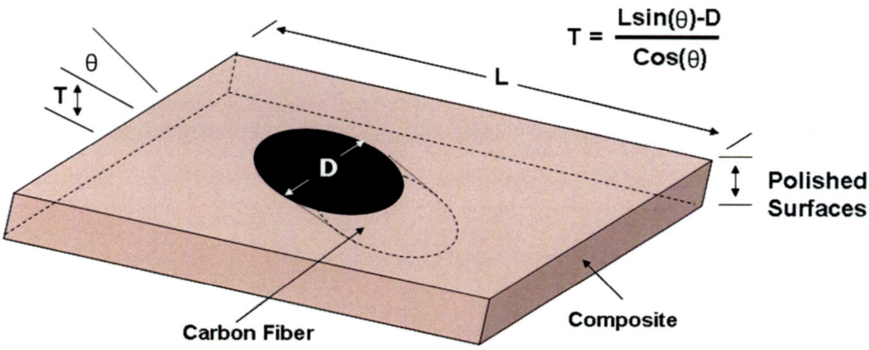

Fig. 12.1 Schematic of a method used to determine the thickness of an ultrathin section

rapidly from its processing temperature (Fig. 12.2c) and not further annealed, only local areas of crystallinity were formed. The fast quenching creates a thermodynamically unfavorable environment for the development of crystals and therefore results in a partially amorphous matrix. As shown in Fig. 12.2(c), there is still some crystallinity due to the inability to quench the material rapidly enough, due to the relatively low thermal conductivity of the composite material as well as the high propensity for this type of thermoplastic to form crystals. Control of the degree of crystallinity and the morphology and size of the crystals is necessary so that mechanical and physical properties can be accurately determined and so they remain consistent (Ref 3 to 5).

In the aforementioned example, the rapid quenching of the thermoplastic-matrix composite did not completely inhibit the formation of crystallinity in the matrix. This was due in part to the rate at which the material was cooled but also the crystallization kinetics of this type of thermoplastic. The effect of cooling on the degree of crystallinity varies for different semicrystalline thermoplastic materials. A comparison of the effect of slow versus fast cooling rates on an engineering thermoplastic glass-fiber-reinforced composite is shown in Fig. 12.3(a and b), respectively. The composite that was subjected to a slow cooling rate formed a high degree of crystallinity in the matrix, while no crystalline regions were found in the matrix of the composite that was rapidly cooled. As a result of the fast cooling rate coupled with the hot-press consolidation, there were areas of residual stresses formed in the part along the fiber-matrix interface, as can be seen in Fig. 12.3(b).

Fiber Nucleation of Spherulite Crystal Growth

The crystallinity in thermoplastic-matrix composites is affected by the fibers, because the fiber surfaces present favorable nucleation sites for crystal growth. Impurities such as dust and free particulates can also nucleate crystals. Figure 12.4(a to c) shows the effect of fiber nucleation on

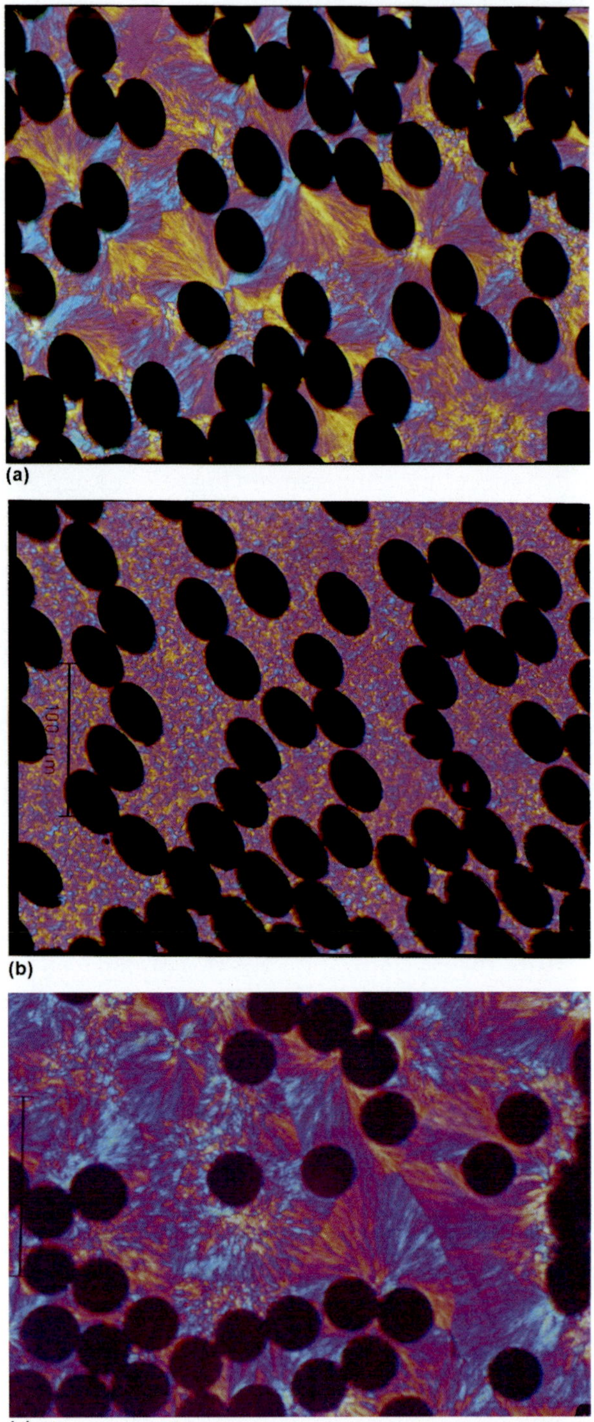

Fig. 12.2 Micrographs taken from ultrathin sections of a high-temperature thermoplastic-matrix carbon fiber composite that were exposed to different cooling rates. (a) Slow cooled. (b) Fast cooled followed by annealing. (c) Fast cooled. Transmitted polarized light, 100× objective

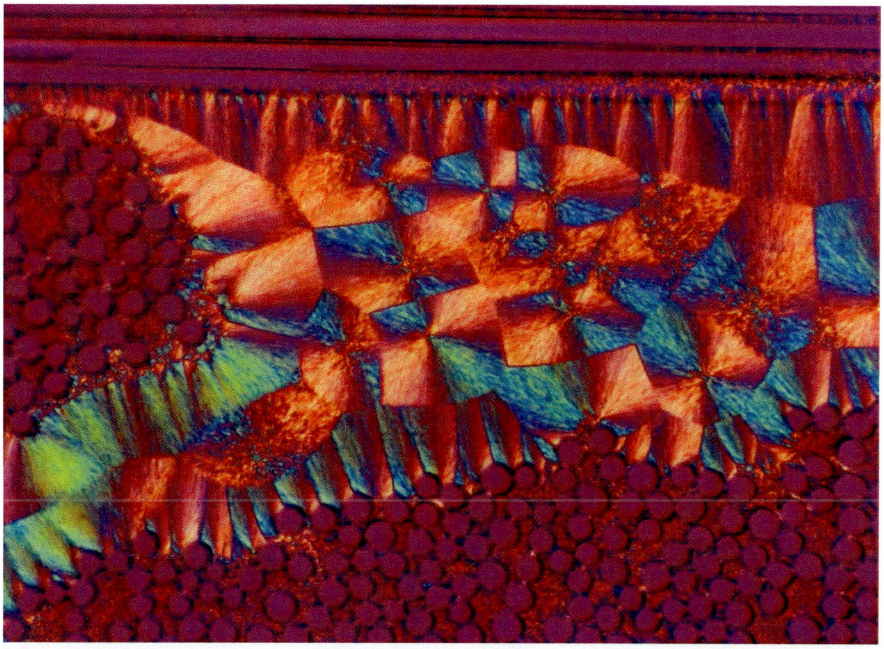

(a)

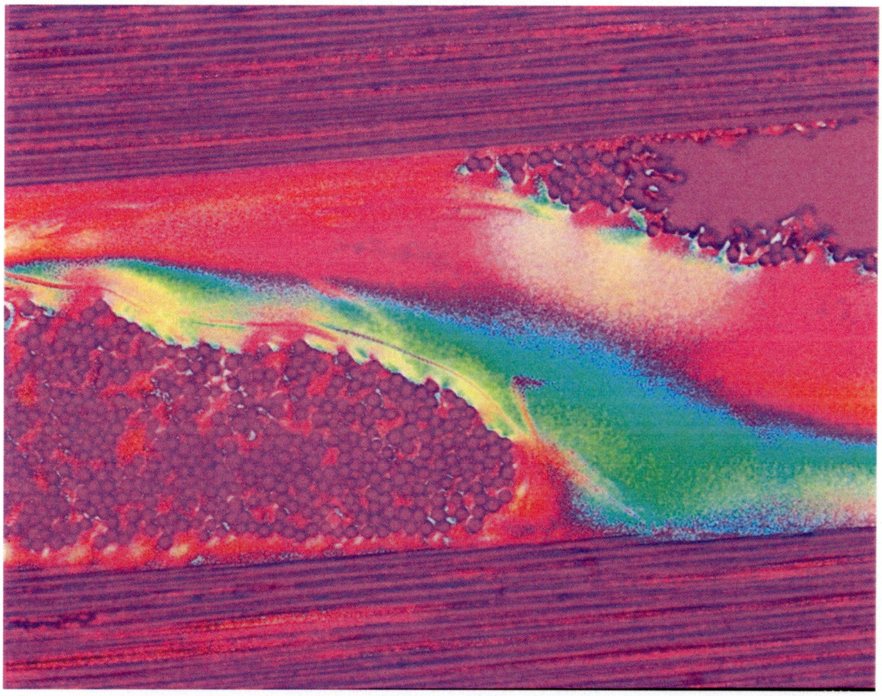

(b)

Fig. 12.3 Matrix morphology differences of an engineering thermoplastic glass fiber composite that was exposed to different cooling rates. (a) Slow cooling rate. (b) Quenched to room temperature. Micrographs were taken from ultrathin sections. Transmitted polarized light, 40× objective

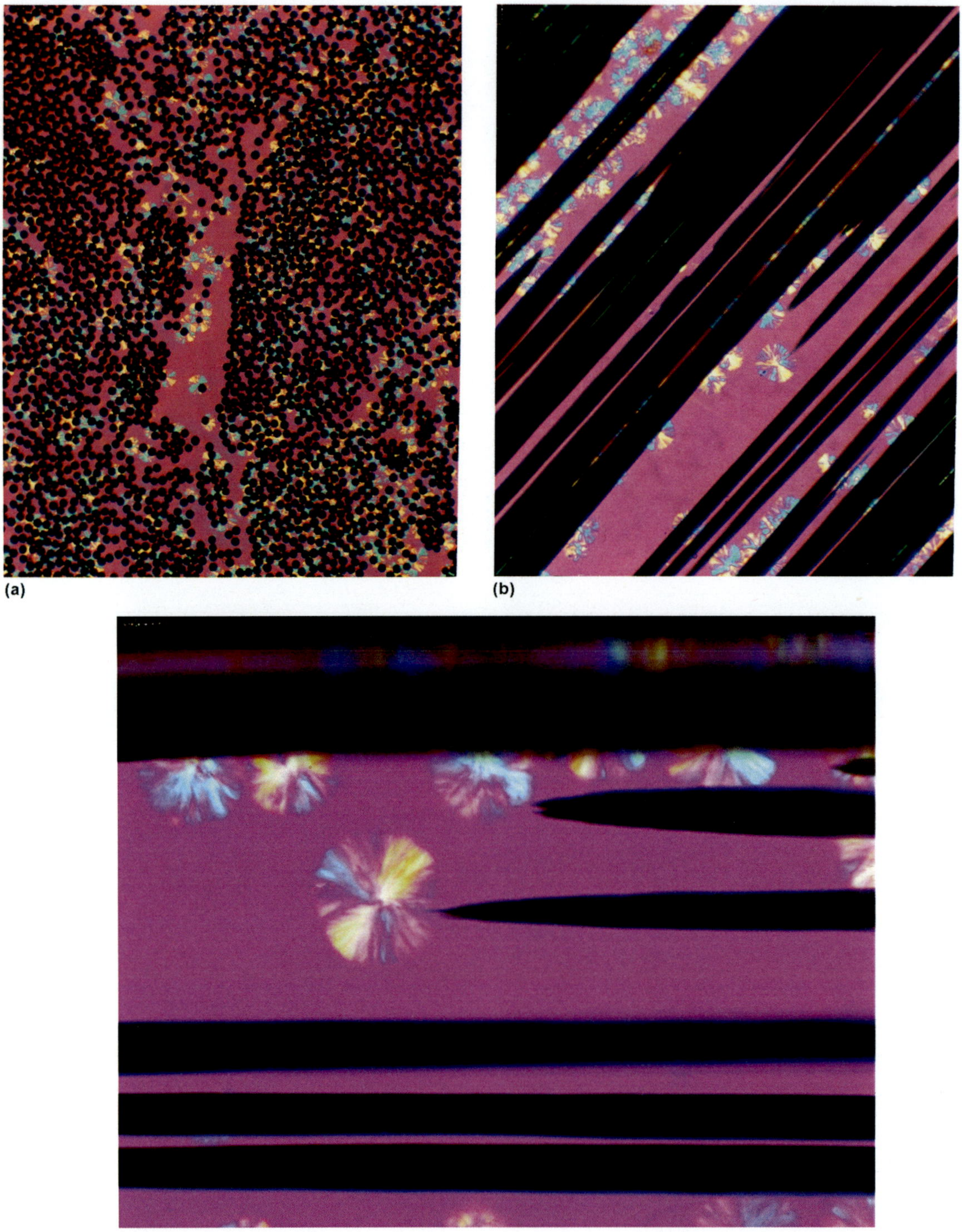

(a) (b) (c)

Fig. 12.4 Fiber nucleation of spherulitic crystal growth in a high-temperature, lightly cross-linked thermoplastic-matrix composite. Micrographs were taken from ultrathin sections of the unidirectional carbon fiber composite. (a) Sectioned through the thickness and perpendicular to the fiber plane. Transmitted polarized light, 10× objective. (b) Sectioned through the thickness and parallel to the fiber plane. Transmitted polarized light, 40× objective. (c) Close-up view of a spherulite that nucleated at the tip of a carbon fiber. Transmitted polarized light, 100× objective

the development of spherulites in a lightly cross-linked, high-temperature thermoplastic-matrix carbon fiber composite. The chemistry of this matrix does not allow for a high degree of crystallinity and therefore provides an excellent example of the nucleation effect of fibers on the development of crystals. Figure 12.4(a) shows a cross section of the composite material taken perpendicular to the fiber plane, as compared to Fig. 12.4(b), which shows a cross section of the composite material taken parallel to the fiber plane. A higher-magnification view of Fig. 12.4(b) is shown in Fig. 12.4(c), highlighting the nucleation of a spherulite at the tip of a carbon fiber. Cross sectioning a composite through the thickness in both the *xz* and *yz* planes of the composite can provide a better understanding of the morphology and location of crystals. If a composite is only sectioned parallel to the fibers through the thickness (Fig. 12.4b), it may be difficult to see if there are spherulites located in the fiber tows and what the distribution of the spherulites is in the composite material. Because the thin section is less than a fiber diameter, the determination of what is in the bulk of the composite is hard to determine. Therefore, sectioning the composite perpendicular to the fiber plane, which provides an end view of the fiber tows, may be necessary to see the distribution of crystals in the composite. Both views are usually required to determine the crystal morphology and location.

Natural Fiber and Resin Composites

The use of fiber-reinforced polymeric composite materials continues to increase throughout the world due to their unique performance attributes. These heterogeneous and anisotropic materials are commonly developed from carbon or glass fibers combined with a petroleum-based polymeric matrix. In recent years, composites made from renewable resources have received increasing attention due to environmental concerns. Rising petroleum costs and consumption has increased the demand for renewable and sustainable alternatives. In response to this effort, the use of natural fibers has been found to be an effective alternative to glass fibers for some composite applications. Some of the most widely used natural fibers for composites are bast and leaf fibers. Bast fibers include hemp, jute, flax, ramie, and kenaf, while leaf fibers include sisal, banana, and pineapple fibers (Ref 6 to 8). As concerns for the environment continue, these types of fibers are finding more application in advanced composites. Recently, the evolution of "green composites" has been taken a step further to include not only natural fibers but high-performance thermosetting matrices made from renewable resources. In this study, a low-temperature-curing matrix was developed from renewable compounds found in sugars, limes, vegetable oils, and nuts. Optical microscopy techniques were used to gain insight into the morphology of this bamboo fiber composite system and the matrix-fiber interactions.

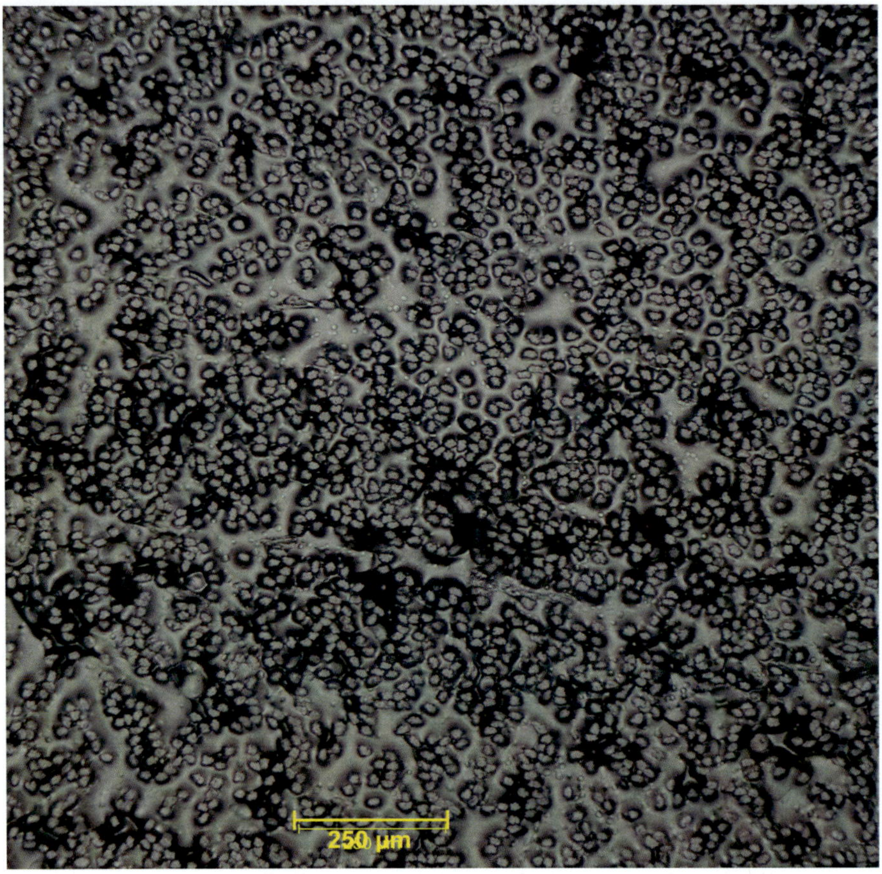

Fig. 12.5 Micrograph of a cross section of a bamboo fiber composite. Bright-field illumination, 10× objective

In Fig. 12.5, there are many areas in the composite that were not impregnated by the resin matrix during the manufacturing process, as seen by dark areas (voids) in the micrograph. The bamboo fibers are randomly dispersed throughout the matrix but are also found clustered in many areas. With higher magnification (Fig. 12.6) and the use of bright-field illumination, the bamboo fibers are found to be irregular in shape. Also in this micrograph, the appearance of a dispersed phase is indicated by light, circular areas. Other features that are even more complex, such as connected phases, are also visible. Dark or shaded areas surrounding the fibers are more easily observed in this higher-magnification micrograph.

The use of phase contrast, which is not commonly performed using reflected-light microscopy, provided more insight into the fiber and matrix microstructure (Fig. 12.7). This contrast technique highlighted the dispersed phase and also more clearly revealed shaded areas around the fibers due to height differences on the sample surface. The height differences on

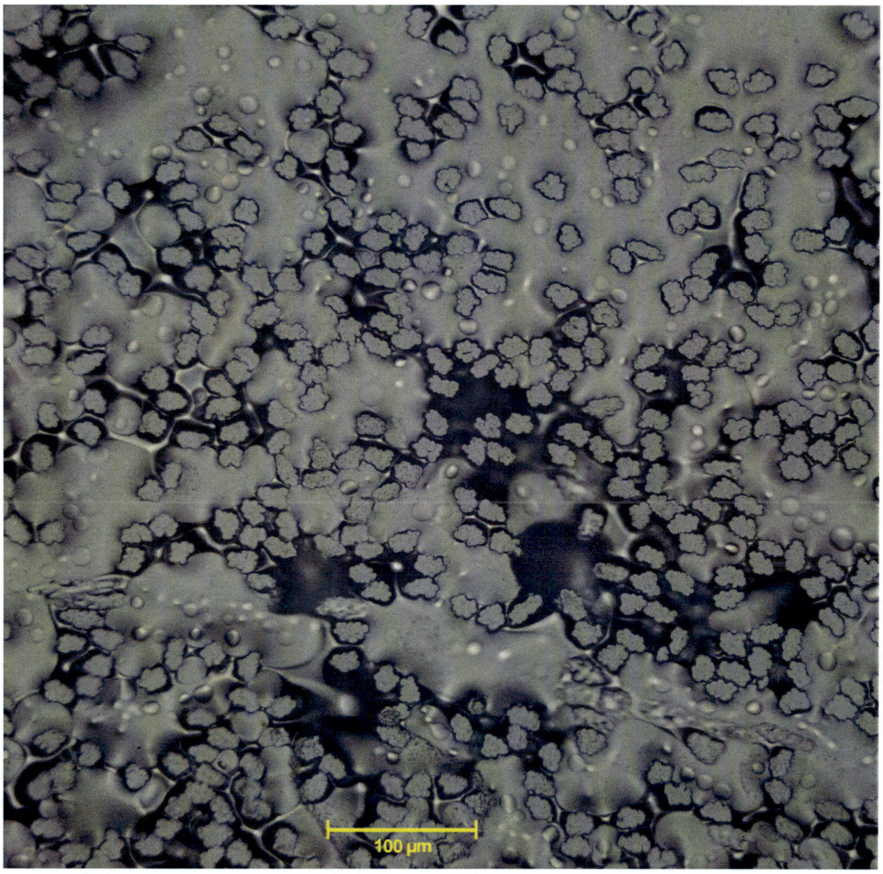

Fig. 12.6 Higher-magnification view of the same area as Fig. 12.5. Bright-field illumination, 25× objective

the sample surface are due to uneven sample removal during the polishing process as a result of different material properties in the interphase area compared to the bulk matrix.

An ultrathin section was prepared from the sample so that transmitted-light optical microscopy could be used to gain further insight into the morphology. By using transmitted-light optical microscopy, the differences in the refractive indices of the phases provide a more detailed analysis of the microstructure. Due to the low contrast of this natural fiber composite, carbon fiber composite laminates were placed at the outer edges of the mount so they could be used as a gage during the thin-section-development procedure. This ultrathin section was less than 1.5 µm thick after it was completed. Transmitted-light optical microscopy revealed that many of the individual fibers were not wetted by the matrix, as shown by the absence of resin in the crevasses on the fiber surface (Fig. 12.8). This appears in the micrograph as individual fibers encircled by resin. Also, lines

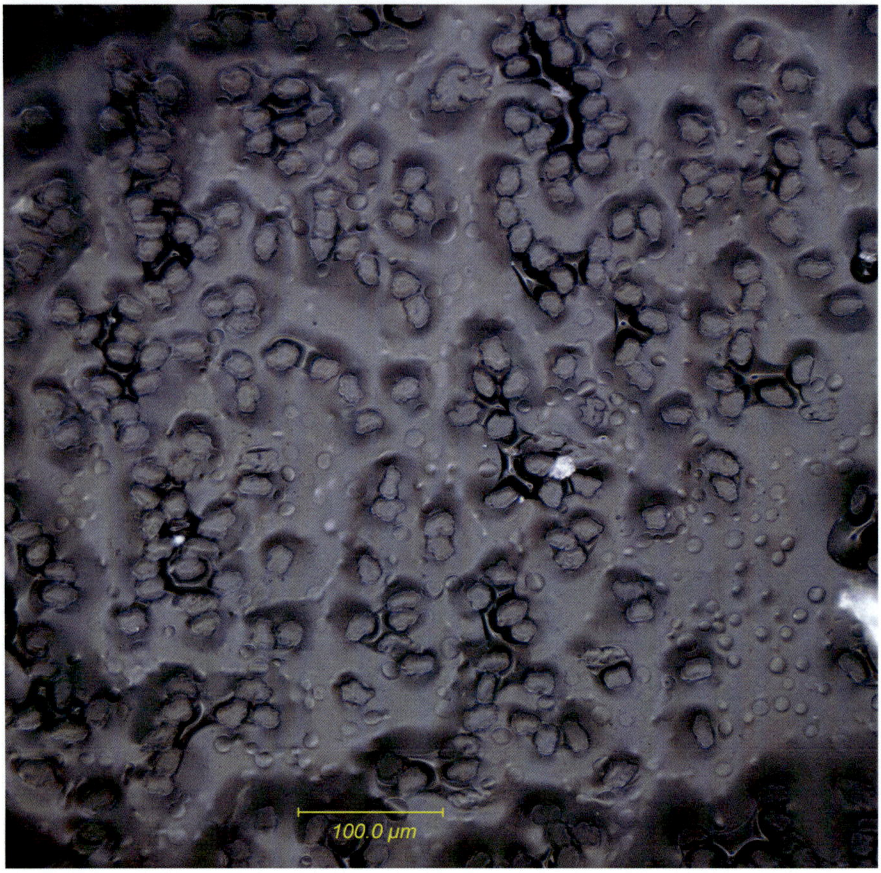

Fig. 12.7 An area in Fig. 12.5 viewed using reflected-light phase contrast with the same magnification as Fig. 12.6. Notice the shading on the outer edges of the micrograph due to this contrast technique.

appear in the matrix that were not detected using reflected-light techniques. The development of the circular dispersed phase appeared to have been affected by these lines (phase) in some areas. In these areas, the resulting circular phases are confined and irregular. Upon further examination of this ultrathin section using Hoffman modulation contrast (Fig. 12.9), it can be seen that the lines are raised, which corresponds to the white lines shown in Fig. 12.8. The raised lines are a harder phase, where the dispersed circular phase is a softer phase. These lines may be a result of one or more of the matrix components curing at different rates or may be due to regions of differing residual stress.

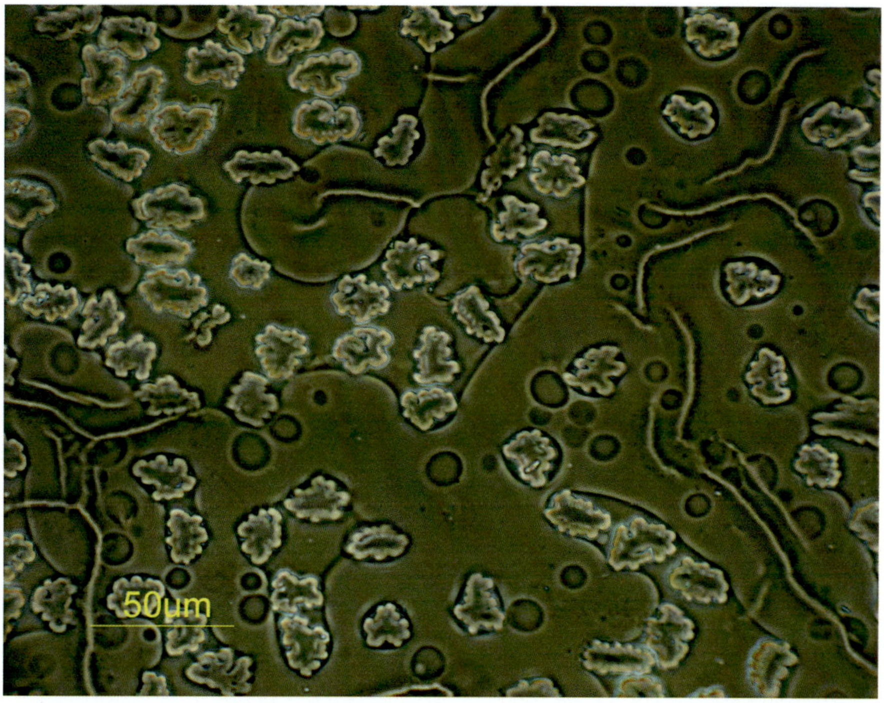

Fig. 12.8 Transmitted-light microscopy of the natural fiber composite ultrathin section. Phase contrast, 40× objective

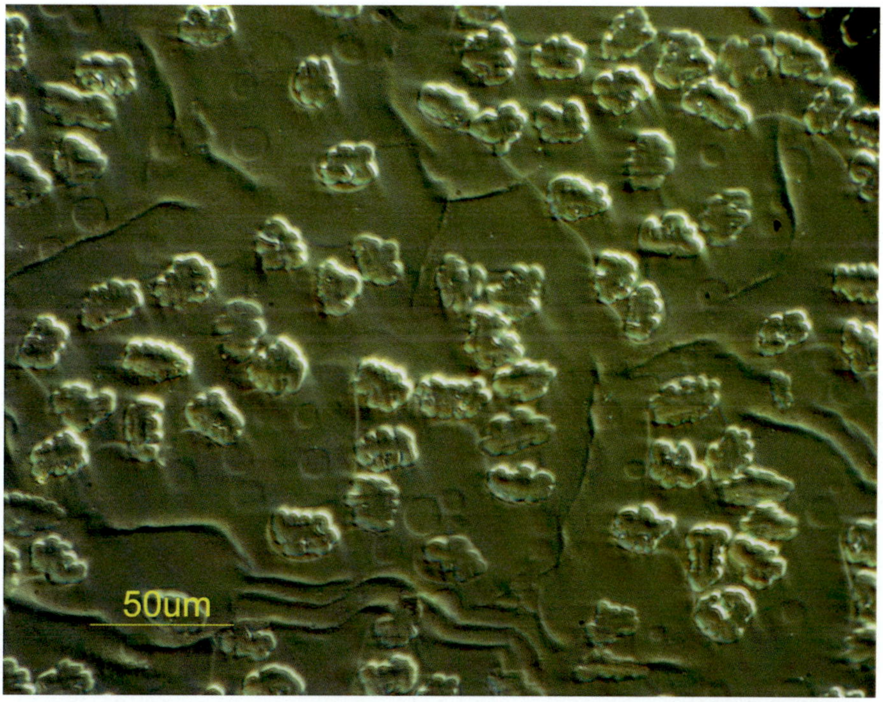

Fig. 12.9 Transmitted-light microscopy of the natural fiber composite ultrathin section. Hoffman modulation contrast, 40× objective

REFERENCES

1. F.W. Billmeyer, Ch 5, Morphology and Order in Crystalline Polymers, *Textbook of Polymer Science,* 2nd ed., John Wiley and Sons, Inc., New York, 1971
2. C. Auer, G. Kalinka, T. Krause, and G. Hinrichsen, Crystallization Kinetics of Pure and Fiber-Reinforced Poly(Phenylene Sulfide), *J. Appl. Polym. Sci.,* Vol 51 (No. 3), 1994, p 407–413
3. S.-L. Gao and J.-K. Kim, Correlation among Crystalline Morphology of PEEK, Interface Bond Strength, and In-Plane Mechanical Properties of Carbon/PEEK Composites, *J. Appl. Polym. Sci.,* Vol 84, 2002, p 1155–1167
4. D. Lu, Y. Yang, G. Zhuang, Y. Zhang, and B. Li, A Study of High-Impact Poly(Phenylene Sulfide), Part 1: The Effect of Its Crystallinity on Its Impact Properties, *Macromol. Chem. Phys.,* Vol 202, 2001, p 734–738
5. J. Cao and L. Chen, Effect of Thermal Cycling on Carbon Fiber-Reinforced PPS Composites, *Polym. Compos.,* Vol 26, 2005, p 713–716
6. G.I. Williams and R.P. Wool, Composites from Natural Fibers and Soy Oil Resins, *Appl. Compos. Mater.,* Vol 7, 2000, p 421–432
7. S. Taj, M.A. Munawar, and S. Khan, Review: Natural Fiber-Reinforced Polymer Composites, *Proc. Pakistan Acad. Sci.,* Vol 44 (No. 2), 2007, p 129–144
8. P.A. Fowler, J.M. Hughes, and R.M. Elias, Review: Biocomposites—Technology, Environmental Credentials and Market Forces, *J. Sci. Food Agric.,* Vol 86, 2006, p 1781–1789

CHAPTER 13

Honeycomb-Cored Sandwich Structure Composites

THE HONEYCOMB SANDWICH STRUCTURE COMPOSITE is a very efficient and complex structure widely used in the aircraft industry (Ref 1, 2). Honeycomb-cored sandwich panels increase part stiffness at a lower weight than monolithic composite materials. By imitating the natural geometric structure of a beehive, the honeycomb core imparts strength and light weight to sandwich panels, while supporting the prepreg skins. The honeycomb sandwich structure composite has high compressive strength in the direction of the cell walls and high shear strength in the plane perpendicular to the cell walls (Ref 3). Perpendicular to the cell walls, the honeycomb core has low strength but is reinforced by the outer composite skins. Commonly, the bond between the honeycomb core and the prepreg skins is created by a film adhesive layer. A surfacing film is often co-cured with the composite sandwich panel to improve the appearance of the part and to provide a smooth, uniform, and nonporous surface.

Film Adhesive/Prepreg Resin Flow and Intermingling

The differences in the morphology of the prepreg, adhesive film, and surfacing film of a honeycomb-cored sandwich structure composite can be observed using optical microscopy. Furthermore, the intermingling of the different resin systems can be identified, which can help with the proper selection of materials and optimization of the processing parameters. Ultimately, this provides more insight into the composite system performance.

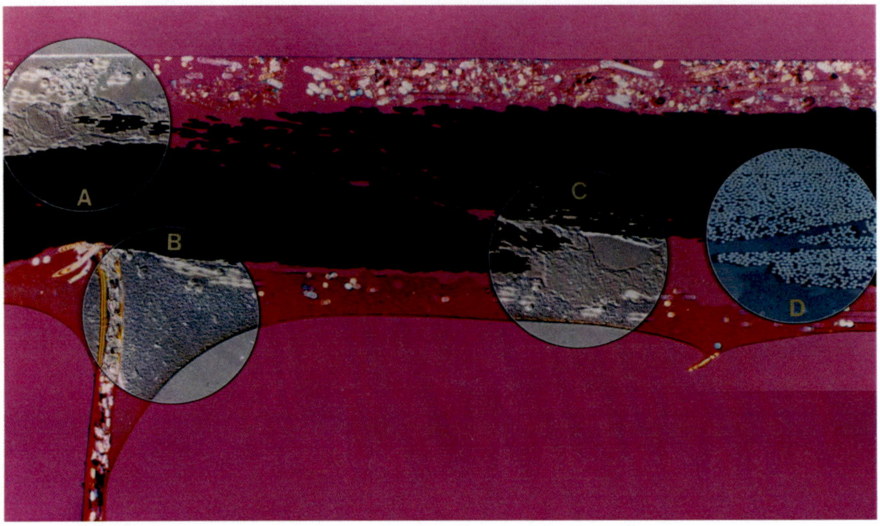

Fig. 13.1 Honeycomb sandwich structure composite cross section (~1 μm ultrathin section) showing differences in the constituents and resin intermingling. Transmitted crossed polarized light with a 530 nm compensator plate. This micrograph and the insets are expanded to 200× magnification. (A–C) Transmitted light, Hoffman modulation contrast. (D) Reflected light, bright-field illumination

To perform this analysis, an ultrathin section (0.5 to 5 μm) must be developed from the honeycomb sandwich structure composite part to be able to use all of the contrast techniques of transmitted-light microscopy.

Figure 13.1 shows an area of a honeycomb-cored sandwich structure composite cross section that is viewed using transmitted polarized light. The 3k-70 plain weave carbon fibers are visible as a black mass or short angled rods, and the epoxy resin as first-order magenta. Birefringent materials are observed throughout the structure, including randomly oriented polyester scrim fibers in the film adhesive and surfacing film, filler materials in the surfacing film, and the Nomex (E.I. du Pont de Nemours and Company) cell wall. Hoffman modulation contrast was used to reveal the subtle morphology differences of the materials used in the honeycomb sandwich structure composite (insets A to C). With this contrast mode, a three-dimensional appearance is developed, indicated by different contrast due to differences in the refractive indices of the resins. By using this technique, the resin flow and intermingling of the various materials was identified. Inset A shows the resin intermingling of the epoxy-based surfacing film with the epoxy matrix in the carbon fiber prepreg. Inset B shows the Nomex core and the phenolic resin coatings on the core cell wall, with the film adhesive fillet that bonds the core to the prepreg skins. Inset C shows the intermingling of the film adhesive with the prepreg resin. To further enhance the appearance of the fibers, a white reflective surface was placed on the back side of the ultrathin section and viewed using reflected-light bright-field illumination (inset D). This provided a three-dimensional view of the fibers and highlighted the fiber structure.

Honeycomb Core Movement and Core Crush

As mentioned previously, the honeycomb core has little strength perpendicular to the cell walls. This can cause problems in manufacturing, leading to such processing defects as honeycomb core crush or honeycomb movement during cure. When autoclave pressure is applied to consolidate the honeycomb sandwich structure composite, the honeycomb core can move or crush due to normal forces acting on chamfered areas of the part. There are three factors, in addition to the processing (autoclave) conditions of temperature ramp rate and applied pressure, that affect honeycomb movement and can result in core crush. These factors include the pressure in the honeycomb core, the combined stiffness of the skin/core before gelation and cure, and the frictional resistance of the prepreg between adjacent plies and on the tooling used for manufacture (Ref 4 to 6). The development and degree of core crush has been documented and successfully measured for different material combinations (Ref 7).

Figure 13.2(a to c) includes micrographs of areas in a honeycomb sandwich structure composite in which the honeycomb core moved during the cure cycle. An ultrathin section was developed, and transmitted light was used to document the area around the honeycomb cell walls. In this figure, the surfacing film, honeycomb cell wall, and film adhesive used to enhance the bond to the core can easily be identified, as described in Fig. 13.1. This figure documents the movement of the core as viewed from the tool side and shows the smooth surface, with surfacing film, against the tool. The deformation of the prepreg plies is evident from the core movement that occurred during the cure cycle. The prepreg plies are wavy in the vicinity of the cell walls, and separation of the prepreg plies is found throughout the thickness. This also caused areas of large voids and cracks in the composite facesheets. In this figure, areas are shown where the controlled-flow prepreg resin spanned some of the regions between the separated prepreg plies. This resin flow occurred before but close to the time when the resin gelled. Figure 13.3(a to c) highlights the areas of the resin between the adjacent prepreg plies, showing that some of these connected areas are intact (Fig. 13.3a), while others have been broken (Fig. 13.3b, c). The broken resin tendon areas are a result of increased ply separation due to additional core movement after the gelation of the resin. During this process, the cure has not progressed far enough for the honeycomb structure to have adequate strength to limit core movement, and therefore, ply separation exceeds the tensile strain of the resin. While this figure documents an extreme case of honeycomb core movement, even slight move-

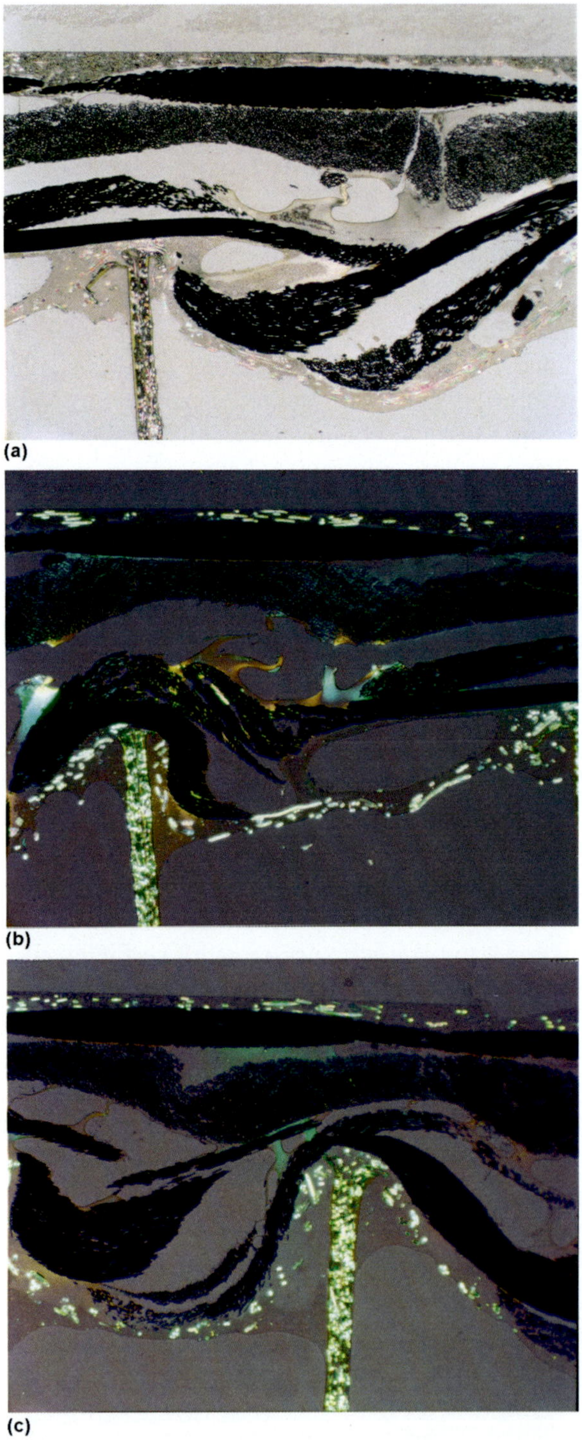

Fig. 13.2 Ultrathin section of an area of a honeycomb sandwich composite structure showing the effects of core movement during manufacturing and the resulting deformation and separation of the prepreg plies. (a) Transmitted crossed polarized light, 20× objective. (b and c) Transmitted polarized light, quarter-wave plate, 20× objective

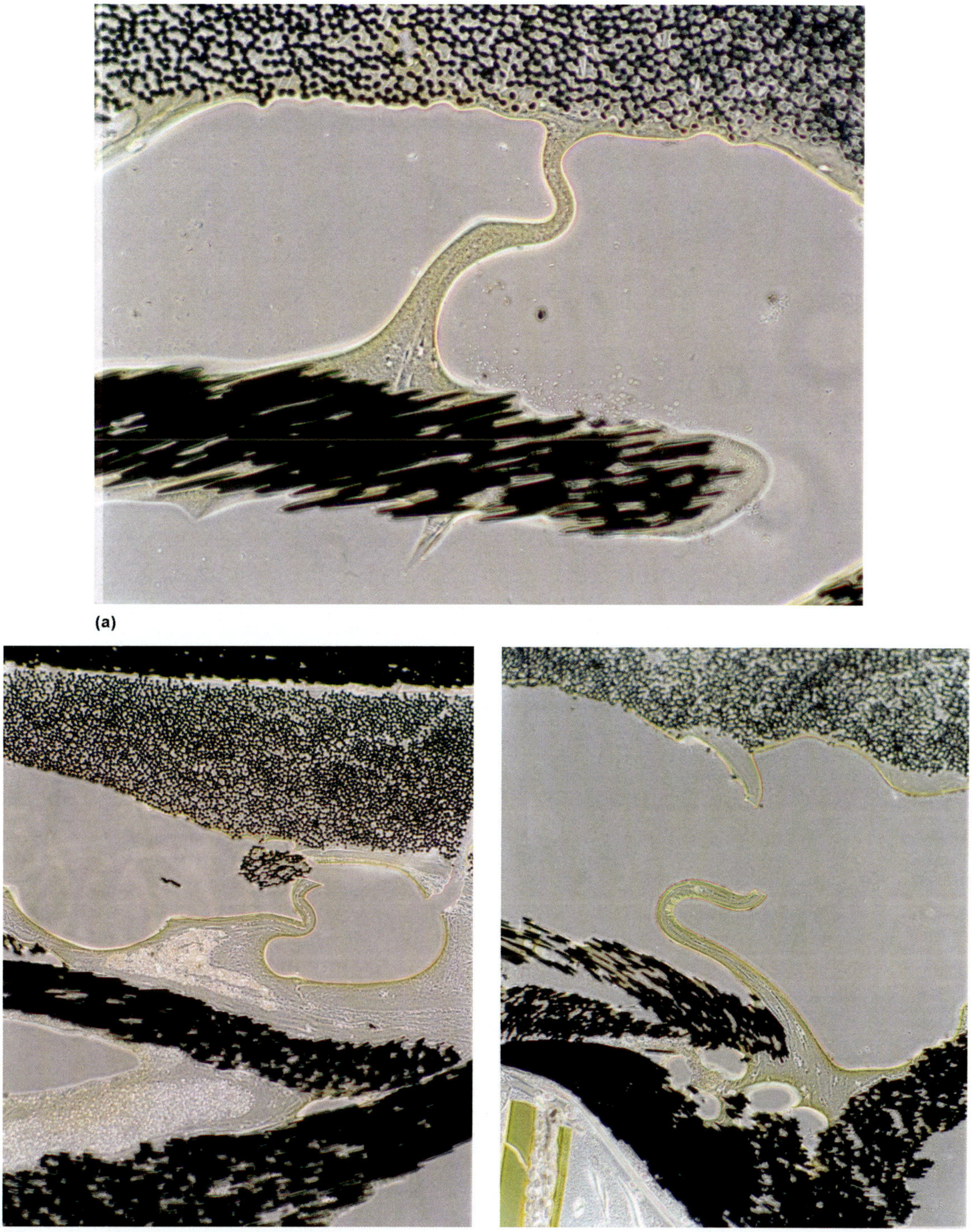

(a) (b) (c)

Fig. 13.3 Ultrathin section of areas of the honeycomb sandwich composite structure where resin was found to span the separated prepreg plies after core movement. (a) Transmitted light, phase contrast, 40× objective. (b and c) Transmitted light, phase contrast, 20× objective

ment can lead to significant alterations to the prepreg skins, reducing the overall strength and integrity of the composite part.

Void Content in Honeycomb Composites

The pressure that can be applied to a honeycomb sandwich structure composite during the cure cycle is limited by the core crush potential of the assembly. Usually, a much lower autoclave pressure is used to cure honeycomb-cored composite structures compared to composite laminates. This lower applied pressure and the nature of the honeycomb core does not allow good translation of autoclave pressure to resin pressure. As a result, the void content may be greater in prepreg skins in a honeycomb structure composite compared to composite laminates. To further aggravate this process, the core often is under vacuum due to the vacuum debulking and compaction procedure. This increases the propensity for voids, especially in the fillet areas where the core may be under vacuum and there is very little compaction pressure. The use of solvent-manufactured prepreg materials is even more susceptible to void formation, because volatile evolution of the residual solvent in the prepreg is more difficult to suppress (Ref 8). Figure 13.4(a) shows a cross section of a self-adhesive solvent-manufactured prepreg system/honeycomb structure composite that has many small voids in the facesheet and fillet area. These small voids are due to the solvent vapor pressure exceeding the resin pressure before gelation and cure (Ref 9, 10). In Fig. 13.4(b and c), large voids are shown in the fillet areas for the same honeycomb sandwich structure composite. The fillet area is very susceptible to void formation if solvents are contained in the prepreg system used against the honeycomb core, because there is little to no resin pressure, and the core is most likely under vacuum.

Honeycomb Core Failure

The skin-to-core bond plays a crucial role in the performance of a honeycomb sandwich composite structure. In some systems, the prepreg matrix has self-adhesive characteristics that enable adequate skin-core bonds, while other prepreg systems require the use of a film adhesive to provide the necessary bond. Some common tests performed on honeycomb composites are climbing drum peel and flatwise tensile tests to characterize the skin-to-core bond strength. After testing the composites, optical microscopy can be employed to observe how and where failure occurred. A plane view of the prepreg/film adhesive interface and separated core can be examined unprepared and can provide insight into the failure. In Fig. 13.5, it can be observed where the Nomex honeycomb core detached from the composite/film adhesive during climbing drum peel testing of the sandwich structure composite. The adhesive surface shows areas where the

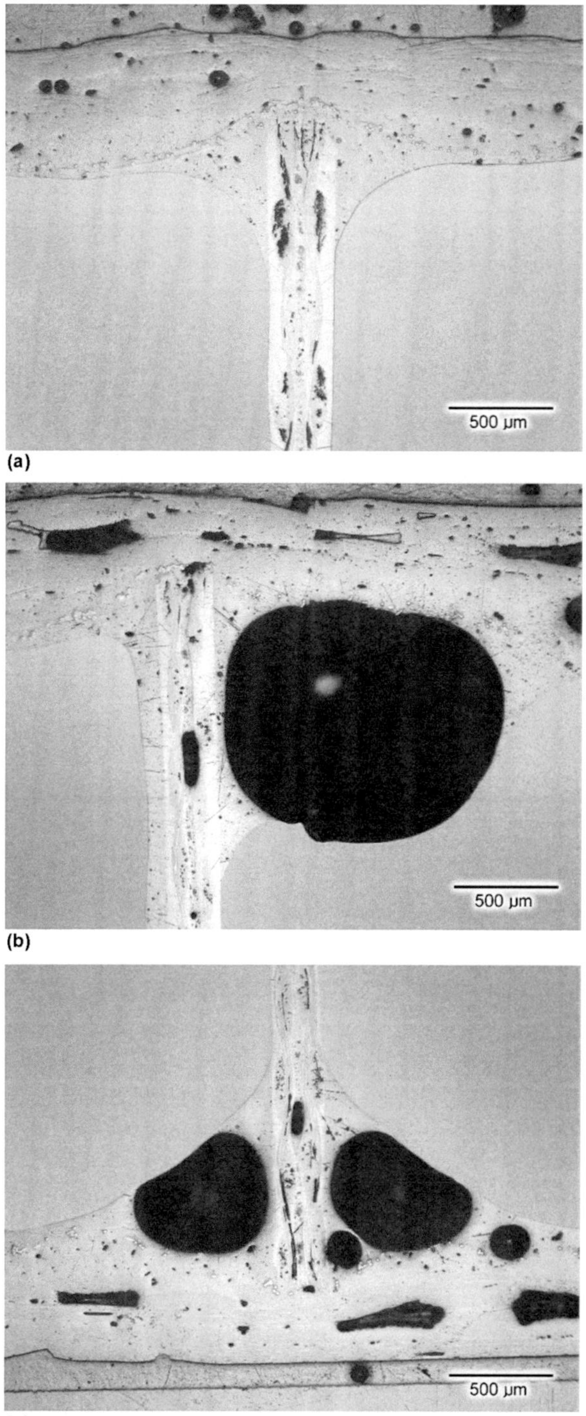

Fig. 13.4 Solvent-generated voids in the prepreg skins and fillet areas of a honeycomb sandwich structure composite. (a and b) Bag side. (c) Tool side. Epi-bright-field illumination, 5× objective. In these micrographs, there is evidence of some scratching on the polished surface. This is due to the high void content that entrapped polishing particles, which are carried to the next step. In some cases, even the use of ultrasonic cleaning cannot remove all of the entrapped polishing particles from the previous step.

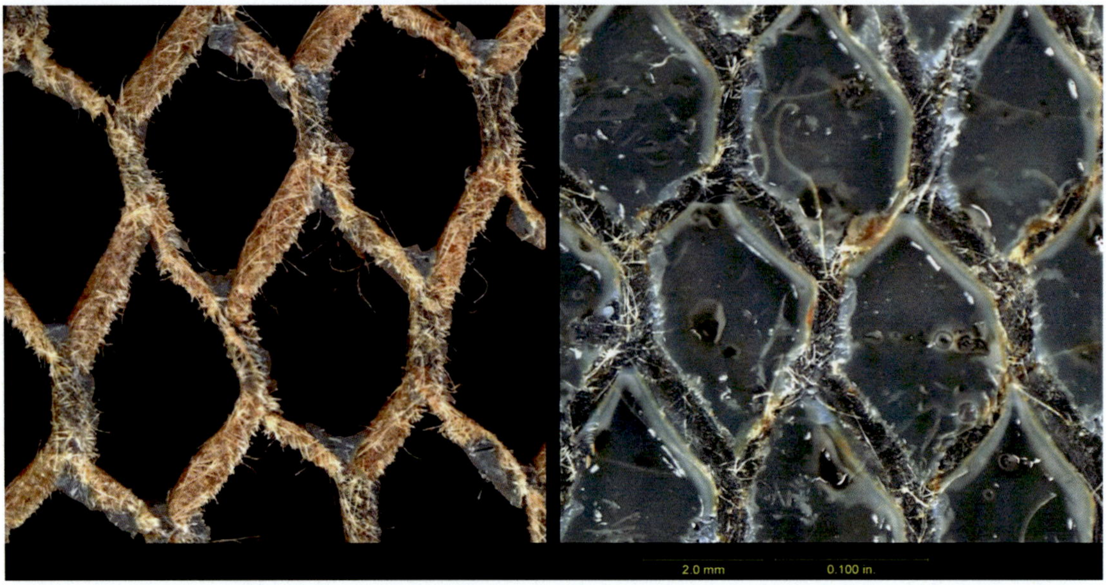

Fig. 13.5 Micrographs of an unprepared honeycomb sandwich composite structure after climbing drum peel testing showing the adhesive surface after failure and the separated honeycomb core. Dark-field illumination, 65 mm macrophotograph

bond was significantly stressed during the peel test. The corresponding image shows areas of the fillets attached to the core walls that were removed during failure.

The failure of a honeycomb sandwich panel with aramid fiber composite facesheets after peel testing is shown in Fig. 13.6. A micrograph of an unprepared honeycomb-cored composite surface after testing is shown in Fig. 13.6(a). Almost complete core failure can be observed in this micrograph. To highlight the microcracks in the aramid fiber facesheet, DYKEM Steel Red dye (Illinois Tool Works, Inc.) was applied on the surface of the composite (Fig. 13.6b) and was shown to penetrate through the facesheet to the core side (Fig. 13.6a). As a result of the dye penetration, the microcracks throughout the thickness of the composite could also be identified. A cross section of the microcracked composite facesheet is shown in Fig. 13.6(c).

A cross section is usually necessary to determine where and how failure occurred in a honeycomb sandwich structure composite. To prepare honeycomb sandwich composites for analysis, the sample must be carefully sectioned and then impregnated with an epoxy casting resin. If the sample is not mounted, the subsequent grinding and polishing steps can cause damage to the honeycomb, and information will be lost. Most of these sections can be mounted without the use of contrast dyes, but a dye may be necessary for glass fiber facesheet/honeycomb-cored composites due to

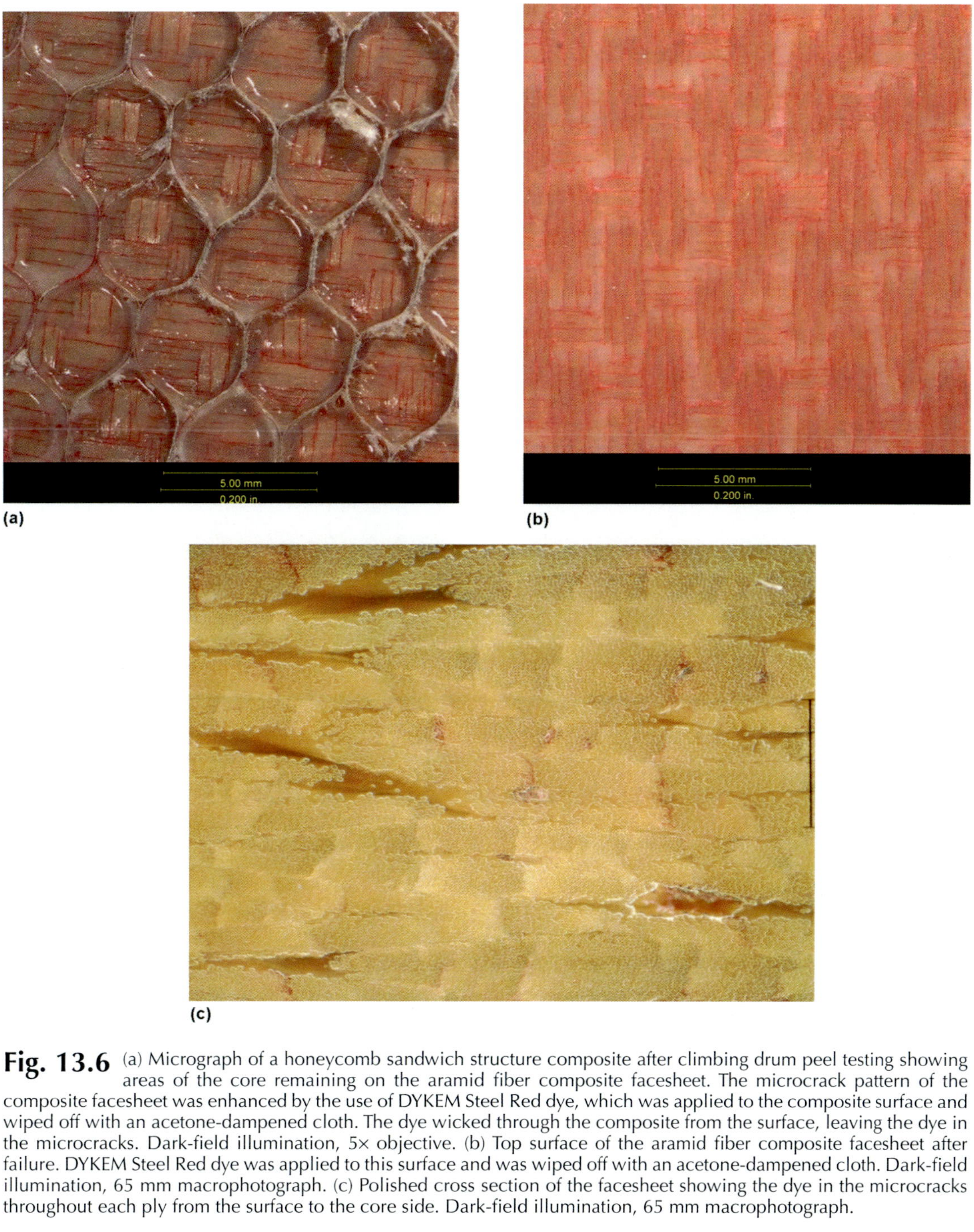

Fig. 13.6 (a) Micrograph of a honeycomb sandwich structure composite after climbing drum peel testing showing areas of the core remaining on the aramid fiber composite facesheet. The microcrack pattern of the composite facesheet was enhanced by the use of DYKEM Steel Red dye, which was applied to the composite surface and wiped off with an acetone-dampened cloth. The dye wicked through the composite from the surface, leaving the dye in the microcracks. Dark-field illumination, 5× objective. (b) Top surface of the aramid fiber composite facesheet after failure. DYKEM Steel Red dye was applied to this surface and was wiped off with an acetone-dampened cloth. Dark-field illumination, 65 mm macrophotograph. (c) Polished cross section of the facesheet showing the dye in the microcracks throughout each ply from the surface to the core side. Dark-field illumination, 65 mm macrophotograph.

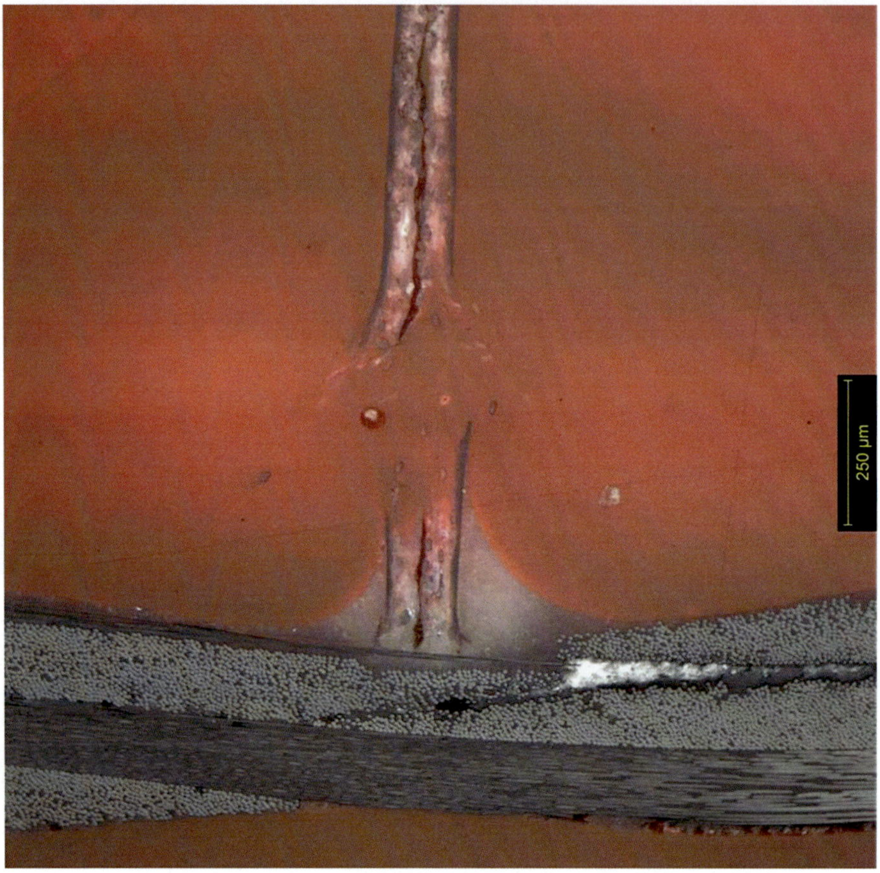

Fig. 13.7 Micrograph of honeycomb core (cell wall) failure. Also shown is a small delaminated area in the carbon fiber plain weave composite near the edge of the fillet due to the stress on the core wall. The sample was mounted in Rhodamine-B-dyed epoxy casting resin. Slightly uncrossed polarized light, 20× objective

the lack of contrast. Core failure is common in honeycomb sandwich structure composites when the formed fillet bond strength exceeds the strength of the core wall (Fig. 13.7). If a film adhesive is used to provide the skin-to-core bond, it must have excellent compatibility with and adhesion to the prepreg composite, or there can be a separation at this interface. These factors and the strength of the core material will determine the nature of the failure. Figure 13.8 shows a series of micrographs of core (cell wall) failure in a honeycomb sandwich structure composite, indicating adequate skin-to-core adhesion. For extremely durable core materials, the failure of the fillet-composite interface may be unavoidable, but this type of failure can also indicate problems at the fillet-composite interface (Fig. 13.9).

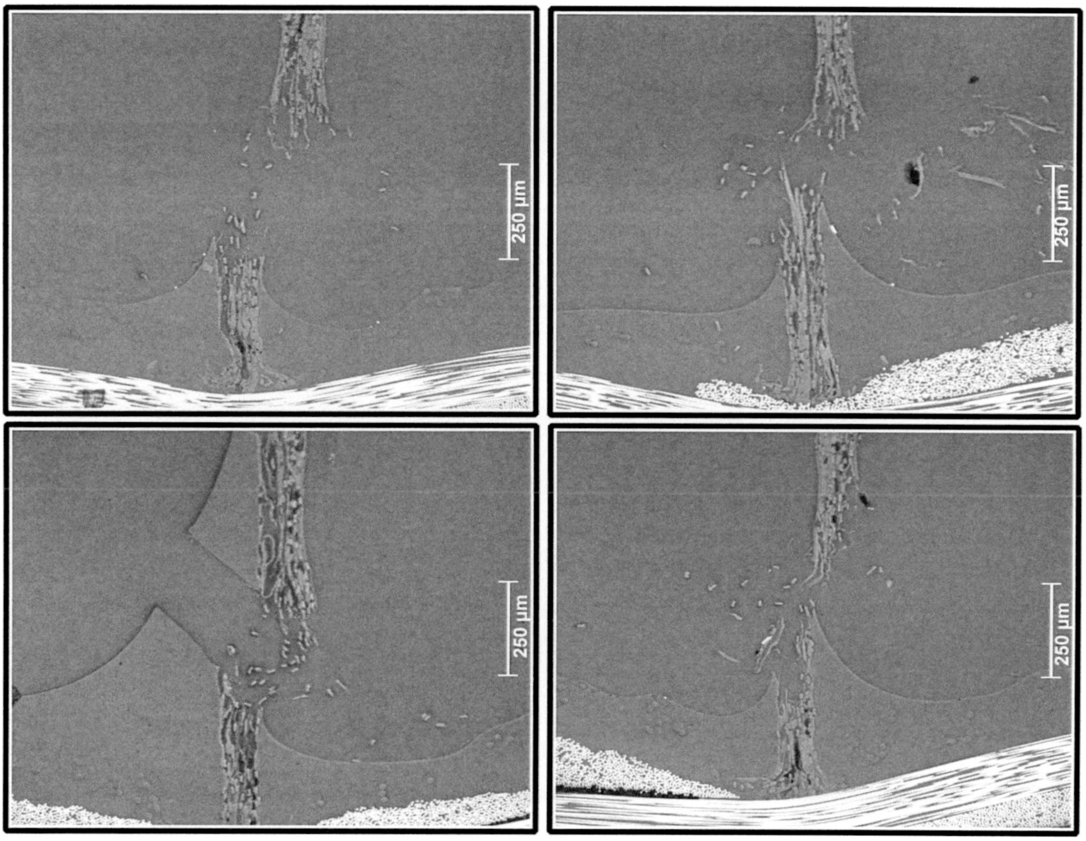

Fig. 13.8 Honeycomb core failure as a result of adequate skin-to-core adhesion. Bright-field illumination, 10× objective

Honeycomb composites that have poor fillet formation are susceptible to multiple failure mechanisms, especially if the prepreg/film adhesive compatibility and bond strength are not adequate. Figure 13.10 shows areas of a honeycomb sandwich structure composite where the film adhesive fillet separated from the prepreg composite. In this material, failure was also found at the adhesive/honeycomb cell wall when there was insufficient or nonexistent fillet formation. No core failure was found in this system.

REFERENCES

1. A.W. Alteneder, D.J. Renn, J.C. Seferis, and R.N. Curran, Processing and Characterization Studies of Honeycomb Composite Structures, *38th International SAMPE Symposium and Exhibition,* 1993, p 1034
2. A.W. Alteneder, Master's thesis, University of Washington, Seattle, 1993

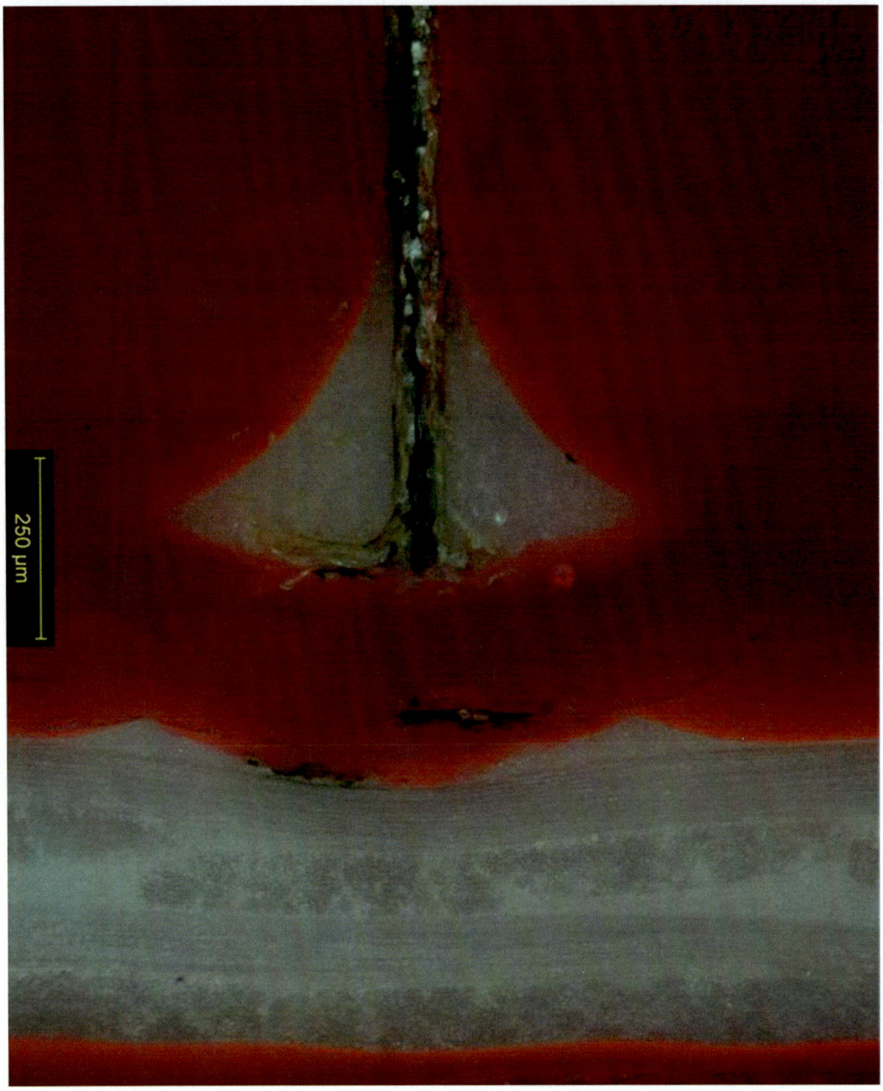

Fig. 13.9 Micrograph of fillet separation from a composite facesheet in a honeycomb sandwich structure composite. The sample was mounted in Rhodamine-B-dyed epoxy casting resin. Slightly uncrossed polarized light, 10× objective

3. A. Marshall, *Handbook of Composites,* G. Lubin, Ed., Van Nostrand Rheinhold Co., New York, 1982
4. T.H. Brayden and D.C. Darrow, Effect of Cure Cycle Parameters on 350 °F Cocured Epoxy Honeycomb Core Panels, *34th International SAMPE Symposium,* 1989, p 861
5. D.J. Renn, T. Tulleau, J.C. Seferis, R.N. Curran, and K.J. Ahn, Composite Honeycomb Core Crush in Relation to Internal Pressure Measurement, *J. Adv. Mater.,* Vol 27 (No. 1), 1995, p 31–40

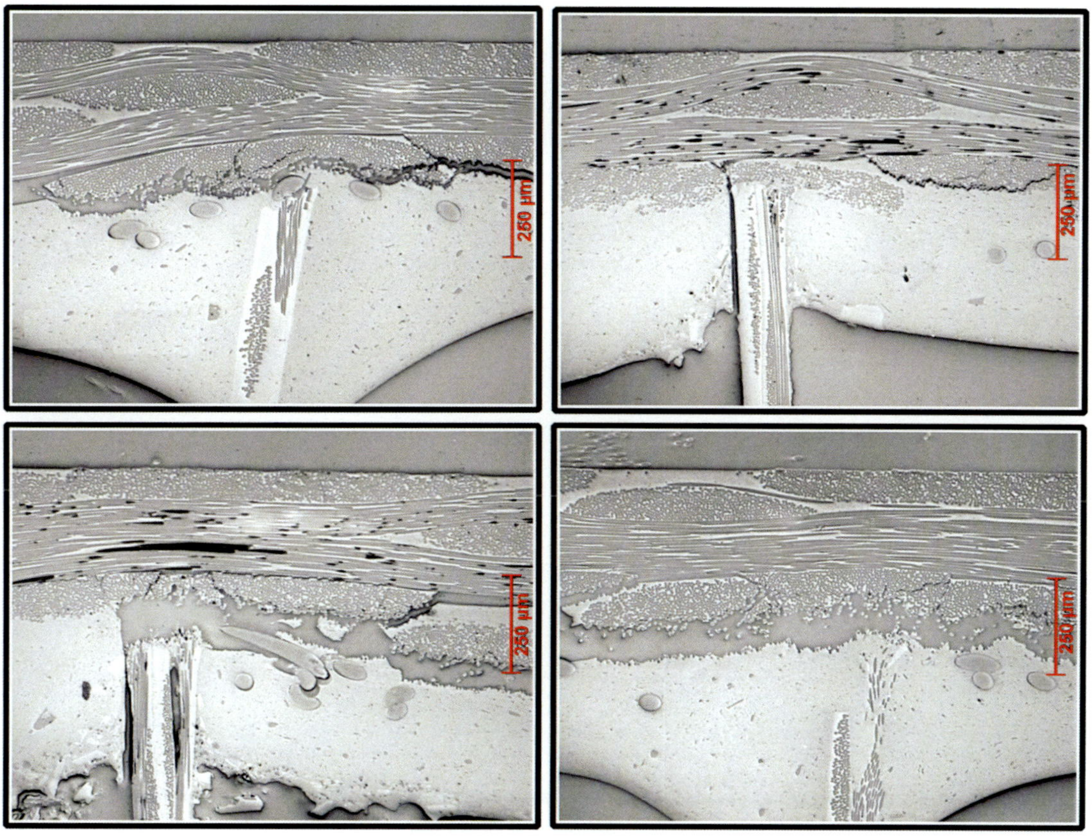

Fig. 13.10 Failure of a honeycomb-cored sandwich structure composite with areas of poor fillet formation and inadequate bond strength between the prepreg and film adhesive. Bright-field illumination, 10× objective

6. C.J. Martin, J.C. Seferis, and M.A. Wilhelm, Frictional Resistance of Thermoset Prepregs and Its Influence on Honeycomb Composite Processing, *Compos. Part A: Appl. Sci. Manuf.*, Vol 21, 1996, p 943–951
7. C.J. Martin, J.W. Putnam, B.S. Hayes, J.C. Seferis, M.J. Turner, and G.E. Green, Effect of Impregnation Conditions on Prepreg Properties and Honeycomb Core Crush, *Polym. Compos.*, Vol 18 (No. 1), 1997, p 90–99
8. B.S. Hayes, J.C. Seferis, and R.R. Edwards, Self-Adhesive Honeycomb Prepreg Systems for Secondary Structural Applications, *Polym. Compos.*, Vol 19 (No. 1), 1998, p 54–64
9. D. Frank-Susich, D.H. Laananen, and D. Ruffner, Cure Cycle Simulation for Thermoset Composites, *Compos. Manuf.*, Vol 4 (No. 3), 1993, p 139–146
10. J.C. Halpin, J.L. Kardos, and M.P. Dudukovic, Processing Science: An Approach for Prepreg Composite Systems, *Pure Appl. Chem.*, Vol 55 (No. 5), 1983, p 893

CHAPTER 14

Surface Degradation of Composites

POLYMER COMPOSITE MATERIALS are subject to degradation if not appropriately protected from the environment. Any undesirable change in properties or appearance of a composite material can be considered a form of degradation. One of the first indicators of degradation is the change in the surface appearance or gloss, followed by a change in the color of the surface resin. The change in the molecular composition may extend only a few nanometers on the surface but can continue to progress into the bulk of the composite matrix. Composite materials having polymeric matrices are susceptible to degradation from heat, sunlight, ozone, atomic oxygen (in space), moisture, solvents (chemicals), fatigue, excessive loading, and combinations of these environmental conditions (Ref 1). The rate and extent of degradation is dependent on the matrix chemistry and fiber reinforcement (Ref 2, 3). The changes in chemical composition at the fiber-matrix interface and interphase region can be more susceptible to degradation than the bulk matrix (Ref 4). Also, fiber orientation is known to influence the degradation behavior (Ref 3, 5).

Heat Effects on Composite Material Surfaces

The proper selection of the matrix material for the environment in which a component is to be used is critical so that long-term durability of the part is maintained. If the temperature to which the composite is exposed in service exceeds the glass transition temperature of the matrix, the material can creep or distort. However, even at lower temperatures than the glass transition temperature, polymeric matrices can degrade, especially if unprotected from the environment (Ref 6, 7). Excessive or prolonged heat can cause polymers to degrade over time, resulting in bond breaking. This degradation in service can lead to embrittlement of the matrix and cause changes in the physical and mechanical properties of a composite part.

Thermo-oxidative degradation begins at the surface and progresses further into a composite part with time. The depth of the oxidative degradation (thermo-oxidation) is a function of the exposure temperature, the length of time of thermal exposure, and the chemistry and constituents of the matrix and composite material. Figure 14.1(a) shows a micrograph of an as-processed epoxy woven glass fabric composite with an epoxy surfacing film. In this image, there is no surface degradation. When the composite was subjected to a temperature of 204 °C (400 °F) for a period of time, it was found that slight oxidation was present on the composite but remained in the surfacing film layer (Fig. 14.1b). As the temperature was further increased to 232 °C (450 °F), the depth of the oxidation was only

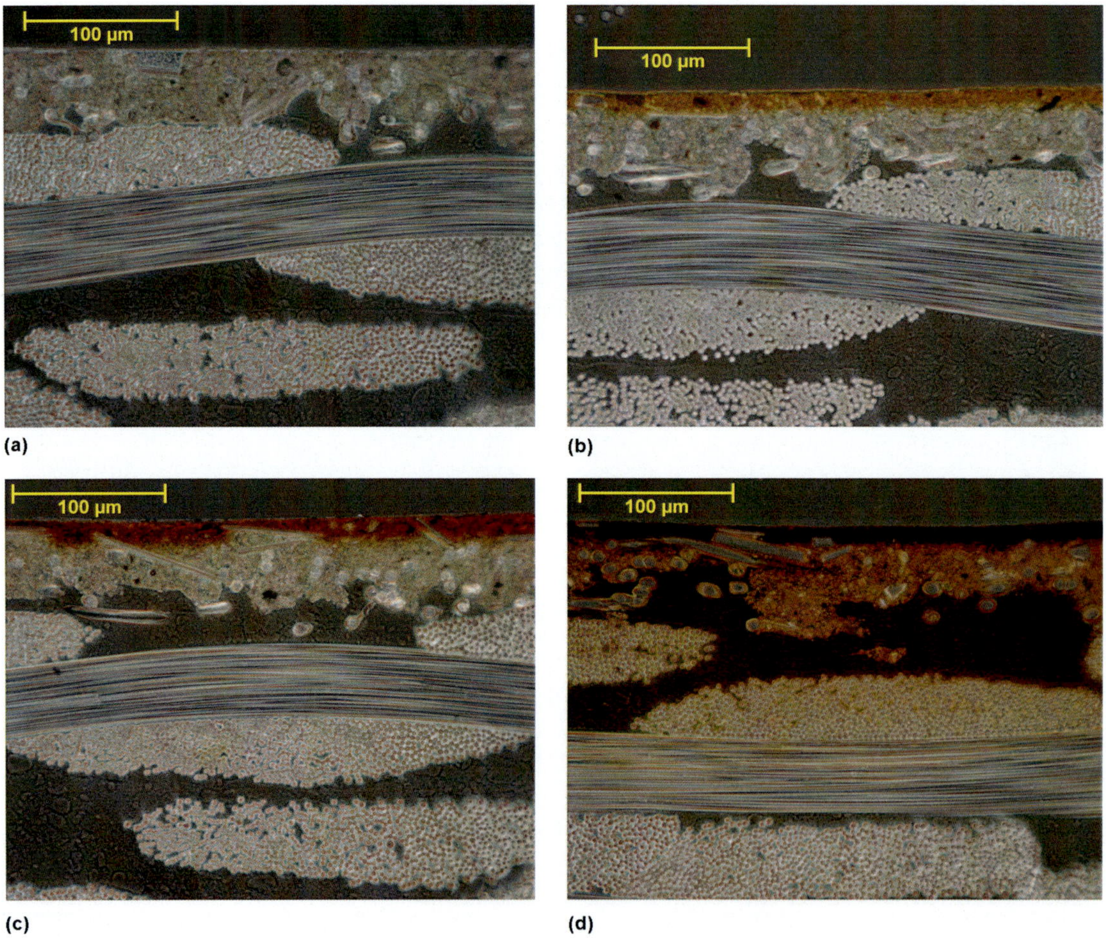

Fig. 14.1 Micrographs showing the effect of temperature on the depth of surface oxidation in a surfacing film/glass fiber composite. (a) As received (no exposure). (b) 204 °C (400 °F) exposure. (c) 232 °C (450 °F) exposure. (d) 260 °C (500 °F) exposure. Transmitted light, phase contrast, 20× objective

slightly greater, but the level of oxidation was increased (Fig. 14.1c). This is noticeable by the darker color of the surface resin. Increasing the temperature to 260 °C (500 °F) further increased the depth of the degradation so that it extended through the first few layers of the composite material (Fig. 14.1d). From analyses such as these, the depth of oxidation can be determined for composite materials as a function of temperature and time.

Ultraviolet-Light Effects on Composite Materials

Matrices in polymeric composites may be subject to degradation by ultraviolet (UV) light (photo-oxidation) if not protected by an appropriate coating. The source of the UV light may be sunlight or one of many types of artificial lights, such as tungsten halogen lights (Ref 8, 9). Composite matrices, as with all polymers, are more or less affected by different wavelengths of light, but the high-energy end of the light spectrum causes more degradation (Ref 10). Figure 14.2 shows how the wavelength of light affected the resin surface oxidation of an epoxy-matrix woven fabric composite. After the 72-hour exposure, it was found that a wavelength of approximately 350 nm caused the most severe surface oxidation. At this high-energy end of the spectrum, the energy exceeds the bond dissociation energies of many covalent bonds (Ref 11). Wavelengths greater than 400 nm showed little visible surface oxidation for this matrix composition at this time interval.

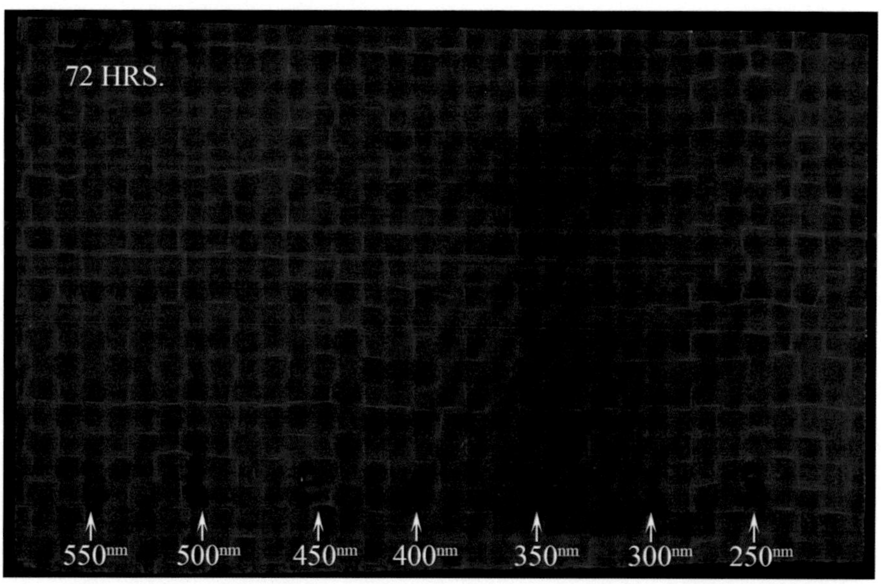

Fig. 14.2 Macroimage of an unprepared woven carbon fabric composite surface showing the extent of surface oxidation after exposure to a range of light wavelengths. 2× magnification photograph of a 10.2 × 15.2 cm (4 × 6 in.) panel

Although surface oxidation may not be visible to the naked eye, cross sectioning a composite and viewing with transmitted light may reveal even slight oxidation. Figure 14.3 shows a micrograph of an ultrathin section that was developed to determine the effect of short-term UV exposure on a carbon-fiber-reinforced composite part. It can be seen that the depth of oxidation in the surfacing film was approximately 5 μm after the UV exposure. With greater exposure to UV light or the presence of other environmental factors, such as heat and humidity, the degradation can be worse (Ref 1). For an unprotected composite, the degradation can continue until only the fibers remain on the surface, further blocking the effect of UV light. Figure 14.4 shows the effect of 10 years of sunlight exposure on a composite material. In this image, it can be seen that the degradation has continued until the first ply of the carbon fibers. Unfortunately for most advanced composite matrices, this type of degradation is possible, and therefore, the composite must be protected from the environment. As with thermo-oxidative degradation, the effect of matrix chemistry and structure does play a significant role in the mechanism and rate of degradation (Ref 12, 13). Usually, the addition of UV absorbers and stabilizers to composite matrices can only slow the onset and rate of degradation and may adversely alter the matrix properties (Ref 14). Ultimately, the UV protection must be accomplished through the appropriate selection of the polymer backbone and matrix chemistry that is relevant to the application and en-

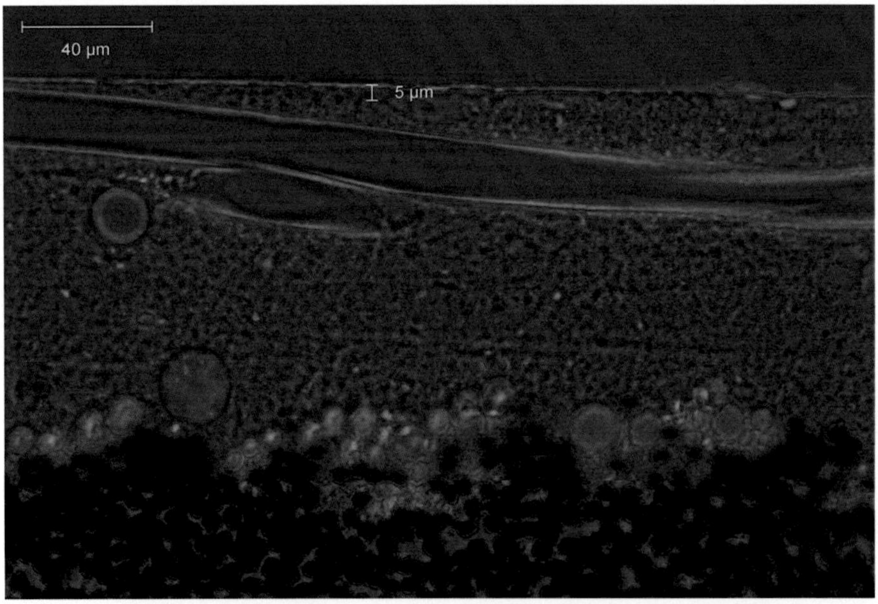

Fig. 14.3 Oxidation on the surface of a woven carbon fabric composite part as a result of short-term ultraviolet-light exposure. The oxidation is found to have penetrated only approximately 5 μm deep into the surfacing film in this time period. Transmitted light, phase contrast, 20× objective

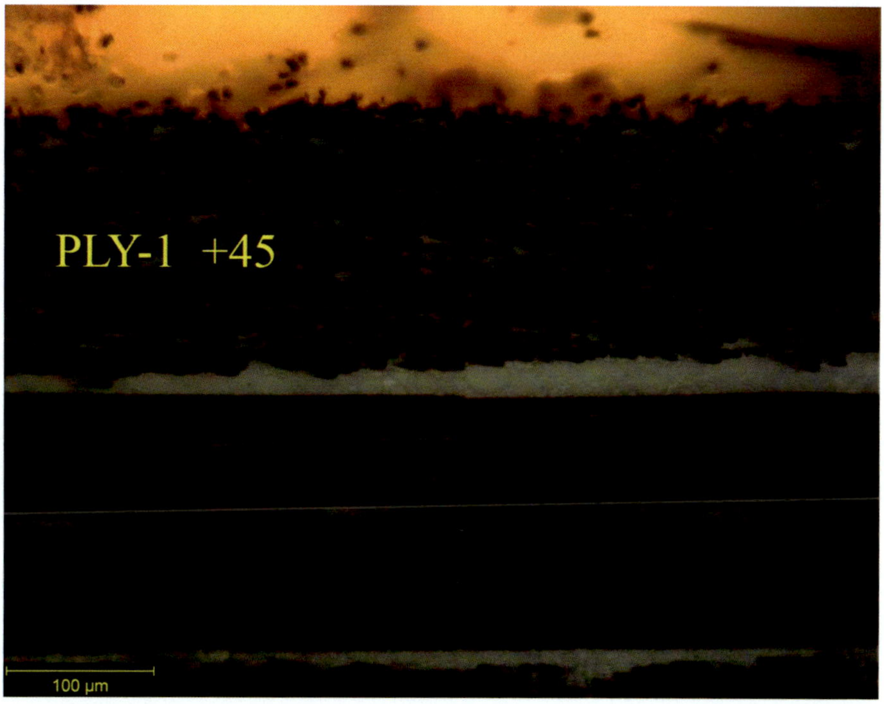

Fig. 14.4 Micrograph showing oxidation on the surface of a woven carbon fabric composite after 10 years of sunlight exposure. Epi-fluorescence, 390–440 nm excitation, 25× objective

vironment. This most likely is in the form of a paint or coating on the composite material.

Atomic Oxygen Effects on Composite Surfaces

The rugged appearance of the surface of the carbon fiber/polyimide-matrix composite specimen shown in Fig. 14.5(a and b) is typical for the polymer composites that flew on the leading edge of the National Aeronautics and Space Administration's (NASA's) Long-Duration Exposure Facility (LDEF). The material specimens and systems on board this 60-ton, experiment-laden satellite were exposed to low Earth orbit for 5.8 years before a last-minute retrieval by the space shuttle Columbia. The jagged peaks and valleys seen in these micrographs were created by the extremely erosive nature of atomic oxygen, the primary atmospheric constituent at the LDEF mission altitudes. In addition to atomic oxygen, the LDEF was bombarded by UV radiation, interstellar dust, micrometeoroids, and a variety of subatomic particles. Although the damage seen here is only a few mils deep, space hardware designers now have a better understanding of what materials can be used in specific locations and for what extent of time as a result of the LDEF.

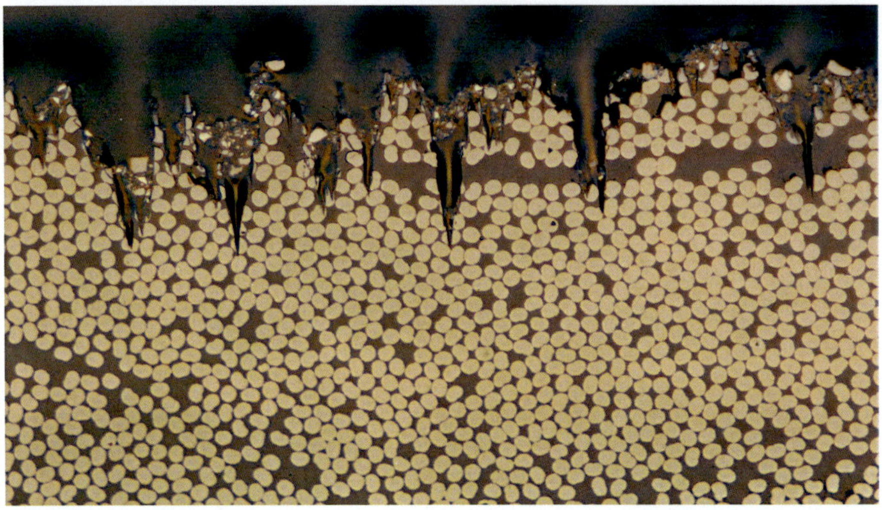

(a)

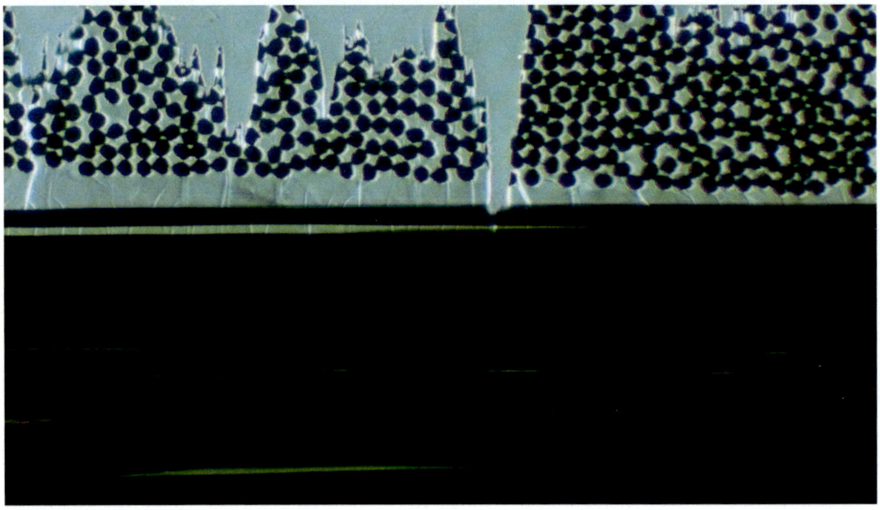

(b)

Fig. 14.5 Degradation of a composite surface after exposure to atomic oxygen. (a) Bright-field illumination, 25× objective. (b) Transmitted light, differential interference contrast, 20× objective

REFERENCES

1. G.A. Luoma and R.D. Rowland, Environmental Degradation of an Epoxy Resin Matrix, *J. Appl. Polym. Sci.,* Vol 32, 1986, p 5777
2. P. Musto et al., Thermal-Oxidative Degradation of Epoxy and Epoxy-Bismaleimide Networks: Kinetics and Mechanism, *Macromol. Chem. Phys.,* Vol 202, 2001, p 3445
3. X. Colin, A. Mavel, C. Marais, and J. Verdu, Interaction Between Cracking and Oxidation in Organic Matrix Composites, *J. Compos. Mater.,* Vol 39 (No. 15), 2005, p 1371–1389

4. K.J. Bowles, M. Madhukar, D.S. Papadopoulos, L. Inghram, and L. McCorkle, The Effects of Fiber Surface Modification and Thermal Aging on Composite Toughness and Its Measurement, *J. Compos. Mater.,* Vol 31 (No. 6), 1997, p 552–579
5. J.D. Nam and J.C. Seferis, Anisotropic Thermo-Oxidative Stability of Carbon Fiber Reinforced Polymeric Composites, *SAMPE Q.,* Vol 24 (No. 1), 1992, p 10–18
6. B.L. Burton, The Thermooxidative Stability of Cured Epoxy Resins, Part 1, *J. Appl. Polym. Sci.,* Vol 47, 1993, p 1821
7. S. Bondzic, J. Hodgkin, J. Krstina, and J. Mardel, Chemistry of Thermal Aging in Aerospace Epoxy Composites, *J. Appl. Polym. Sci.,* Vol 100, 2006, p 2210–2219
8. "Ultraviolet Radiation from Fluorescent Lamps," Lamp Section, National Electrical Manufacturers Association, Rosslyn, VA, May 4, 1999
9. B.L. Patkus, Technical leaflet, "The Environment," Section 2, Leaflet 4, Northeast Document Conservation Center, Andover, MA
10. A. Rivaton et al., Photo-Oxidation of Phenoxy Resins at Long and Short Wavelengths, Part I: Identification of the Photoproducts, *Polym. Degradation Stab.,* Vol 58, 1997, p 321
11. F. Rodriguez, *Principles of Polymer Systems,* 3rd ed., Hemisphere Publishing Co., New York, 1989
12. B. Mailhot et al., Study of the Degradation of an Epoxy/Amine Resin, Part 1: Photo- and Thermo-Chemical Mechanisms, *Macromol. Chem. Phys.,* Vol 206, 2005, p 575
13. V. Bellenger and J. Verdu, Photooxidation of Amine Crosslinked Epoxies, Part 2: Influence of Structure, *J. Appl. Polym. Sci.,* Vol 28, 1983, p 2677
14. G.A. George et al., Photo-Oxidation and Photoprotection of the Surface Resin of a Glass Fiber-Epoxy Composite, *J. Appl. Polym. Sci.,*

CHAPTER 15

The Effects of Lightning Strikes on Polymeric Composites

AS GREATER PORTIONS of new commercial airplane designs are constructed of polymer composites, it has become increasingly important to design these structures to mitigate the effects of lightning strikes (Ref 1). This design consideration is especially vital when composite structures contain jet fuel. In designing composite aircraft structures, the cost and efficiency of repairs and maintenance must be considered as well as the initial component cost. Assessing the extent of damage in a polymer composite structure after a lightning strike is a key element in developing future aircraft composite parts that are affordable, repairable, and maintainable (Ref 2).

Metal aircraft structures have relatively high electrical conductivity, which permits lightning currents to readily pass through to ground. The high electrical current of the lightning arc can cause remelting and rapid quenching of the metal alloy, resulting in changes in microstructure and properties. Composite conductivity, however, is not as homogeneous as in metals. In carbon-fiber-reinforced composites, the fibers are electrically conductive, but these are surrounded by an electrically insulating polymer matrix. As a result, the electrical current arcs between carbon fibers and damages the polymer-matrix resin, both within the composite laminate plies (intraply arcing) and between plies (interply arcing). To minimize damage from a lightning strike, metal foil, expanded metal foil (metal foil that is slit and drawn in two directions to produce a grid structure), or interwoven wire fabric prepreg is placed on the exterior surface of the composite part. The metal foil or wire-modified fabric conducts much of the electrical arc current to ground, reducing or eliminating damage to the underlying composite.

Lightning damage in polymer composites, as in metal structures, is manifested by damage at both the macroscopic or visual level and within the material microstructure. In addition to visual damage assessment, nondestructive inspection techniques are employed to detect damage within the composite part. These inspection techniques include ultrasound and x-ray analysis. It is also important to assess damage that may have occurred at the microstructural level, and special specimen-preparation techniques have been developed to study this microscopic damage.

Assessment of Microstructural Damage in Composites

The macroeffects of a lightning strike on composites are usually apparent and can be documented with macrophotography, but the microeffects can only be observed with a clean cross section of the damaged area. The zone in and around the strike is very fragile, due to the vaporized and degraded matrix (Ref 3). Care must be taken in cross sectioning and mounting in order to preserve as much of the critical information as possible. Mounting and sectioning fiber-reinforced composite specimens that have been subjected to lightning strikes is best accomplished with a two-stage mount. It is advisable on the primary mount to add a dye to the mounting epoxy, such as Rhodamine B laser dye (see Chapter 2, "Sample Preparation and Mounting," in this book). The dye allows the mounting resin to be distinguished from the original composite specimen and helps to highlight damage features. The best method for creating an artifact-free specimen is to first impregnate the strike area under vacuum using an epoxy casting resin, followed by the application of pressure during the cure. This encapsulated area will allow sectioning through the center of the strike and thus hold the fragile material in place, minimizing artifacts. The specimen can be remounted to provide adequate handling of the microscopic sample.

Polished cross-sectional mounts can be examined with a variety of microscopy techniques, including polarized light, bright- and dark-field illumination, and epi-fluorescence. The damage characteristics depend on many factors, including the type of lightning strike protection material on the surface (expanded foil or interwoven wire), the polymer matrix, the fiber type and volume, the lay-up of the composite, the thickness and characteristics of the surface paint layers, and the energy level of the lightning strike (Ref 4). The microstructure of fiber-reinforced polymer-matrix composites that have been exposed to high-energy arcing is quite different and can vary widely depending on the constituents. The strike area in fiber-reinforced thermoset-matrix composites usually consists of a central hot-spot zone, where the fibers are degraded, surrounded by a zone where the polymer is vaporized.

Laboratory-generated lightning strikes and the effects of high currents traveling through the composite and the transition zone, caused by high-energy arcing, are shown in the following figures (Ref 2). Figure 15.1 shows common surface features of lightning strike damage in a carbon fiber fabric polymer-matrix composite part. Carbon fibers can be seen protruding from the center of the strike zone. In the strike zone, the matrix is vaporized on the surface, along with some of the carbon fibers. This composite material has no lightning strike protection on the surface and therefore has significant damage. Figure 15.2 is a cross-sectional montage of the edge of the strike zone, which shows areas where there was interply arcing. Figure 15.3(a) was photographed slightly below the main region of damage at the strike center near a Kevlar (E.I. du Pont de Nemours and

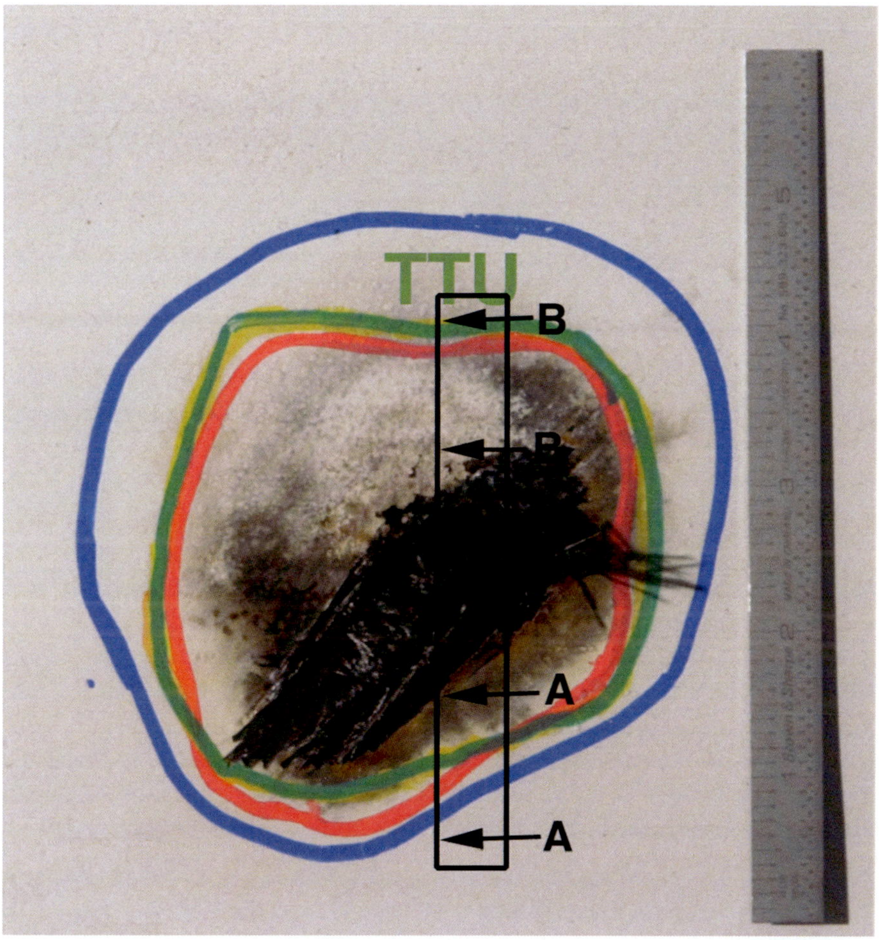

Fig. 15.1 Photograph of a painted carbon-fiber-reinforced composite part surface after a zone 1A lab-induced lightning strike. Two cross sections were taken from this area after impregnation with a casting resin. The cross sections for microscopic analysis are labeled "A" and "B," and the direction of the arrows shows the section plane.

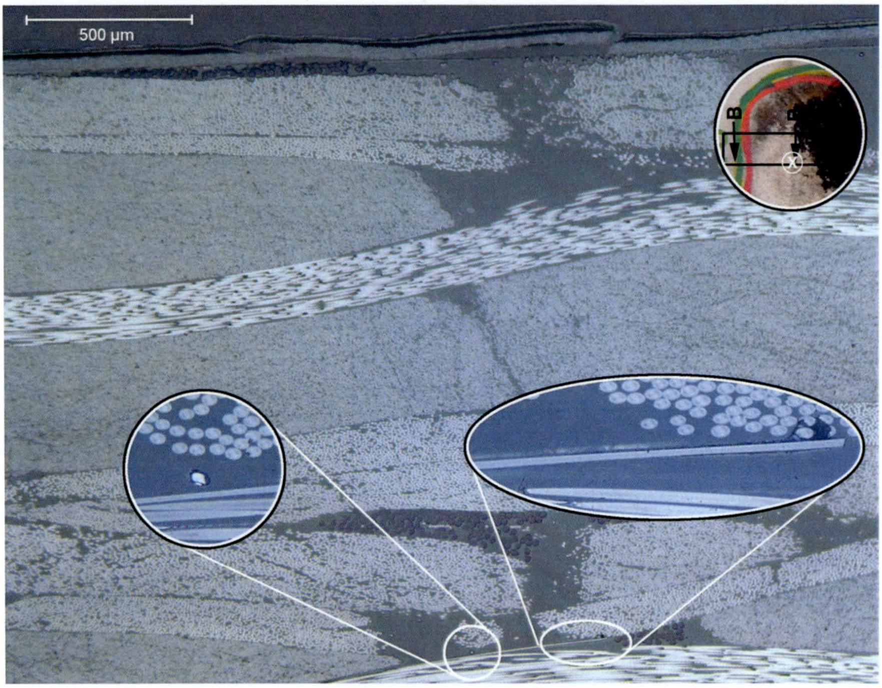

Fig. 15.2 Montage of the edge of the lightning strike zone that corresponds to section plane B in Fig. 15.1. Slightly uncrossed polarized light, 10× objective (25× objective insets)

Company) stitch. It is difficult to see the extent of damage from the interply arcing in this micrograph, but with the use of epi-fluorescence, the detail of the arcing event is much more visible. Figure 15.3(b) shows the damage from interply arcing between carbon fibers and microcracking in the Kevlar stitch. In the lightning damage zone, there was significant microcracking and delamination, as can be seen in Fig. 15.4. Figure 15.5 shows an area under the lightning strike zone that exhibited interply arcing, microcracking, and delamination.

Figure 15.6 shows the result of a lightning strike on a panel consisting of expanded aluminum foil over a honeycomb sandwich carbon fiber composite. When compared to Fig. 15.1, it can be seen that the use of a conductive foil on the surface of the composite part significantly reduced the damage. To further analyze the area under the strike zone, the damaged area was impregnated under vacuum with a low-viscosity epoxy casting resin. No contrast dye was added to the casting resin. The first layer of expanded aluminum foil was vaporized, and some of the outer carbon fibers degraded (Fig. 15.7). The conductive layer protected the rest of the carbon fibers and composite structure. Only a few localized microcracks were created in the composite. Figure 15.8 is a section taken underneath the lightning strike attachment point, showing that the mechanical force from the strike impact buckled the honeycomb cell wall.

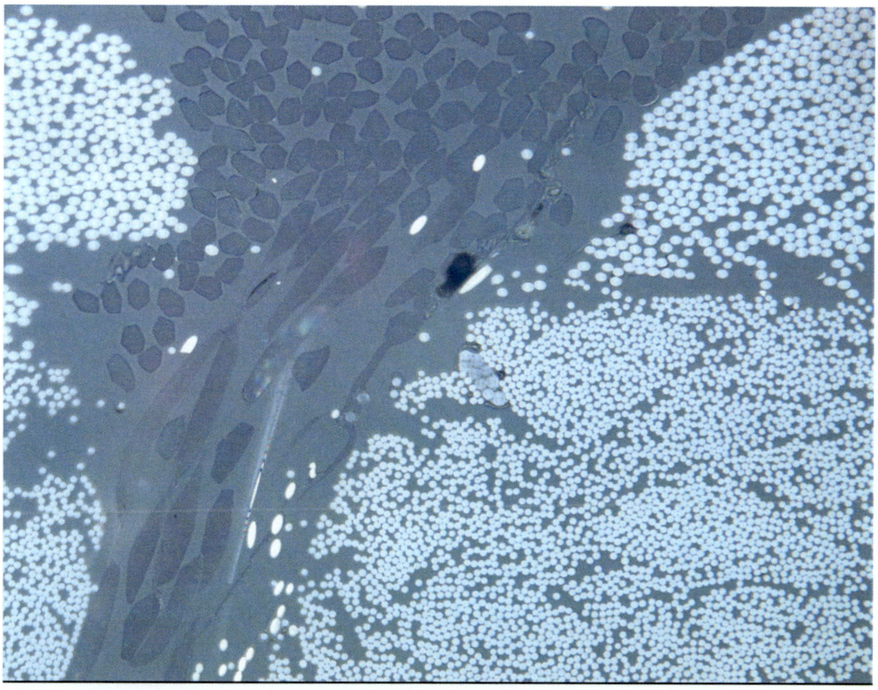

(a)

(b)

Fig. 15.3 Cross section taken at the edge of the strike zone corresponding to section plane A in Fig. 15.1. The micrographs are taken away from the main damage area to show the effects of interply arcing. A Kevlar stitch is shown between the carbon fibers. (a) Slightly uncrossed polarized light, 25× objective. (b) Epi-fluorescence, 390–440 nm excitation, 25× objective

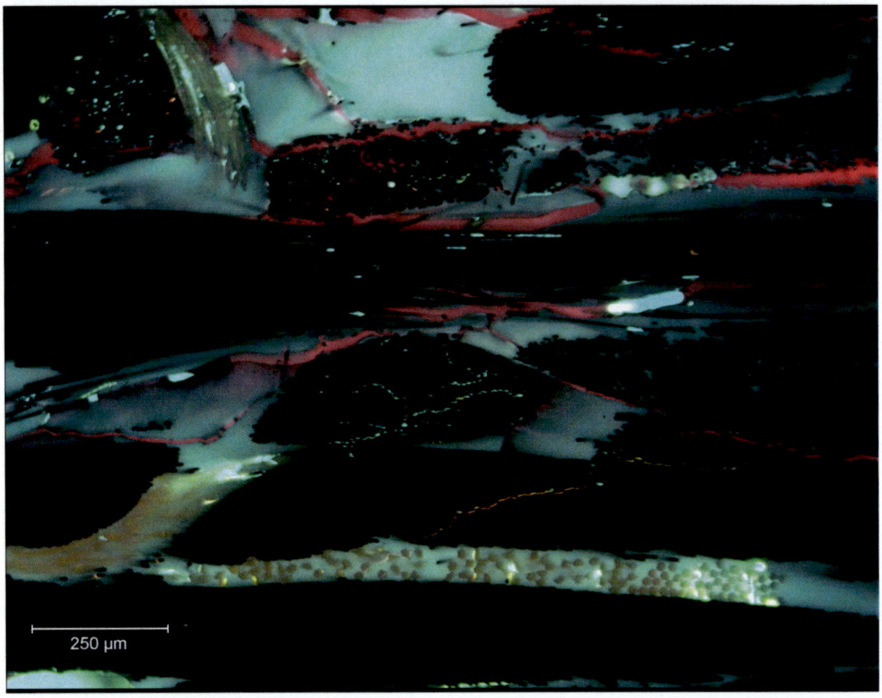

Fig. 15.4 Micrograph taken at an area under the lightning strike zone showing delamination and microcracking in the composite. Epi-fluorescence, 390–440 nm excitation, 10× objective

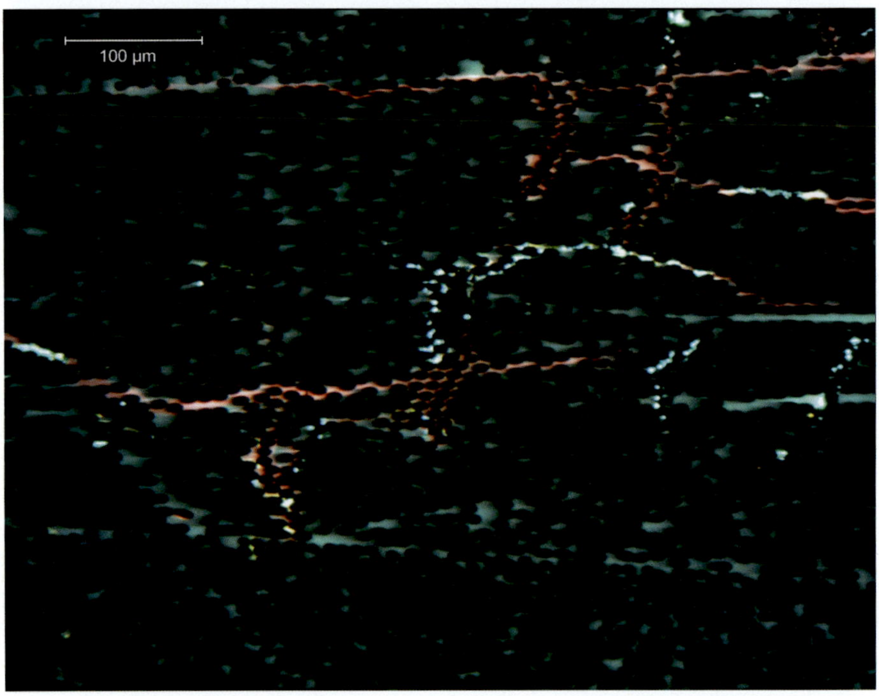

Fig. 15.5 Micrograph taken under the strike zone showing damage induced by the lightning strike. Epi-fluorescence, 390–440 nm excitation, 25× objective

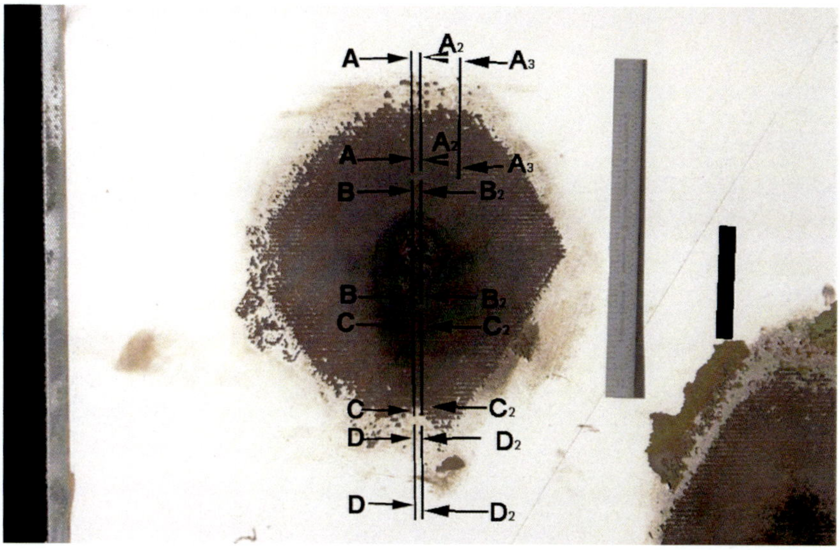

Fig. 15.6 Photograph of a painted composite surface, protected with expanded aluminum foil, after a zone 1A lab-induced lightning strike. A cross-sectional map is superimposed over the lightning-strike-damaged area.

Fig. 15.7 Area under the lightning strike zone where the expanded aluminum foil was vaporized. The surface shows only slight damage. Slightly uncrossed polarized light, 10× objective

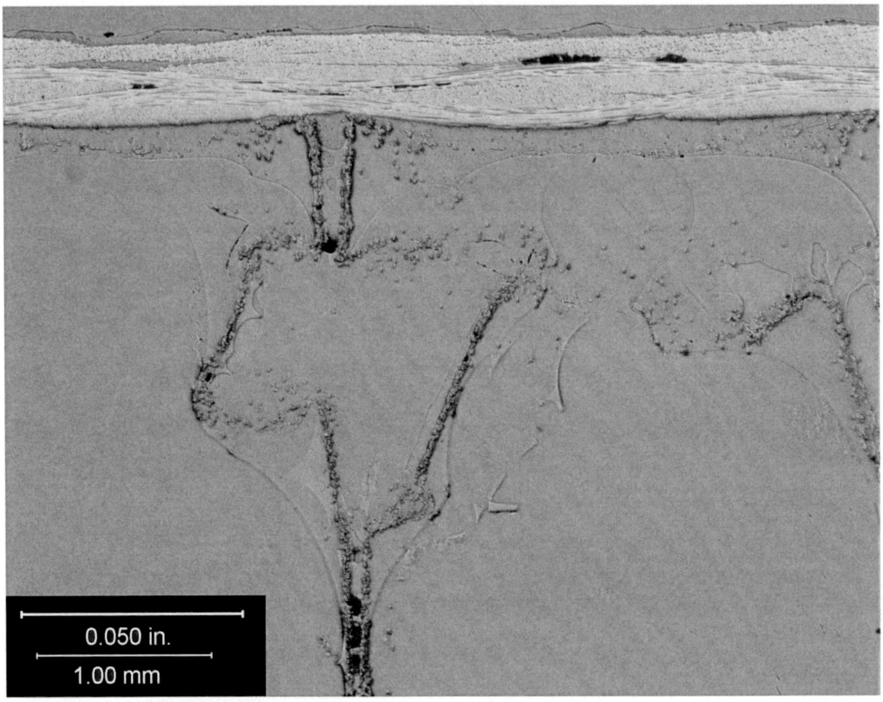

Fig. 15.8 Macroimage of an area under the lightning strike attachment point showing crushing of a honeycomb cell wall. Bright-field illumination, 65mm macrophotograph

The following figures were developed from a composite lap joint that was subjected to a lightning strike. The damaged lap joint was impregnated with Rhodamine-B-dyed epoxy resin. Prior to the final polishing step, the specimen was dried, and Magnaflux Zyglo (Magnaflux Corp.) penetrant was applied before completing the sample preparation with the final polish. Figure 15.9 shows the effect of arcing between the titanium fastener and carbon fiber composite. This arcing caused the edges of the fastener to melt and areas of the composite to degrade, which created an irregular gap between the two dissimilar materials. In the location of the composite edge near the fastener (Fig. 15.10), it can be seen that areas of the matrix were completely vaporized. This is shown by impregnation of the Rhodamine-B-dyed casting epoxy into the carbon fibers, as highlighted by the inset micrograph. Also shown in the figure are heat-affected fibers. Figure 15.11(a and b) is of the same location as illustrated in Fig. 15.10 but taken 1.5 mm (0.06 in.) from the interface, showing the effects of interply arcing as observed using different contrast methods. The use of epi-fluorescence may be a necessary technique to observe the full extent of the damage from a lightning strike. This is further illustrated in Fig. 15.12, which shows the same material that was subjected to

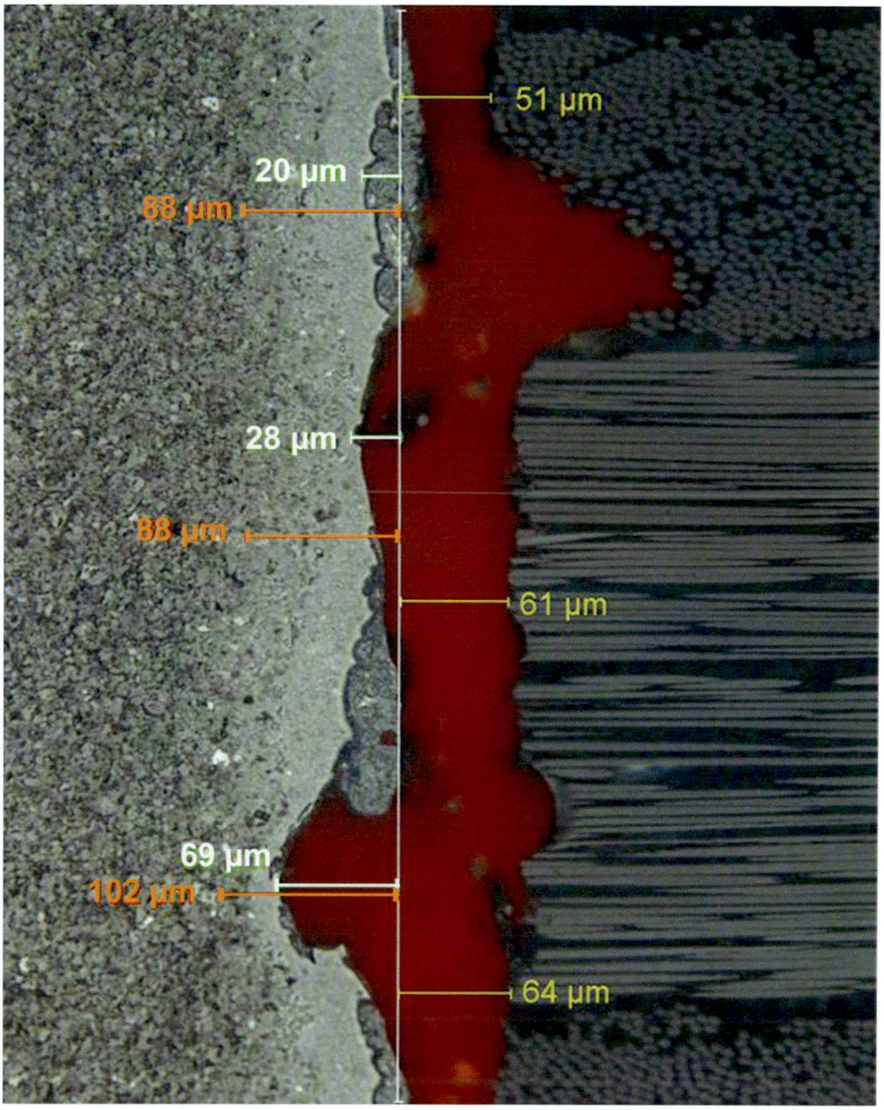

Fig. 15.9 Micrograph showing the effect of electrical arcing between a titanium fastener and carbon fiber composite. Slightly uncrossed polarized light, 25× objective

a lightning strike, revealing both interply and intraply arcing in the composite.

To protect a composite surface from the damage inflicted by a lightning strike, there have been many different types of woven wires, meshes, and screens, all of which proved more or less effective. An alternative method for protecting a composite surface from lightning strike damage is to bond a monolithic sheet of conductive foil to the surface of the composite. Fig-

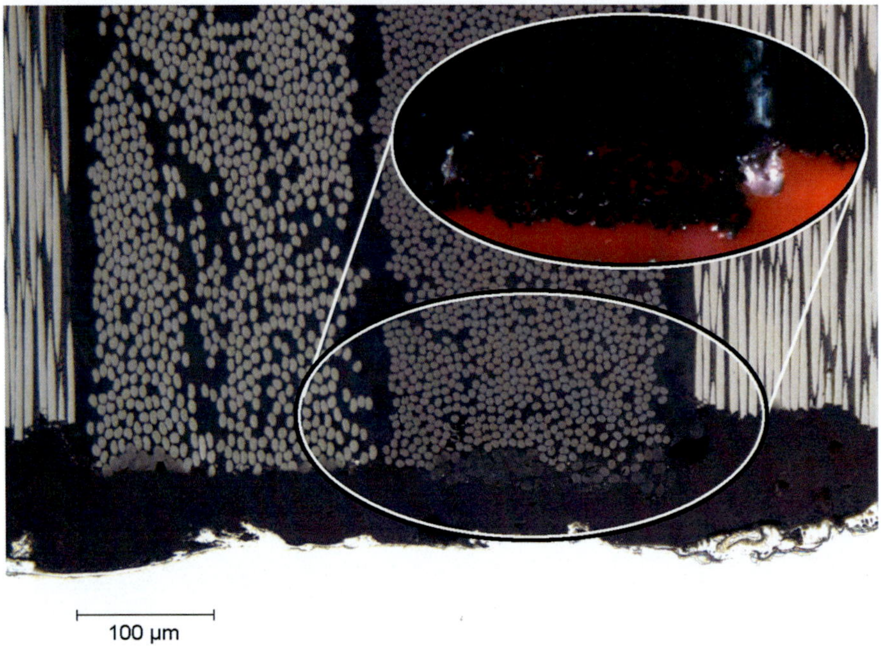

Fig. 15.10 Micrograph showing damage at the interface of a carbon fiber composite from high-current arcing between a titanium fastener. Slightly uncrossed polarized light, 25× objective

ure 15.13(a and b) shows the effect of a lightning strike on a composite laminate that had a metal foil top surface. In the area directly under the strike, the metal foil was vaporized. At the outer edge of the strike, a nugget of melted metal was found to remain attached to thc foil. Further preparation of the reflected-light sample to an ultrathin-section sample revealed more information about the heat generation created by the lightning strike. The vaporized and degraded matrix in the first ply can easily be seen by using transmitted-light optical microscopy. Deeper under the damage zone, areas of matrix oxidation and microcracking were found in the composite plies.

In composites that have been struck by lightning, with or without surface protection, there are usually areas that have heat damage. This damage is commonly found near the surface but can extend throughout the composite, depending on the constituents. As discussed previously, heat can degrade and vaporize both fibers and matrix. Figure 15.14(a to c) shows areas in a composite material where there was heat damage under the strike zone. In Fig. 15.14(a and b), large hot spots are evident on the surface, with fiber degradation at the surface of the strike. Figure 15.14(b) shows that the high impact energy in this localized area under the lightning strike zone caused crazing to occur in the interlayer area of the composite. A single hot spot is shown in the carbon fiber composite in Fig. 15.14(c).

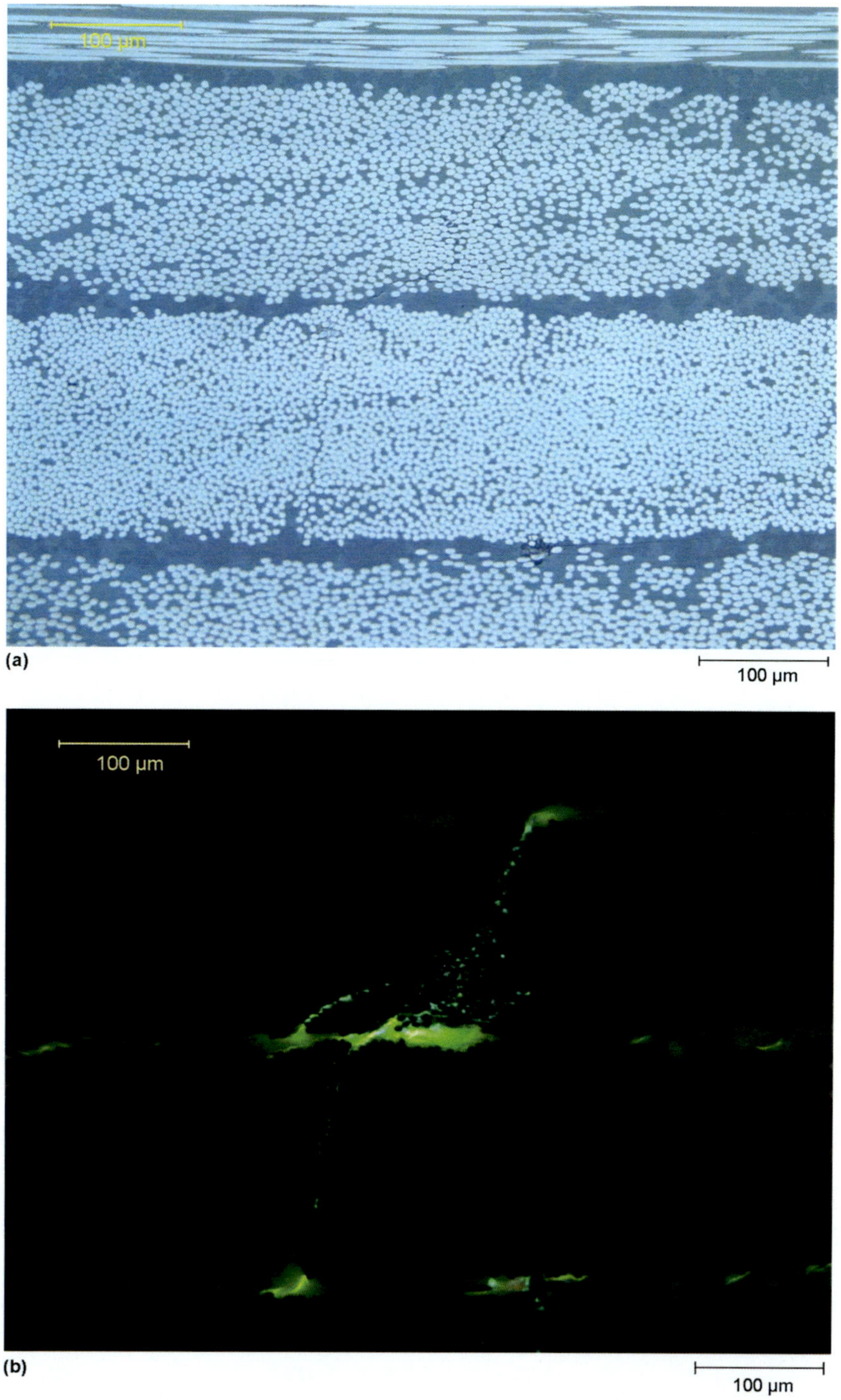

Fig. 15.11 Area sectioned 1.5 mm (0.06 in.) from the titanium fastener/composite interface showing the effects of interply arcing in the composite. (a) Bright-field illumination, 25× objective. (b) Epi-fluorescence, 390–440 nm excitation, 25× objective

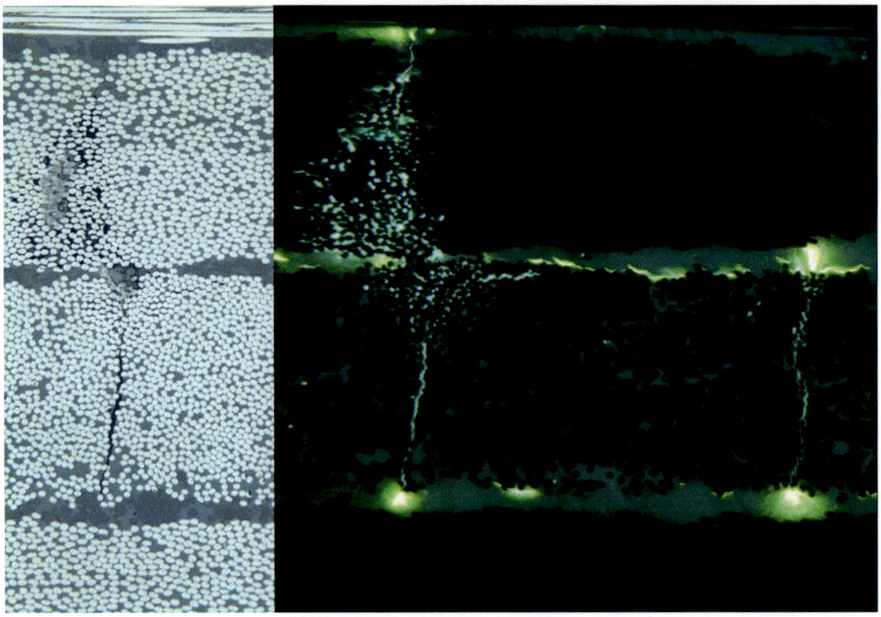

Fig. 15.12 Interply and intraply arcing events that occurred in a carbon fiber composite after being subjected to a lab-induced lightning strike. Comparison of different contrast methods: bright-field illumination and epi-fluorescence, 390–440 nm excitation, 25× objective

The effects of lightning strike damage in a composite structure are summarized in Fig. 15.15. In this figure, there are areas of vaporized and degraded matrix and fibers as well as matrix microcracking.

These micrographs illustrate the variety of lightning damage features that are being studied in carbon-fiber-reinforced polymer composites. The preceding examples have shown that the analysis of lightning strike damage of composite materials requires the use of many different sample-preparation techniques and the full extent of the optical microscope.

REFERENCES

1. L.M. Gammon, N. Baxter, C.A. Meyr, H. Oberson, E.R. Stack, and A. Falcone, The Effects of Lightning Strikes and High Current on Aerospace Materials, *Microsc. Microanal.,* Vol 15 (Suppl. 2), 2009, p 28
2. L.M. Gammon and A. Falcone, Lightning Strike Damage in Polymer Composites, *Adv. Mater. Process.,* Aug 2003, p 61–62
3. L.M. Gammon and B.S. Hayes, The Effects of Lightning Strike and High Current on Polymer Composites, *35th Annual International Metallographic Society,* Aug 2002 (Quebec City)
4. E. Rupke, *Lightning Direct Effects Handbook,* AGATE-WP3.1-031027--043-Design Guideline, Work Package Title: WBS3.0 Integrated Design and Manufacturing, Lightning Technologies Inc., Pittsfield, MA, March 1, 2002

(a)

(b)

Fig. 15.13 Lightning strike damage in a carbon fiber composite laminate having metal foil on the surface for protection. (a) Slightly uncrossed polarized light, 4× objective. (b) Transmitted light (ultrathin section), circular polarized light, 4× objective. The impregnation outline of the epoxy casting resin can be seen surrounding the metal nugget.

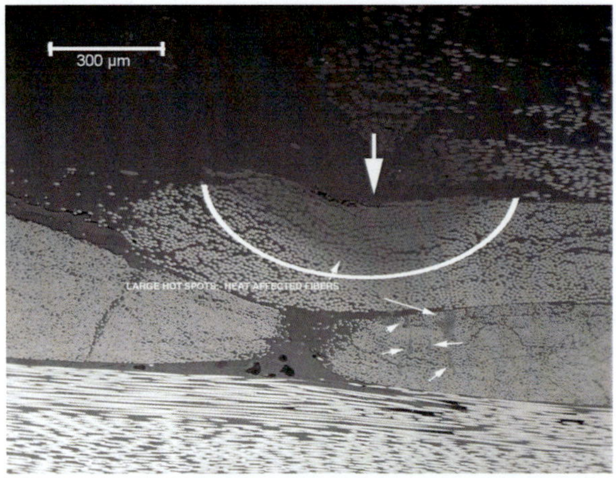

(a)

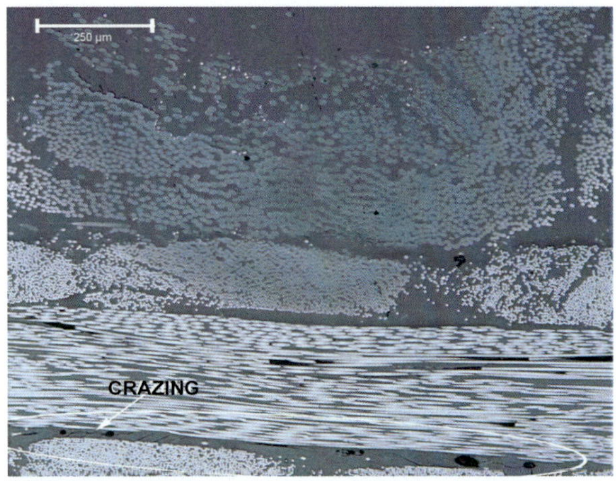

(b)

(c)

Fig. 15.14 Carbon fiber composite cross sections showing heat damage from lab-induced lightning strikes. (a) Section showing heat-affected fibers. Bright-field illumination, 10× objective. (b) Area under the strike zone showing matrix crazing due to the impact. Bright-field illumination, 10× objective. (c) Local hot spot in a composite. Bright-field illumination, 10× objective

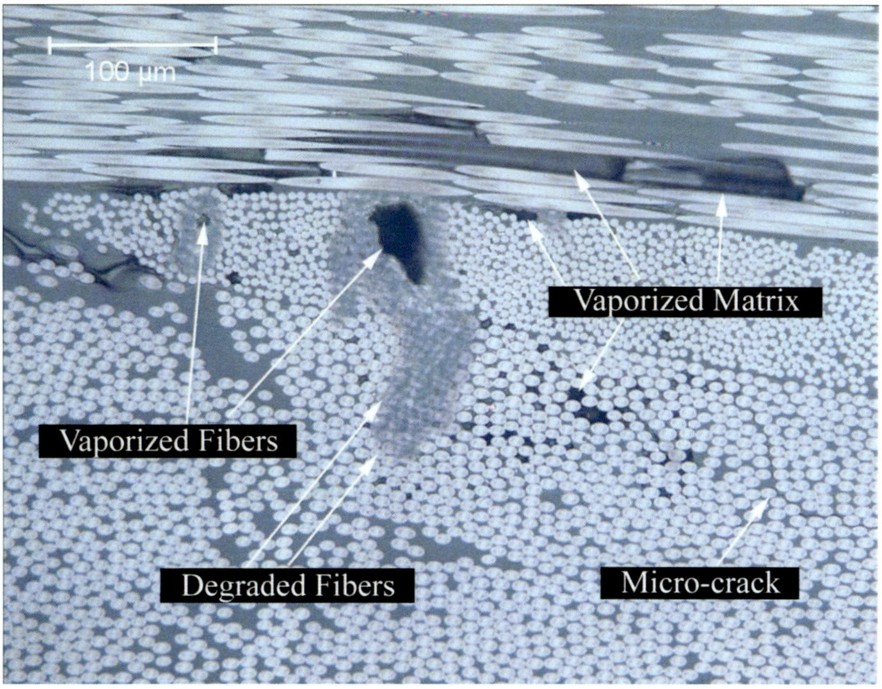

Fig. 15.15 Lightning strike damage in a carbon fiber composite material showing fiber and matrix vaporization and degradation as well as microcracking. Bright-field illumination, 25× objective

Index

D

E

F

G

H

I

J

K

L

S

T

U

V

FSC
www.fsc.org
MIX
Papier | Fördert
gute Waldnutzung
FSC® C083411

Kolibri 360 GmbH
Ferdinand-Jühlke-Straße 7
99095 Erfurt, Deutschland
produktsicherheit@kolibri360.de